VOLUME THIRTY THREE

THE ENZYMES

Inhibitors of the Ras Superfamily G-proteins, Part A

VOLUME THIRTY THREE

THE ENZYMES

Inhibitors of the Ras Superfamily G-proteins, Part A

Edited by

FUYUHIKO TAMANOI

Department of Microbiology, Immunology, and Molecular Genetics, Molecular Biology Institute
University of California
Los Angeles, CA 90095, USA

AMSTERDAM • BOSTON • HEIDELBERG • LONDON
NEW YORK • OXFORD • PARIS • SAN DIEGO
SAN FRANCISCO • SINGAPORE • SYDNEY • TOKYO
Academic Press is an imprint of Elsevier

Academic Press is an imprint of Elsevier
225 Wyman Street, Waltham, MA 02451, USA
525 B Street, Suite 1800, San Diego, CA 92101-4495, USA
Radarweg 29, PO Box 211, 1000 AE Amsterdam, The Netherlands
The Boulevard, Langford Lane, Kidlington, Oxford, OX5 1GB, UK
32 Jamestown Road, London NW1 7BY, UK

First edition 2013

ISBN: 978-0-12-416749-0
ISSN: 1874-6047

For information on all Academic Press publications visit our website at store.elsevier.com

Printed and bound by CPI Group (UK) Ltd, Croydon, CR0 4YY

Transferred to digital print 2012

CONTENTS

PREFACE

The Ras superfamily G-proteins function as a molecular switch that regulates a variety of biochemical reactions in the cell. These proteins play important roles in cell growth, proliferation, morphology, motility, and others. Activation of these proteins is associated with human cancer and diseases. Over the years, a variety of approaches have been taken to develop small molecule inhibitors of the Ras superfamily G-proteins. This and the next volume (Volumes 33 and 34) provide the first publications of comprehensive overview of these studies.

I am very grateful to the authors for their effort in providing exciting chapters that detail their original approach to the problem. I also thank Mary Ann Zimmerman of Elsevier for her guidance, Helene Kabes of Elsevier for her support, and Phoebe Phan of UCLA for her communication and assistance during the preparation of this volume.

Fuyuhiko Tamanoi
May 2013

CHAPTER ONE

The Ras Superfamily G-Proteins

Ashley L. Tetlow, Fuyuhiko Tamanoi[1]
Department of Microbiology, Immunology and Molecular Genetics, Molecular Biology Institute, Jonsson Comprehensive Cancer Center, University of California, Los Angeles, Los Angeles, California, USA
[1]Corresponding author: e-mail address: fuyut@microbio.ucla.edu

Contents

Abstract

The Ras superfamily G-proteins are monomeric proteins of approximately 21 kDa that act as a molecular switch to regulate a variety of cellular processes. The structure of the Ras superfamily G-proteins, their regulators as well as posttranslational modification of these proteins leading to their membrane association have been elucidated. The Ras superfamily G-proteins interact at their effector domains with their downstream effectors via protein–protein interactions. Mutational activation or overexpression of the Ras superfamily G-proteins has been observed in a number of human cancer cases. Over the years, a variety of approaches to inhibit the Ras superfamily G-proteins have been developed. These different approaches are discussed in this volume.

1. THE Ras SUPERFAMILY G-PROTEINS

The Ras superfamily G-proteins comprise more than 150 monomeric proteins of approximately 21 kDa (Fig. 1.1A) [1]. These proteins act as a molecular switch to regulate a variety of biochemical reactions by alternating between an active GTP-bound form and an inactive GDP-bound form. In its active conformation, these proteins bind their downstream effector

The Enzymes, Volume 33
ISSN 1874-6047
http://dx.doi.org/10.1016/B978-0-12-416749-0.00001-4

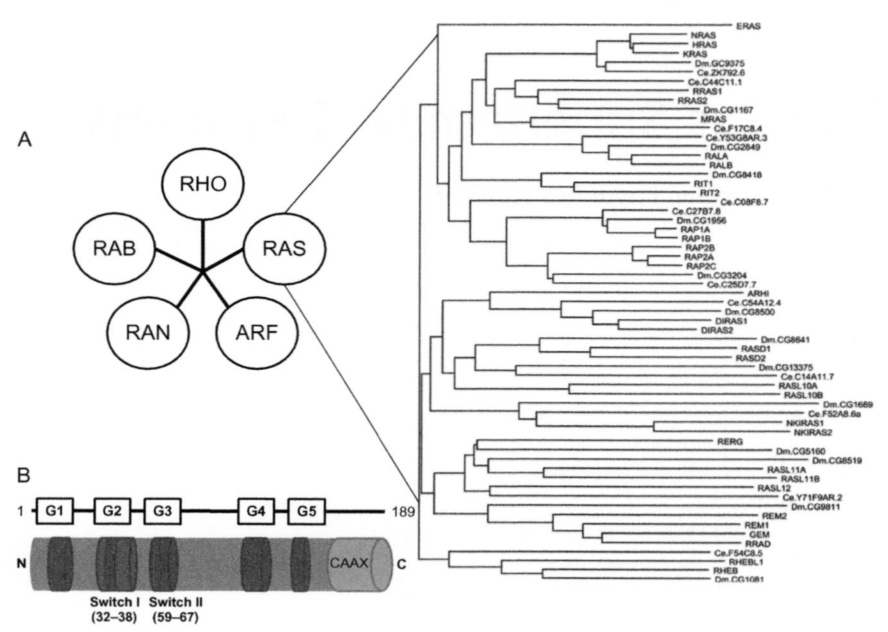

Figure 1.1 (A) Expanded dendrogram of RAS subfamily members from *H. sapiens*, *D. melanogaster* (Dm), and *C. elegans* (Ce). Human protein names are in uppercase letters. Branch lengths are directly proportional to the number of differences between sequences compared. *Modified from Ref. [52].* (B) RAS family domain architecture depicting G-motifs 1–5 (blue), core effector domain (purple), and membrane targeting sequence (orange). *Modified from Ref. [1,52].* (See color plate.)

proteins leading to cellular effects. It is believed that these G-proteins exist mainly in a GDP-bound form under starvation conditions and that the percentage of GTP-bound form increases upon stimulation. For example, in a study that examined the percentage of GTP-bound Ras in rat pheochromocytoma (PC12) cells, it was found that 6% of Ras was complexed with GTP in serum-starved culture. However, the percentage of GTP-bound Ras was rapidly increased to 24% after stimulation with NGF and this high level was maintained for at least 16 h [2]. A similar result was reported by Muroya et al. [3].

This superfamily is divided into subfamilies. While there are overlaps in the cellular processes they affect, general comments can be made concerning processes that each subfamily proteins affect. The Ras subfamily proteins are involved in proliferation, growth, and gene expression and include three isoforms of Ras: K-Ras, N-Ras, and H-Ras [1]. Rheb proteins also belong to the Ras subfamily and they are involved in the mTOR signaling pathway leading to the regulation of protein synthesis, growth, and cell cycle [4]. The Rho subfamily proteins include Rho, Rac, and Cdc42 that have important roles in regulating cytoskeletal organization, motility, and invasion in addition to gene expression [5,6]. The Rab and Arf subfamily proteins regulate membrane trafficking and vesicular transport processes [1,7]. Finally, the Ran subfamily G-proteins are involved in protein transport into the nucleus [8].

Conserved motifs among the Ras superfamily proteins are depicted in Fig. 1.1B, which shows the H-Ras protein. Five short stretches of conserved sequences are identified and they are called the G-motifs [9]. Sequences in the G-1 and G-3 motifs include residues such as Gly-12 and Gln-61 that play critical roles in the binding and hydrolysis of guanine nucleotides. The G-1 motif contains a conserved sequence $GX_4GK(S/T)$. G-2 motif encompasses residues 32–40 that define a so-called effector domain, a site where the interaction with downstream effector proteins takes place. G-3 motif contains the conserved sequence DX_2G. The invariant aspartate-57 binds the catalytic Mg^{2+} mediated by a water molecule. G-4 motif contains a sequence motif (N/T)(K/Q)XD and G-5 motif contains residues 144–146 that interact with the guanine nucleotides indirectly. Finally, many of these proteins share a conserved C-terminal motif [10]. In the case of the Ras or Rho family G-proteins, they end with the so-called CAAX motif where C is cysteine, A is an aliphatic amino acid, and X is the C-terminal residue. Rab proteins end with the CC or CXC motif. The C-terminal motifs are recognized by protein prenyltransferases that initiate a series of C-terminal modification events that are critical for their membrane association (discussed below).

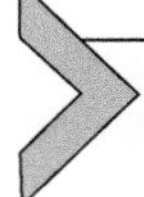

2. OVERVIEW OF STRUCTURE, ACTIVITY, AND REGULATORY PROTEINS

2.1. Three-dimensional structure

Many proteins in the Ras superfamily G-proteins are crystallized and their structures have been elucidated [11,12]. Figure 1.2 shows the structure of H-Ras protein which is remarkably similar to that of the guanine nucleotide-binding domain of bacterial EF-Tu. The Ras protein consists of five α-helices (A1–A5), six β-strands (B1–B6), and five polypeptide loops (G1–G5). Binding of GTP involves a variety of key interactions with amino

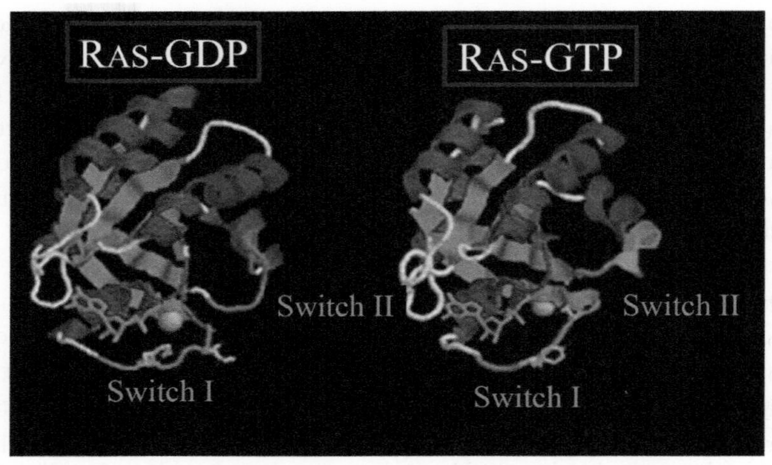

Figure 1.2 The structure of GTP-binding-protein RAS in GTP-bound active form (right) and GDP-bound inactive form (left). GTP analog is depicted in sticks. Ras-GTP analog (PDB ID: 5P21). Ras-GDP (PDB ID: 4Q21). *Modified from Ref. [12,53].* (See color plate.)

acid side chains. Coordination of Mg^{2+} with oxygens of the phosphates of GTP as well as to the hydroxyls of Thr-35 and Ser-17 was elucidated. Since the determination of the structure of Ras, structures of a variety of the Ras superfamily G-proteins have been determined. They include Rho, Cdc42, Ran, Rab, Arf, Ran, Rheb, RAD, and Rap [13–20].

Comparison of the GTP-bound form and the GDP-bound form of Ras revealed that there are mainly two regions of the protein that exhibit conformational change [21]. One of these regions (switch I) is the so-called effector domain where interactions with downstream effector proteins take place. The other region, called switch II, is in the region encompassing residues 59–67. This region is involved in the interaction with an activator called GEF (GDP/GTP exchange factor). Thus, structural studies provided mechanistic basis of the action of the Ras superfamily G-proteins.

NMR analyses of the Ras superfamily G-proteins have also been carried out. Solution structure of full-length farnesylated and nonfarnesylated H-ras proteins has been elucidated [22]. NMR has also been utilized to obtain detailed mechanistic data on Ras cycling as well as defects conferred by oncogenic mutations or pathogenic conditions (RASopathy) caused by mutations in the regulatory proteins [23]. Solution structure and dynamics of RalB have been characterized by NMR [24].

2.2. GEF and GAP regulate the activities of the Ras superfamily G-proteins

The Ras superfamily G-proteins have intrinsic activities that are stimulated by regulatory proteins as shown in Fig. 1.3. In the case of Ras, its intrinsic GTPase activity hydrolyzes GTP, but this reaction is very slow; taking more than an hour to hydrolyze GTP bound to the protein. This slow intrinsic GTPase is dramatically enhanced by a family of proteins called GAP (GTPase activating protein). This family of proteins includes RasGAP, NF1, and Gap1. NF1 is the product of neurofibromatosis type I, a gene whose mutations lead to genetic disorder characterized by the appearance of benign tumors of the peripheral nervous system. RasGAP enhances the intrinsic activity of Ras through the stabilization of glutamine 61, which coordinates a water molecule to perform a nucleophilic attack. RapGAP

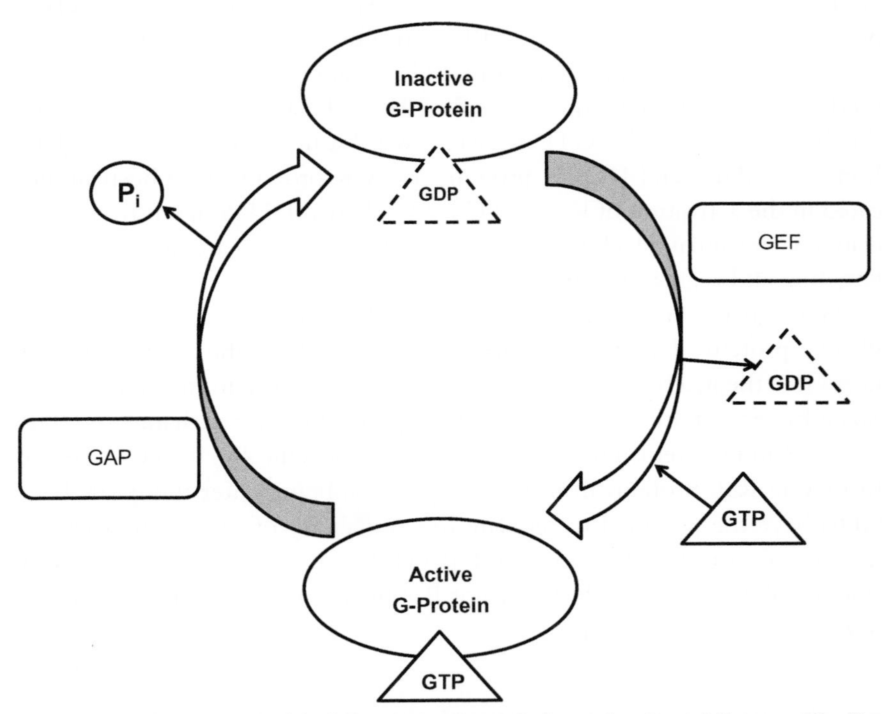

Figure 1.3 Schematic model of the molecular switch mechanism of Ras superfamily class GTPases in continuous cycle between active and inactive states. The proteins GAP and GEF aid in the transition of GTPases. GEF (guanine nucleotide exchange factors) removes GDP from inactive G-proteins. GAP (GTPase activating protein) hydrolizes GTP and an inorganic phosphate is released.

contains an arginine at 789 that is able to stabilize the transition state during the catalysis of GTP [25]. RapGAPs are specific for Rap proteins and utilize a threonine residue in switch II instead of the typical glutamine. RapGAPs also have an asparagine at residue 290, replacing the Gln61 seen in Ras, Rho, or Ran GAPs [26]. In the case of RabGAPs, the TBC (Tre-2, Bub2, and Cdc16) domain supplies two catalytic residues in *trans*, an arginine finger analogous to Ras/Rho family GAPs and a glutamine finger that substitutes for the glutamine in the DxxGQ motif of the GTPase [27]. RanGAP is an interesting case because its action does not utilize an arginine finger [28].

The bound GDP can be exchanged with GTP, but this intrinsic exchange activity is very slow. This is stimulated by a group of regulatory proteins called GEF. GEF interacts with GDP-bound form and exchanges bound GDP with fresh GTP. The Cdc25 family acts on Ras, Ral, and Rap and includes Cdc25, SOS, and EPAC. The crystal structure of SOS/Ras complex has been elucidated [29]. This interaction with GEF takes place between residues 32 and 67. GEF sterically occludes the magnesium-binding site, causing the GDP-bound G-protein to lose affinity for GDP. GDP is released and subsequently replaced by GTP and GEF loses its affinity for the G-protein. Rho GTPases have two distinct families of GEFs: Dbl homology (DH) and DOCK proteins. DOCK proteins have been implicated in the activation of Rac and Cdc42, whereas DH domains have associated C-terminal pleckstrin homology domains that are implicated in targeting and regulatory functions [30,31].

GDI (guanine nucleotide-dissociation inhibitor) is another type of regulatory protein. This protein interacts with GDP-bound G-protein and keeps it in the inactive form. GDI removes small G-proteins from the membrane by sequestration of their lipid tails, preventing the exchange of GDP for GTP and the return to an active state [32]. Specifically, Rab GDI serves to recycle Rab proteins from their target membranes after transport. Rab GDIs have the potential to regulate the availability of certain intracellular transport factors via inhibition of GDP/GTP exchange. Rab3A-GDI is also able to solubilize Rab9GDP, but not Rab9GTP, further enhancing transport factor regulation [33].

2.3. Posttranslational modification and membrane association

Many of the Ras superfamily G-proteins are posttranslationally modified by a series of C-terminal modification events that are depicted in Fig. 1.4 for Ras [34]. In the first step, these proteins are modified by the addition of an

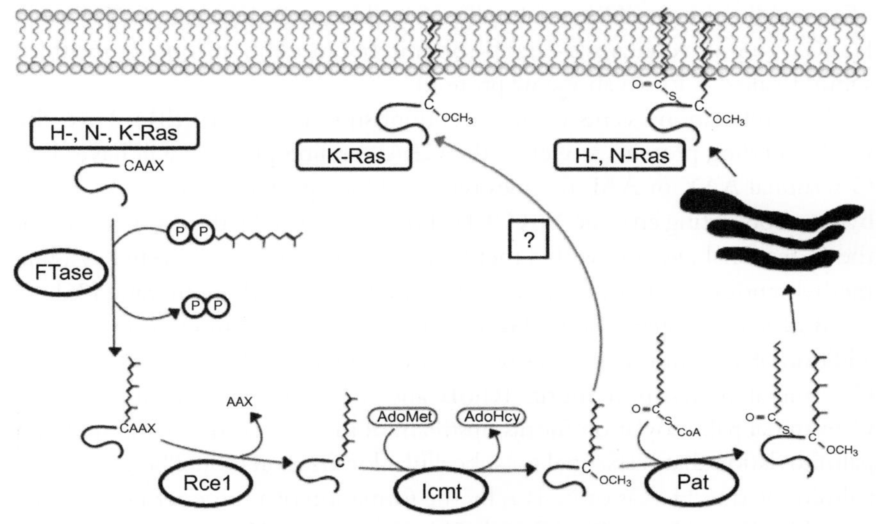

Figure 1.4 Diagram showing posttranslational modification of H-, N-, and K-Ras. RAS is synthesized on free ribosomes as a globular hydrophilic protein. Nascent RAS encounters FTase in the cytosol and, after farnesylation, it gains affinity for, and is transported to, membranes of the ER where it encounters the subsequent CAAX-processing enzymes RCE1 and ICMT. Following CAAX processing, K-RAS deviates from the path of the palmitoylated RAS isoforms and proceeds directly to the plasma membrane via a poorly understood pathway that may involve cytosolic chaperones. N-RAS and H-RAS proceed to the cytosolic face of the Golgi apparatus, where they are palmitoylated by palmitoyltransferase (PAT) and then traffic via vesicles to the plasma membrane. *Modified from Ref. [53].*

isoprenoid. In the case of proteins ending with the CAAX motif, a farnesyl group is covalently attached to a cysteine via thioether linkage. This occurs with proteins such as Ras and Rheb. With proteins ending with the CAAL motif such as Rho, Rac, and Cdc42, they are modified by the addition of a geranylgeranyl group [34]. Rab proteins end with CC or CXC motif and are modified by the addition of two geranylgeranyl moieties. Three types of enzymes catalyze these modification events. Protein farnesyltransferase (FTase) catalyzes the transfer of a farnesyl group from farnesyl pyrophosphate, an intermediate in cholesterol biosynthesis, to a cysteine in the CAAX motif. On the other hand, protein geranylgeranyltransferase I (GGTase-I) catalyzes the addition of a geranylgeranyl group to the cysteine in the CAAL motif. Two cysteines at the end of Rab proteins are both modified by the addition of a geranylgeranyl group that is catalyzed by Rab geranylgeranyltransferase (RabGGTase). FTase and GGTase-I are heterodimeric enzymes consisting of α- and β-subunits. The α-subunit is shared between FTase and GGTase-I. RabGGTase also contains α- and β-subunits but in addition contains the third subunit called REP (Rab escort protein).

Postprenylation events occur in endoplasmic reticulum (ER). First, the CAAX ending proteins as well as the CAAL ending proteins will have their C-terminal AAX or AAL residues removed by a proteolytic event catalyzed by Ras-converting enzyme 1 (RCE1). This is followed by the modification of the C-terminal cysteine with a methyl group catalyzed by isoprenyl cysteine methyltransferases (ICMT). Both are located on the ER membrane [35].

With some proteins such as H-Ras and N-Ras, further modification by the addition of a palmitic acid takes place at cysteines located just upstream of the C-terminal prenylation motif. RhoB and TC10 G-proteins also contain C-terminal palmitoylation motifs upstream of the CAAX motif and undergo palmitoylation that is essential for subcellular localization [36]. The addition of palmitic acid to H-Ras or N-Ras by the formation of a thioester bond is catalyzed by Ras palmitoyltransferase. This enzyme is a Golgi-resident protein that consists of two subunits DHHC9 and GCP16. Since a thioester linkage is reversible, this modification provides dynamic recycling of these proteins. On the other hand, K-Ras contains a polybasic domain located N-terminus to the farnesylated cysteine that provides increased affinity to membrane lipids. This transport process is currently under active investigation [37].

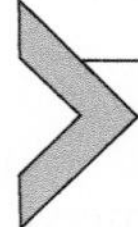

3. UPSTREAM AND DOWNSTREAM SIGNALING PATHWAYS

Upstream events initiating from the activation of receptors on cell surface lead to the activation of Ras superfamily G-proteins (Fig. 1.5). For example, in the case of Ras proteins, EGFR activation causes tyrosine phosphorylation of the receptor facilitating the binding of Grb2 through its SH2 domain. This protein also has SH3 domains and one of these, N-terminal SH3, binds SOS that acts as a RasGEF. This interaction is mediated by the SH3–Proline-rich domain interaction. SOS then activates Ras by facilitating GDP/GTP exchange [38]. As for Rho, Rac, and Cdc42, a variety of GEFs including the DH domain-containing GEFs: βPIX, Sos1, Trio (DH1), and Tiam-1 are involved in their activation [39–41].

Ras superfamily G-proteins activate downstream signaling pathways by interacting with effector proteins (Fig. 1.5). For Ras proteins, three main

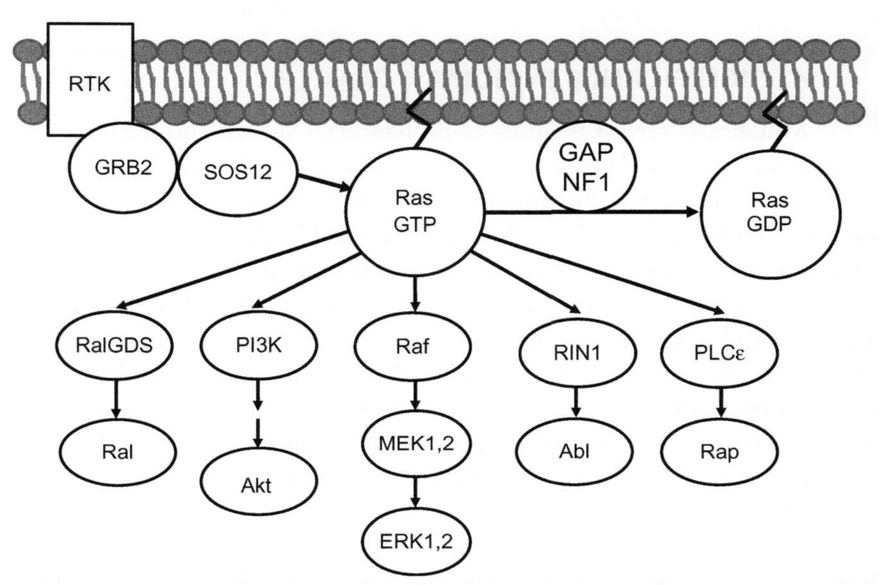

Figure 1.5 Diagram of RAS-GTP downstream effector proteins: RalGDS, PI3K, Raf, and RIN1. The diagram also shows RTK activation of membrane bound Ras, triggered by growth factors, cytokines, neurotransmitters, and hormones. (For color version of this figure, the reader is referred to the online version of this chapter.)

pathways have been extensively studied. First, Ras has been shown to interact with the N-terminal regulatory domain of Raf resulting in the activation of Raf kinase. This then leads to the activation of MEK kinase and ERK kinase. Another downstream effector is phosphatidylinositide 3-kinase (PI3K) that causes conversion of PIP2 to PIP3, resulting in the activation of AKT. Finally, Ras also activates RalGDS (Ral guanine nucleotide-dissociation stimulator), leading to the activation of Ral G-protein. Rac family proteins interact with PAK kinase and activate this enzyme. Rho proteins have multiple downstream effectors including Rho kinase that activates a number of reactions including phosphorylation of myosin light chain and regulation of actin cytoarchitecture and mDia that also aids this process. Rho proteins modulate the activity of many transcription factors including SRF, NFκB, c/EBPβ, Stat3, Stat5, FHL-2, PAX6, GATA-4, E2F, estrogen receptor α/β, CREB, and transcription factors that depend on the JNK and p38 MAP kinase pathways [42].

4. SIGNIFICANCE IN HUMAN CANCER

Mutations of the Ras family G-proteins have been identified in a variety of human cancer. The frequency is overall 20–30%. In pancreatic cancer, activation of K-ras was identified in more than 60% of cases. K-ras is the most frequently mutated Ras followed by N-ras [43]. These mutations occur mainly in the residue 12, 13, or 61, residues that are involved in the hydrolysis of GTP [9]. Thus, mutant proteins have decreased inherent GTPase activity. In addition, these mutant proteins are defective in GAP stimulation of GTPase activity. As a consequence, the proteins remain in the GTP-bound form leading to constitutive activation.

In contrast to Ras family proteins, no activating mutations of the Rho family proteins have been identified. Instead, overexpression of these proteins has been observed in a number of cases. Rho overexpression is known to cause mitosis-associated detachment from epithelial sheets, which is thought to aid in cancer dissemination and metastasis [44]. Gain-of-function mutations of GEF proteins have also been shown to be oncogenic, such as in the *VAV1* oncogene [45]. Overexpressed Rho GEF-H1 is transcriptionally activated by hPTTG1 (human pituitary tumor-transforming gene 1), thereby promoting breast cancer metastasis [46].

Perturbations of the signaling pathways upstream of Ras could lead to Ras activation. Mutations that lead to EGFR overexpression or overactivity have been associated with a number of cancers, including lung cancer and glioblastoma multiforme [47,48]. In this latter case, a more or less specific mutation of EGFR, called EGFRvIII, is often observed. Mutations, amplifications, or misregulations of EGFR or family members are implicated in about 30% of all epithelial cancers.

Activation of Ras has emerged as a mechanism of drug resistance. In the case of melanoma, recent development of B-Raf inhibitors, such as vemurafenib, has provided dramatic advances in clinical treatment [49]. Fifty percent of melanomas harbor activating (V600) mutations in the serine–threonine protein kinase B-Raf. However, resistance to B-Raf inhibitors has emerged as a major problem [50]. A number of the resistance mechanisms involve activation or upregulation of Ras proteins including mutations in N-Ras. In the case of lung cancer, inhibition of EGFR by the use of monoclonal antibody or small molecule inhibitors has been a promising approach. However, lung cancer harboring K-ras mutations is resistant to this therapy. Presence of K-ras mutation has been a predictive marker for the approach involving EGFR inhibition [51]. Recent studies on the Ras/Raf/MEK/ERK and PI3K/PTEN/AKT signaling cascades have shown that these signaling pathways play functional roles in malignant transformation and drug resistance of hematopoietic, breast, and prostate cancer cells.

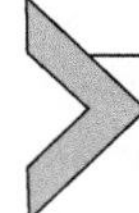

5. STRATEGIES TO INHIBIT THE Ras SUPERFAMILY G-PROTEINS

A variety of strategies have been developed to inhibit the Ras superfamily G-proteins. This is the topic of this and the following volume, including detailed discussions in each chapter of each approach. Here, we will discuss different inhibition methods briefly. The approaches can be grouped into four different categories.

The first approach can be termed "direct binders." Compounds that directly target Ras superfamily G-proteins have been identified. Using NMR-based screens, compounds have been identified that bind Ras. Deeper understanding of the structure and function of the Ras superfamily G-proteins using NMR led to the finding that there are two forms of GTP-bound Ras and that there is a dynamic equilibrium between the two forms. Compounds that shift the equilibrium have been identified and shown to inhibit the Ras signaling. The second approach is to block the activation of Ras by interfering with the action of GEF. Some of the compounds that are found to bind directly to G-proteins interact with the protein at the site where GEF interacts. Screens to find G-protein/GEF interaction inhibitors led to the identification of novel compounds. The third approach is to target protein–protein interactions between G-protein and its downstream target. Finally, approaches to inhibit membrane association of the Ras superfamily G-proteins have been explored. This approach was prompted by the discovery that a set of enzymes, prenyltransferases, catalyze posttranslational modification events that result in increased affinity for membranes. Small molecule inhibitors of protein prenyltransferases have been developed. This approach was taken in the 1990s that led to the development of farnesyltransferase inhibitors (FTIs) that exhibited promising activity in animal studies and were evaluated in clinical trials. However, there was a shortcoming in this approach, as FTIs were incapable of inhibiting prenylation and membrane association of K-Ras and N-Ras proteins due to their alternative prenylation; these proteins can be geranylgeranylated in the presence of FTIs. Because a major portion of Ras-driven cancer involves K-Ras and N-Ras, FTIs are not effective for the therapy of human cancer. To overcome this problem, small molecule inhibitors of protein GGTIs have been developed. GGTIs can be combined with FTIs to inhibit N-Ras and K-Ras. Another strategy was to inhibit postprenylation events that include proteolytic removal of three C-terminal amino acids as well as methylation of C-terminal cysteine. Inhibitors of the enzymes catalyzing these events have been developed. In addition, further studies on membrane association of Ras family proteins are ongoing. High throughput screening for inhibitors of membrane association of K-Ras protein led to the identification of small molecule inhibitors. In addition, proteins acting as chaperons for Ras have been characterized.

ACKNOWLEDGMENTS

We thank members of Tamanoi Lab for discussion. This work is supported by NIH grant CA41996.

REFERENCES

[1] K. Wennerberg, K.L. Rossman, C.J. Der, The Ras superfamily at a glance, J. Cell Sci. 118 (2005) 843–846.
[2] M. Nakafuku, T. Satoh, Y. Kaziro, Differentiation factors, including nerve growth-factor, fibroblast growth-factor, and interleukin-6, induce an accumulation of an active ras. Gtp complex in rat pheochromocytoma PC12 cells, J. Biol. Chem. 267 (1992) 19448–19454.
[3] K. Muroya, S. Hattori, S. Nakamura, Nerve growth factor induces rapid accumulation of the GTP-bound form of p21ras in rat pheochromocytoma PC12 cells, Oncogene 7 (1992) 277–281.
[4] P.J. Aspuria, F. Tamanoi, The Rheb family of GTP-binding proteins, Cell. Signal. 16 (2004) 1105–1112.
[5] S.J. Heasman, A.J. Ridley, Mammalian Rho GTPases: new insights into their functions from in vivo studies, Nat. Rev. Mol. Cell Biol. 9 (2008) 690–701.
[6] K. Schlessinger, A. Hall, N. Tolwinski, Wnt signaling pathways meet Rho GTPases, Genes Dev. 23 (2009) 265–277.
[7] M. Zerial, H. McBride, Rab proteins as membrane organizers, Nat. Rev. Mol. Cell Biol. 2 (2001) 107–117.
[8] K. Scheffzek, C. Klebe, K. Fritzwolf, W. Kabsch, A. Wittinghofer, Crystal-structure of the nuclear Ras-related protein Ran in its Gdp-bound form, Nature 374 (1995) 378–381.
[9] T.E. Dever, M.J. Glynias, W.C. Merrick, GTP-binding domain: three consensus sequence elements with distinct spacing, Proc. Natl. Acad. Sci. U.S.A. 84 (1987) 1814–1818.
[10] K. Del Villar, D. Dorin, I. Sattler, J. Urano, P. Poullet, et al., C-terminal motifs found in Ras-superfamily G-proteins: CAAX and C-seven motifs, Biochem. Soc. Trans. 24 (1996) 709–713.
[11] H.R. Bourne, D.A. Sanders, F. McCormick, The GTPase superfamily: conserved structure and molecular mechanism, Nature 349 (1991) 117–127.
[12] E.F. Pai, U. Krengel, G.A. Petsko, R.S. Goody, W. Kabsch, et al., Refined crystal-structure of the triphosphate conformation of H-Ras P21 at 1.35 a resolution—implications for the mechanism of Gtp hydrolysis, EMBO J. 9 (1990) 2351–2359.

[13] Y. Wei, Y. Zhang, U. Derewenda, X. Liu, W. Minor, et al., Crystal structure of RhoA-GDP and its functional implications, Nat. Struct. Biol. 4 (1997) 699–703.
[14] K. Rittinger, P.A. Walker, J.F. Eccleston, K. Nurmahomed, D. Owen, et al., Crystal structure of a small G protein in complex with the GTPase-activating protein rhoGAP, Nature 388 (1997) 693–697.
[15] Y. Zhu, L. Hu, Y. Zhou, Q. Yao, L. Liu, et al., Structural mechanism of host Rab1 activation by the bifunctional Legionella type IV effector SidM/DrrA, Proc. Natl. Acad. Sci. U.S.A. 107 (2010) 4699–4704.
[16] S.E. Greasley, H. Jhoti, C. Teahan, R. Solari, A. Fensome, et al., The structure of rat ADP-ribosylation factor-1 (ARF-1) complexed to GDP determined from two different crystal forms, Nat. Struct. Biol. 2 (1995) 797–806.
[17] M. Stewart, H.M. Kent, A.J. McCoy, Structural basis for molecular recognition between nuclear transport factor 2 (NTF2) and the GDP-bound form of the Ras-family GTPase Ran, J. Mol. Biol. 277 (1998) 635–646.
[18] Y. Yu, S. Li, X. Xu, Y. Li, K. Guan, et al., Structural basis for the unique biological function of small GTPase RHEB, J. Biol. Chem. 280 (2005) 17093–17100.
[19] A. Yanuar, S. Sakurai, K. Kitano, T. Hakoshima, Crystal structure of human Rad GTPase of the RGK-family, Genes Cells 11 (2006) 961–968.
[20] N. Nassar, G. Horn, C. Herrmann, A. Scherer, F. McCormick, et al., The 2.2 A crystal structure of the Ras-binding domain of the serine/threonine kinase c-Raf1 in complex with Rap1A and a GTP analogue, Nature 375 (1995) 554–560.
[21] M. Paduch, F. Jelen, J. Otlewski, Structure of small G proteins and their regulators, Acta Biochim. Pol. 48 (2001) 829–850.
[22] R. Thapar, J.G. Williams, S.L. Campbell, NMR characterization of full-length farnesylated and non-farnesylated H-Ras and its implications for Raf activation, J. Mol. Biol. 343 (2004) 1391–1408.
[23] M.J. Smith, B.G. Neel, M. Ikura, NMR-based functional profiling of RASopathies and oncogenic RAS mutations, Proc. Natl. Acad. Sci. U.S.A. 110 (2013) 4574–4579.
[24] R.B. Fenwick, S. Prasannan, L.J. Campbell, D. Nietlispach, K.A. Evetts, et al., Solution structure and dynamics of the small GTPase RalB in its active conformation: significance for effector protein binding, Biochemistry 48 (2009) 2192–2206.
[25] J.L. Bos, H. Rehmann, A. Wittinghofer, GEFs and GAPs: critical elements in the control of small G proteins, Cell 129 (2007) 865–877.
[26] A. Scrima, C. Thomas, D. Deaconescu, A. Wittinghofer, The Rap-RapGAP complex: GTP hydrolysis without catalytic glutamine and arginine residues, EMBO J. 27 (2008) 1145–1153.
[27] X. Pan, S. Eathiraj, M. Munson, D.G. Lambright, TBC-domain GAPs for Rab GTPases accelerate GTP hydrolysis by a dual-finger mechanism, Nature 442 (2006) 303–306.
[28] M.J. Seewald, C. Korner, A. Wittinghofer, I.R. Vetter, RanGAP mediates GTP hydrolysis without an arginine finger, Nature 415 (2002) 662–666.
[29] H. Sondermann, S.M. Soisson, S. Boykevisch, S.S. Yang, D. Bar-Sagi, et al., Structural analysis of autoinhibition in the Ras activator Son of sevenless, Cell 119 (2004) 393–405.
[30] J. Yang, Z. Zhang, S.M. Roe, C.J. Marshall, D. Barford, Activation of Rho GTPases by DOCK exchange factors is mediated by a nucleotide sensor, Science 325 (2009) 1398–1402.
[31] S.M. Soisson, A.S. Nimnual, M. Uy, D. Bar-Sagi, J. Kuriyan, Crystal structure of the Dbl and pleckstrin homology domains from the human Son of sevenless protein, Cell 95 (1998) 259–268.
[32] A.B. DiracSvejstrup, T. Sumizawa, S.R. Pfeffer, Identification of a GDI displacement factor that releases endosomal Rab GTPases from Rab-GDI, EMBO J. 16 (1997) 465–472.

[33] T. Soldati, M.A. Riederer, S.R. Pfeffer, Rab GDI: a solubilizing and recycling factor for rab9 protein, Mol. Biol. Cell 4 (1993) 425–434.
[34] I.M. Ahearn, K. Haigis, D. Bar-Sagi, M.R. Philips, Regulating the regulator: post-translational modification of RAS, Nat. Rev. Mol. Cell Biol. 13 (2012) 39–51.
[35] M.H. Gelb, L. Brunsveld, C.A. Hrycyna, S. Michaelis, F. Tamanoi, et al., Therapeutic intervention based on protein prenylation and associated modifications, Nat. Chem. Biol. 2 (2006) 518–528.
[36] D. Michaelson, J. Silletti, G. Murphy, P. D'Eustachio, M. Rush, et al., Differential localization of Rho GTPases in live cells: regulation by hypervariable regions and RhoGDI binding, J. Cell Biol. 152 (2001) 111–126.
[37] P.A. Konstantinopoulos, M.V. Karamouzis, A.G. Papavassiliou, Post-translational modifications and regulation of the RAS superfamily of GTPases as anticancer targets, Nat. Rev. Drug Discov. 6 (2007) 540–555.
[38] A.B. Vojtek, C.J. Der, Increasing complexity of the Ras signaling pathway, J. Biol. Chem. 273 (1998) 19925–19928.
[39] B. Aghazadeh, K. Zhu, T.J. Kubiseski, G.A. Liu, T. Pawson, et al., Structure and mutagenesis of the Dbl homology domain, Nat. Struct. Biol. 5 (1998) 1098–1107.
[40] X.H. Liu, H. Wang, M. Eberstadt, A. Schnuchel, E.T. Olejniczak, et al., NMR structure and mutagenesis of the N-terminal Dbl homology domain of the nucleotide exchange factor trio, Cell 95 (1998) 269–277.
[41] D.K. Worthylake, K.L. Rossman, J. Sondek, Crystal structure of Rac1 in complex with the guanine nucleotide exchange region of Tiam1, Nature 408 (2000) 682–688.
[42] S. Aznar, J.C. Lacal, Rho signals to cell growth and apoptosis, Cancer Lett. 165 (2001) 1–10.
[43] I.A. Prior, P.D. Lewis, C. Mattos, A comprehensive survey of Ras mutations in cancer, Cancer Res. 72 (2012) 2457–2467.
[44] J.M. Vasiliev, T. Omelchenko, I.M. Gelfand, H.H. Feder, E.M. Bonder, Rho overexpression leads to mitosis-associated detachment of cells from epithelial sheets: a link to the mechanism of cancer dissemination, Proc. Natl. Acad. Sci. U.S.A. 101 (2004) 12526–12530.
[45] J. Han, K. Luby-Phelps, B. Das, X. Shu, Y. Xia, et al., Role of substrates and products of PI 3-kinase in regulating activation of Rac-related guanosine triphosphatases by Vav, Science 279 (1998) 558–560.
[46] Y.C. Liao, J.W. Ruan, I. Lua, M.H. Li, W.L. Chen, et al., Overexpressed hPTTG1 promotes breast cancer cell invasion and metastasis by regulating GEF-H1/RhoA signalling, Oncogene 31 (2012) 3086–3097.
[47] Y. Yatabe, T. Takahashi, T. Mitsudomi, Epidermal growth factor receptor gene amplication is acquired in association with tumor progression of EGFR-mutated lung cancer, Cancer Res. 68 (2008) 2106–2111.
[48] C.T. Kuan, C.J. Wikstrand, D.D. Bigner, EGF mutant receptor vIII as a molecular target in cancer therapy, Endocr. Relat. Cancer 8 (2001) 83–96.
[49] G. Bollag, P. Hirth, J. Tsai, J. Zhang, P.N. Ibrahim, et al., Clinical efficacy of a RAF inhibitor needs broad target blockade in BRAF-mutant melanoma, Nature 467 (2010) 596–599.
[50] R. Nazarian, H. Shi, Q. Wang, X. Kong, R.C. Koya, et al., Melanomas acquire resistance to B-RAF(V600E) inhibition by RTK or N-RAS upregulation, Nature 468 (2010) 973–977.
[51] R. Califano, L. Landi, F. Cappuzzo, Prognostic and predictive value of K-RAS mutations in non-small cell lung cancer, Drugs 72 (Suppl 1) (2012) 28–36.
[52] J. Colicelli, Human RAS superfamily proteins and related GTPases, Sci. STKE 2004 (2004) RE13.
[53] J.F. Hancock, Ras proteins: different signals from different locations, Nat. Rev. Mol. Cell Biol. 4 (2003) 373–384.

CHAPTER TWO

NMR Study to Identify a Ligand-Binding Pocket in Ras

Till Maurer[1], Weiru Wang[1]
Department of Structural Biology, Genentech Inc., One DNA Way, South San Francisco, California, USA
[1]Corresponding authors: e-mail address: maurer.till@gene.com; wang.weiru@gene.com

Contents

Abstract

Despite decades of intense drug discovery efforts, to date no small molecules have been described that directly bind to Ras protein and effectively antagonize its function. In order to identify and characterize small-molecule binders to KRas, we carried out a fragment-based lead discovery effort. A ligand-detected primary nuclear magnetic resonance (NMR) screen identified 266 fragments from a library of 3285 diverse compounds. Protein-detected NMR using isotopically labeled KRas protein was applied for hit validation and binding site characterization. An area on the KRas surface emerged as a consensus site of fragment binding. X-ray crystallography studies on a subset of the hits elucidated atomic details of the ligand–protein interactions, and revealed that the consensus site comprises a shallow hydrophobic pocket. Comparison among the crystal structures indicated that the ligand-binding pocket is flexible and can be expanded upon ligand binding. The identified ligand-binding pocket is proximal to the protein–protein interface and therefore has the potential to mediate functional effects. Indeed,

The Enzymes, Volume 33
ISSN 1874-6047
http://dx.doi.org/10.1016/B978-0-12-416749-0.00002-6

some ligands inhibited SOS1-dependent nucleotide exchange, although with weak potency. Several Ras ligands have been published in literature, the majority of which were discovered using NMR-based methods. Mapping of the ligand-binding sites revealed five areas on Ras with a high propensity for ligand binding and the potential of modulating Ras activity.

1. KRAS AS A DRUG TARGET

Ever since Ras was identified as a human oncogene more than 30 years ago [1–8], it has remained on the center stage of cancer research (reviews [9,10]). This is primarily due to the prevalence of Ras mutations found in human cancers. Over 20% of tumors express some kind of oncogenic Ras (Sanger website, http://www.sanger.ac.uk/genetics/CGP/cosmic). Among the three Ras isoforms (H, N, and K), KRas mutations constitute the most common aberrations and occur in three of the top four most lethal neoplasms in the United States: lung, colon, and pancreatic cancer [11]. Ras-driven cancers are associated with poor prognosis [12]. Supporting evidence comes from experiments in animal models where inactivation of oncogenic Ras results in tumor shrinkage [13–15]. Thus, inhibition of oncogenic Ras is considered a promising strategy for pharmacological intervention [16,17].

Ras is a key regulator in multiple cellular processes leading to survival, proliferation, and differentiation [12,18]. Through posttranslational modifications [17], Ras proteins are associated with the cell membrane [19,20], where they facilitate the complex interplay between extracellular stimuli and intracellular signaling cascades. Upon activation, Ras attracts effector proteins, such as Raf [21–24] and PI3K [25], to the inner leaf of plasma membrane where they can transmit signal downstream. Withdrawal of extracellular growth hormone stimulation results in deactivation of Ras.

Ras is the founding member of the guanine nucleotide-binding protein (G-protein) superfamily [26–28]. It alternates between the GTP-bound "on" state and GDP-bound "off" state. The nucleotide exchange is predominately a passive process and is controlled by cofactor proteins [18,29]. The Ras^{GDP} to Ras^{GTP} switch is catalyzed by guanine nucleotide exchange factors (GEFs), notably SOS1 [30]. The Ras^{GTP} to Ras^{GDP} switch is achieved by hydrolysis of the bound GTP. Ras itself possesses a low level of intrinsic GTPase activity [31], not sufficiently fast to facilitate *in vivo* switching. The GTPase activity is enhanced by a factor of 300 through interaction with GTPase-activating proteins (GAPs), such as GAP1 [32]. Oncogenic mutations tend to impair the intrinsic GTPase activity [33–37] and make Ras insensitive to GAP stimulation, thereby prolonging the activated state.

The natural substrate-binding characteristics dictate that Ras lacks the classical Pauling-type ligand [38,39] as GTP, the transition state, and the hydrolysis product GDP all bind Ras with picomolar affinity [18]. Furthermore, the concentration of GTP in the living cell would necessitate a low femto- to attomolar binder to compete with endogenous nucleotide. That ligand would also have to mimic the nonactive state of Ras while being selective to the constantly activated, oncogenic form of the protein. Directly competing against GTP being an infeasible path, an alternate strategy is to identify inhibitors that will bind in the regions outside the nucleotide-binding site of Ras and either modulate Ras activity utilizing an allosteric mechanism or interfering with protein–protein interactions (PPI) by binding to the regions involved with complex formation. Those regions appear relatively flat, and surface mapping indicates a lack of preexistent pockets for small-molecule binding. It is therefore challenging to find small molecules that bind Ras and can serve as a starting point for medicinal chemistry elaboration by traditional activity-based screening methods. As several protein players are involved in Ras signaling cascades, PPI is an essential part of Ras function. Hence, we think it is a more accurate description to classify Ras as a PPI target, and as such fragment-based lead discovery (FBLD) seems to be a viable path forward as successful examples have shown in this methodology [40].

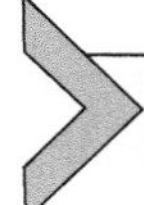

2. NUCLEAR MAGNETIC RESONANCE FRAGMENT SCREEN

2.1. Introduction

FBLD [41] has emerged as an alternate path to small-molecule inhibitors of proteins implicated in the cause and development of diseases. This is largely facilitated by our rapidly growing knowledge of the molecular basis of disease but also due to technical advances in the field of structural biology including X-ray crystallography and nuclear magnetic resonance (NMR) spectroscopy. To date, the modern drug discovery paradigm still heavily depends on high-throughput screening (HTS) as a primary method to find active compounds as starting points [42,43]. Using these methods, the pharmaceutical industry has been particularly successful on a group of functionally relevant proteins with well-defined substrate-binding sites, such as kinases and proteases. Detailed structural understanding of the ligand–protein interactions, primarily through X-ray crystallography, facilitates chemistry elaboration and rapid increase in potency and selectivity. The sensitivity of HTS, however, is limited because it generally employs a biochemical assay, for instance using substrate turnover as readout, to identify molecules that interfere with protein function. Also, an optimal HTS library is dominated by lead-like molecules with molecular weights above 300 Da.

Another path to the discovery of small-molecule inhibitors of functionally important proteins is to engage purely biophysically driven screening methods. Here, the objective is not necessarily to find "active" compounds *per se*, but to identify possible binding sites on the protein surface together with the corresponding binders. The question of activity would be addressed later at the fragment hit-to-lead (FH2L) stages. FBLD focuses on weakly binding fragment compounds [44–46] with molecular weights (~120–250 Da) around half that of traditional HTS screening library compounds. These small chemical entities can exploit less well-defined binding sites on the protein surface and have potential in finding allosteric sites or targeting PPIs. Biophysical methods such as NMR [47] or surface plasmon resonance [48] offer high sensitivity for the detection of fragment binding. Aside from the identification of ligands for a known binding site using ligand-based NMR methods, a fragment-based screen can also reveal novel binding sites in proteins when protein-based methods can be applied. Combined with the structure-based design, successful drug candidates can be derived. Examples include the Bcl-2 [49] and BACE1 [50] inhibitors.

Those enzymes that appear to lack any small-molecule-binding pocket that can be utilized for functional inhibition present a challenge to small-molecule drug discovery. Strategies of using broader biophysical methods to find binders first and then engage a mode of action or enzymatic mechanism later in the process are well suited to tackle such targets. NMR is a very sensitive means to monitor molecular interactions between proteins and small-molecule ligands. Ligand-based NMR methods rely on through-space interactions of the ligand with the target protein. When the ligand is in close proximity (under 6 Å), transfer of polarization between ligand and target protein can take place. When one of the interaction partners, in our case the protein, is saturated by selective irradiation, the magnitude of the ligand's polarization is modulated. Comparing the protein-saturated spectrum with the nonprotein-saturated spectrum, the saturation transfer difference (STD) [51] can be used to differentiate binders from nonbinders. The underlying process is comparable to other energy transfer methods such as fluorescence resonance energy transfer [52], but NMR offers superior spectral resolution with each chemically equivalent ligand atom, in our case the protons, resolved and leading to a separate NMR signal with a characteristic chemical shift. In the difference spectrum, this results in a signal that is proportional to the distance from the ligand proton to the protein surface. The resolution together with the knowledge of ligand-specific chemical shift pattern allows simultaneous measurement of multiple compounds and the identification of binders from a mixture of compounds. A similar method, water-LOGSY (ligand observed gradient spectroscopy) [53], uses an indirect transfer via the bulk solvent in the sample.

2.2. The design of NMR fragment-screening cascade against KRas4B G12D mutant

NMR fragment-screening cascade, in general, starts with ligand-based methods, such as STD, water-LOGSY, or a combination of them. These methods can detect ligand binding (or better ligand-target proximity) with affinity ranging from tens of nanomolar to hundreds of millimolar. Such a large dynamic range is far beyond that of biochemical assays and most other biophysical methods. The only drawback in using ligand-detected NMR is that the signal is only about compound binding without any discrimination of binding site, a question that is addressed in the confirmation step using isotopically labeled protein.

KRas^{GTP} and KRas^{GDP} are in distinct structural states. No prior knowledge could suggest whether the primary screen should bias toward either one. In addition, fragments binding to either of the two states would be of interest at this stage. We decided to screen the 1:1 mixture of diphosphate and triphosphate analogue-loaded forms of KRas protein. The primary screen will most likely identify both specific binders to particular epitopes of KRas as well as nonspecific binders that will transiently interact with multiple sites without a consistent orientation. The number of primary binders tends to be high (in the order of 30% of the library) and a focus on specific binding sites is necessary in the triage process.

In order to identify binders to a specific site on the target protein, we chose to use protein-based NMR methods such as 2D ^{15}N heteronuclear single quantum coherence (HSQC) [54] with isotopically labeled protein. These methods allowed us to measure changes in the "chemical environment" of NMR active nuclei (e.g., ^{15}N in the backbone amide group). Those changes can be the consequence of interactions with other molecules, such as a fragment compound. Therefore, following the chemical shift changes provides a sensitive means of monitoring molecular interactions with spatial resolution on the protein side. With the assignments of KRas available in the public domain (entry no. 17785, BMRB [55]), the ligand-binding site could be identified by mapping the residues with observed changes in the 2D NMR spectrum onto a KRas structure.

2.3. The fragment library

A high-quality fragment library is the most important component for a fragment screen. The fragment-screening library was designed to represent a diverse collection of low molecular-weight compounds from commercially available sources. The compounds were selected to contain 16 or fewer non-hydrogen atoms as the main exclusion criterion. The consequence of this was that the compounds had a molecular weight of less than 300. The second filter was for the molecules to have a partitioning coefficient (clogP) of less than 3.0 (to minimize artifacts from poor solubility) and to contain at least one ring and three or fewer rotatable bonds. Further criteria that were applied included the sum of all nitrogen and oxygen atoms greater than 2 and less than 5, a polar surface area of less than 80, no more than 2 hydrogen bond donors, and no more than 4 hydrogen bond acceptors, bicycles, single- or bicyclic rings that were connected by either 1, 2, or 3 bonds. These criteria are similar in principal to those defined by the "rule of three" proposed by Congreve and colleagues [56]. The library was then triaged experimentally in two steps. Firstly, the structural integrity of each compound was confirmed using classical NMR methods. Secondly, we assessed the solubility and aggregation behavior at increasing ligand concentrations using 1D NMR and STD spectra. As a rule, all measured STD intensities should be well under 15% as defined by the ratio between on- and off-resonance 1D spectra [57]. This step eliminated promiscuous binding and aggregated fragments. It also improved the library quality for subsequent X-ray crystallography, where high compound concentration is required. The final step was to formulate the compound mixtures for primary screening. The above-described properties used for the selection of each compound in the fragment library were used to calculate the three top principal components for these properties in the fragment set. A compound identified being close to the center of the principal component analysis plot was defined as a reference compound. The Diverse Subset algorithm with MACCS Structural Keys fingerprints as implemented in the program MOE (Molecular Operating Environment; Chemical Computing Group, Canada) was then applied using the reference compound as a seed. In this way, we obtained a sorted list. Each fragment in a given mixture had a large chemical diversity and in consequence a low probability of having an overlapping NMR spectrum. A spread of diversity across the screening library also allows estimates of the overall hit rate from the first mixture experiments.

Mixtures of five fragments were made from 100 mM d6-DMSO stock solutions and stored in 384-well plates. Reference 1D as well as STD spectra were recorded for each mixture without addition of protein. This step was necessary because although the single compounds were tested for aggregation, we found mixtures that appear to have formed aggregates based on the composition of that mixture. Any form of compound aggregation leads to intense STD signals and thus would produce a false positive in the primary screen.

2.4. The ligand-based primary screen

The screen was carried out following the workflow depicted in Fig. 2.1. The fragment library of 3285 compounds was screened in mixtures of five compounds per measurement. Each compound in the mixture was at a concentration of 250 μM. The ligand-based spectra were recorded in the presence of 10 μM KRas (1–169) half (5 μM) of which was loaded with GDP ($KRas^{GDP}$) and the other half (5 μM) loaded with GMPPCP ($KRas^{GTP}$)

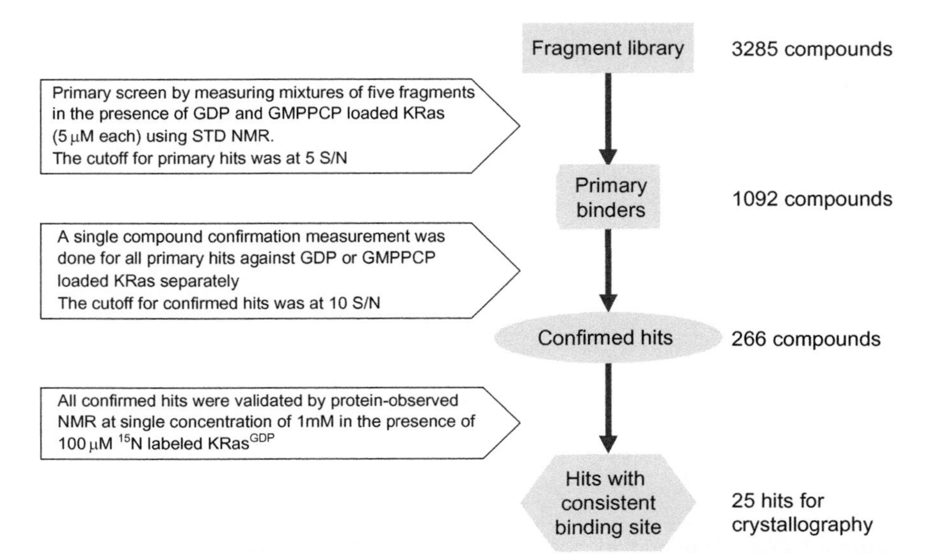

Figure 2.1 The NMR screen workflow in the discovery of KRas binders. *Adapted from Ref. [58].* (For color version of this figure, the reader is referred to the online version of this chapter.)

using the nucleotide exchange protocol published previously [58]. The sample buffer contains 150 mM NaCl, 20 mM d6-TRIS, and 10 mM $MgCl_2$ in 99% $2H_2O$ at pH 8.0. The d6-DMSO concentration was maintained at ca. 5%.

The samples were automatically generated by "just-in-time" sample preparation protocol using a TECAN EVO pipetting robot with a custom-made 1 mm pipetting needle. Samples in 3 mm NMR tubes were transported to a 600 MHz Bruker spectrometer by a Bruker Sample Rail system. Fragment binding was monitored using STD NMR spectroscopy by measuring 256 on- and 256 off-resonance transients interleaved at 284 K in 3 mm sample tubes with a sample volume of 180 μL. Figure 2.2 shows an example set of binding spectrum and reference data from a typical 5-compound mixture.

Binders were identified by their NMR chemical shifts in the STD spectra by comparison to the 1D reference spectra recorded under similar buffer

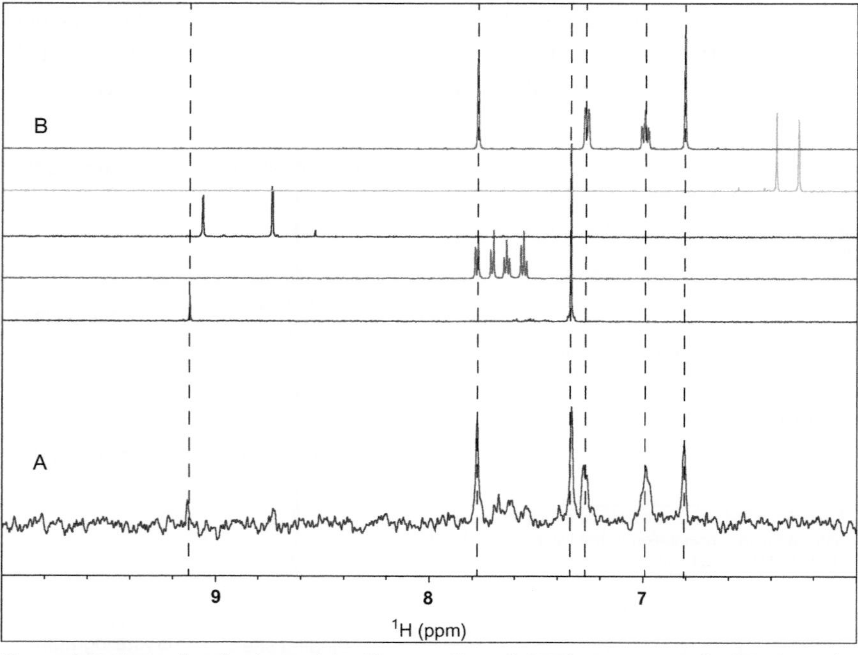

Figure 2.2 Example of primary hits. The top five 1D NMR spectra marked B show the reference data of the mixture. The bottom trace marked A shows the STD difference spectrum. The binders from this mixture are identified based on the chemical shifts of the ligand protons (dotted lines). *Adapted from Ref. [58].* (See color plate.)

conditions (the fragment library was characterized in PBS at pH 7.2). Compounds that showed a signal-to-noise ratio greater than 5 in the STD spectrum were selected as primary hits. This resulted in the sum of 1092 fragments identified as primary binders corresponding to a hit rate of 36.4%. All the primary hits were subsequently remeasured on their own in two separate samples containing either di- or triphosphate nucleotide loaded KRas. We applied a more stringent cutoff at this stage and selected those compounds that showed signal to noise of greater than 10 in the STD spectrum. This resulted in 266 confirmed hits or a hit rate 8.8% of the primary library.

2.5. Hit validation using protein-observed NMR

In the next step of the screening cascade, all 266 primary hits were subjected to further validation in a protein-observed experiment. (Figure 2.3 shows a comparison of the reference with the spectra of three different compounds.) We measured amide–nitrogen–proton correlations in a ^{15}N HSQC spectrum [59]. The objective of applying this process was to look for changes in the ^{15}N HSQC spectrum as an indicator of ligand binding. In the first round of experiments, the fragments were added at a single concentration, usually 1 mM, into a sample containing 100 μM ^{15}N labeled KRas^{GDP}. Kraulis et al. showed that the KRas structure becomes more dynamic upon GTP loading, which leads to extensive line broadening [60]. We noted that the line broadening leads to disappearance of signals particularly in the important switch regions and in the ligand-binding site. Therefore, KRas^{GDP} was used here for its higher quality ^{15}N HSQC spectra compared to that of KRas^{GTP}.

We assigned the backbone amide resonances of KRas based on data from the published assignments on HRas [61]. For the nonidentical amino acid residues in KRas, we confirmed the assignment by triple resonance experiments HNCA and HN(CO)CACB using ^{13}C/^{15}N labeled protein and ^{15}N NOESY and 2D HSQC spectra. Our assignments were later confirmed by the KRas assignments with the BMRB entry number 17785.

Hit validation data were analyzed by comparing the chemical shifts of the ^{15}N HSQC spectra in the presence and absence of ligand. The ligand-imparted chemical shift perturbations were grouped by their propensity to shift certain residues and the magnitude of the shift scaled to the base frequencies of the corresponding dimension [62]. Interestingly, most primary hits perturbed only very few signals and were not followed up. Twenty-five

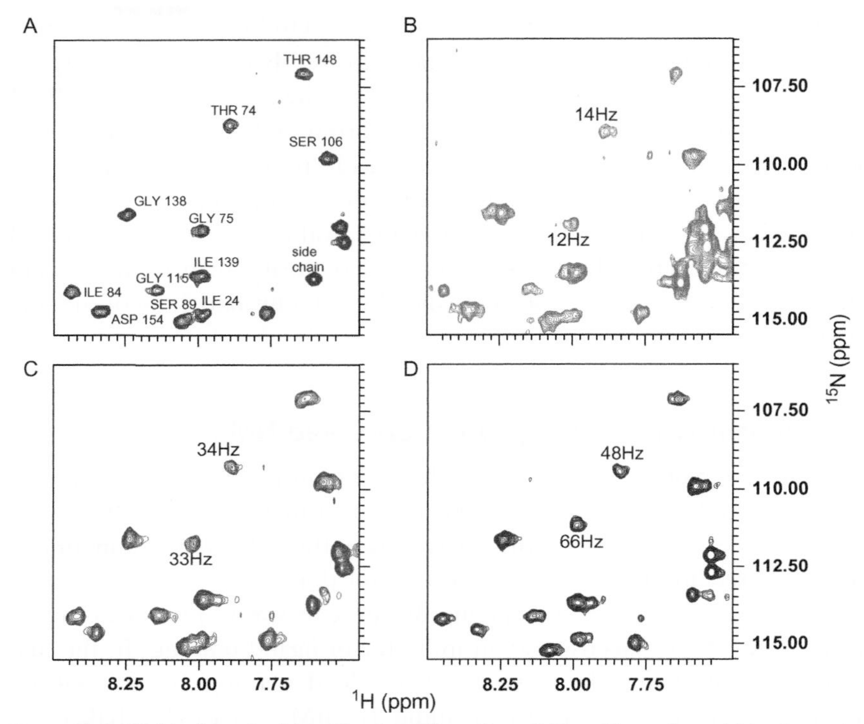

Figure 2.3 2D binding site identification of confirmed hits. All confirmed hits from the proton-based ligand-detected experiments are measured against 100 μM ^{15}N-labeled KRas. Shifts indicated that some of these bind to an epitope around residues T74/G75 and L56/D57. For clarity, the region showing the amide resonance peaks of T74/G75 is shown. Panel (A) is the ligand-free reference spectrum, (B) 2 mM BZDM, (C) 2 mM BZIM, and (D) 2 mM DCAI represent three different primary fragment hits. The magnitude of the shift perturbations is given in Hz. *Adapted from Ref. [58].* (See color plate.)

compounds shifted a similar set of resonances. These were the amide proton/nitrogen resonance signals of residues T74, G75, L56, and D57 (Fig. 2.3). Although other shift patterns were observed with various fragments (e.g. D124 and H95), they could not be mapped to a contiguous surface of the crystal structure. In contrast, the four most prevalently shifted residues flanked a shallow but distinct pocket in the surface opposite to the switch region depicted in Fig. 2.4. In our original analysis using the crystal structure coordinates of the GDP-bound form of KRas (PDB code 3GFT), this pocket was occupied by the side chain of Y71 and was only present in the PDB entry 1CRP, indicating that this side chain is mobile. It is also involved in a hydrogen bond upon complex formation with SOS1.

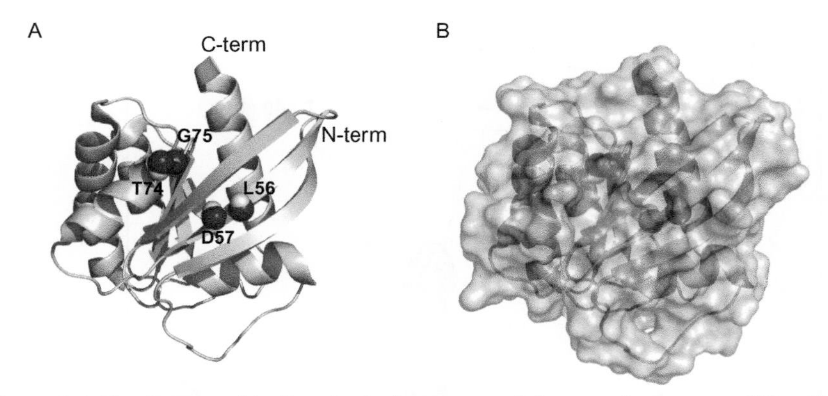

Figure 2.4 The fragment binding site in the context of the overall structure of KRas. The cross peaks stemming from the amide groups are mapped onto the X-ray structure of KRas. Panel (A) shows their position with respect to the backbone. Panel (B) has the protein surface superimposed and clearly shows that there is a pocket centered around the observed shifts. (See color plate.)

A mutant KRas with this residue substituted for Ala did not show modulated behavior in the KRas/SOS1 interaction assay (data not shown), leaving room to speculate on the functional importance of this residue and the position of its side chain.

2.6. FH2L using K_d determinations based on protein-observed NMR

The final NMR step in the workflow entailed K_d determinations by a 5-point compound titration and monitoring the magnitude of chemical shift changes. The plot of these changes induced by compound addition against the ligand concentration and a nonlinear regression curve fit of the shift values using a single-site tight-binding model allowed the calculation of K_d values and errors. The binding model considered compound depletion. The depiction of an example of compound concentration versus the combined ^{1}H and ^{15}N chemical shift change to a single-site binding function resulting in the binding K_d value of the fragment is shown in Fig. 2.5. As all followed-up fragments shifted multiple residues, this analysis could be done with two to five resonance signals, and an average K_d could be reported. The K_d determinations were important in the hit-to-lead process, as at this time no biochemical readout was available for compounds with the low binding affinities of between 10 and 1000 μM associated with fragment ligands. With an upper limit due to ligand solubility and a lower limit due to

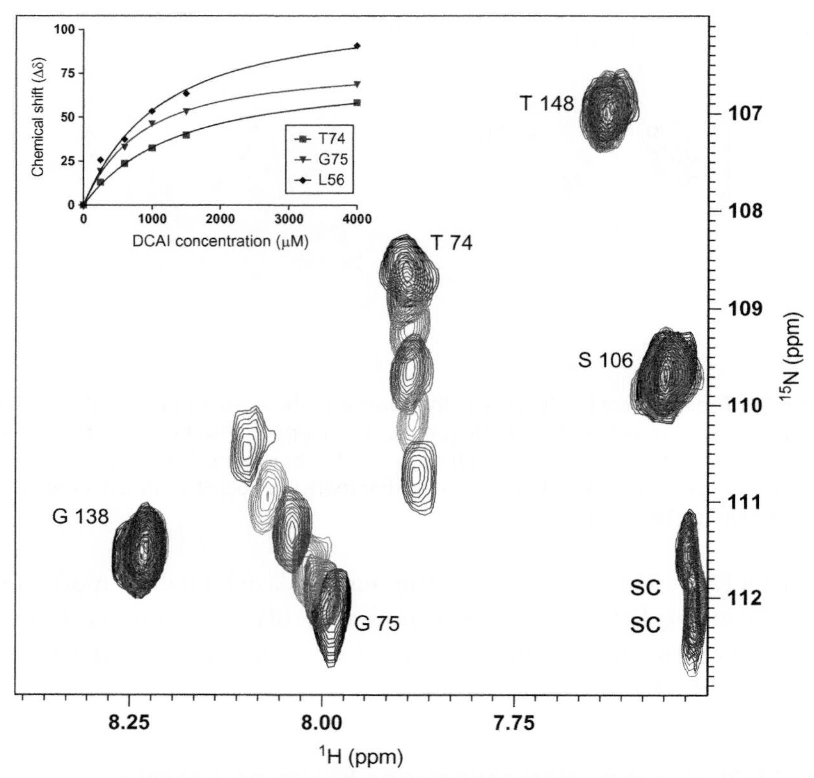

Figure 2.5 K_d measurements. K_d values for the fragment hits are determined by titration of increasing fragment concentration (0.125, 0.25, 0.5, 1.0, and 2.0 mM) into ^{15}N-labeled KRas at 100 μM. K_d values are calculated by a nonlinear fit using a single-site tight-binding model in GraphPad Prism. *Adapted from Ref. [58].* (See color plate.)

intermediate exchange, this process was applicable for an affinity range from single-digit millimolar to double-digit micromolar. These values are often not covered in the affinity range of a typical HTS-based assay but very well matched for fragment-based discovery. This final workflow together with chemical modification and X-ray crystallography brings the primary fragment hit to affinity values where compounds start to show activity in a biochemical assay often with a threshold value of around 30 μM and where the "normal" hit-to-lead machinery would take over.

The FH2L process is generally guided by the information from co-crystal structures. Alternate scaffolds are consequently either synthesized or found at a commercial source to determine SAR and identify compounds with greater binding affinity. As a first pass, all repurchased or newly synthesized compounds were measured for solubility and aggregation using NMR. Compounds identified as soluble above 500 μM were subsequently measured at 1 mM in the presence of 100 μM ^{15}N-labeled KRas. As shown in Fig. 2.3, the magnitude of the chemical shift perturbations obviously changed in different compound series. Although not applicable in a quantitative fashion, this was used as a criterion to prioritize compounds for analysis in the last NMR step of the workflow, the 5-point ligand titration. Additional residues beyond the residues T74, G75, L56, and D57 started to show chemical shift perturbations, for example, Asp 54 and Leu 53. This was seen as an indication of additional interactions and thus a potential increase in binding affinity, in particular, because NMR observe amide groups are directly involved in polar interactions. Both shifts magnitude and additional shift were used to prioritize affinity determinations, which were in turn used as a criterion for compound prioritization in X-ray crystallography. The K_d value of DCIA was determined to be approximately 1.1 mM ± 0.2 mM.

3. PROTEIN LIGAND CO-STRUCTURES

3.1. The crystal structure of KRas4B^{G12D} (KRas$_m$)

Co-crystal structure determination is a critical step in fragment-based screens. It provides the ultimate confirmation ofligand binding to the target protein in a specific mode. The atomic resolution details of the ligand–protein interactions form the basis for chemistry expansion and potentially result in rapid increase in binding affinity. The structures will reveal the ligand-binding site on the protein and any conformational changes concurrent with the binding. The most efficient way of evaluating large number of fragment hits crystallographically is by soaking hit compounds into a previously formed crystal of the target protein in its unliganded form. To this end, we determined the crystal structures offragments in complex with KRas4B^{G12D} (will be referred to as KRas$_m$ later) in either GDP or GTP states. The GTP state was obtained by incorporating the nonhydrolyzable GTP analogue GMPPCP for its prolonged stability that allows time for crystallization.

Crystal structures of Ras isoforms have been previously described in the literature [63–68] and were shown to share a high degree of similarity. All Ras isoforms adopt a common α/β-fold. The core structure is composed of a 5-stranded parallel β-sheet. In addition, a special β-strand (β2) is added to the edge of the core sheet in an antiparallel orientation. Five α-helices flank on both sides of the β-core (Fig. 2.6). β2 and the preceding loop, usually

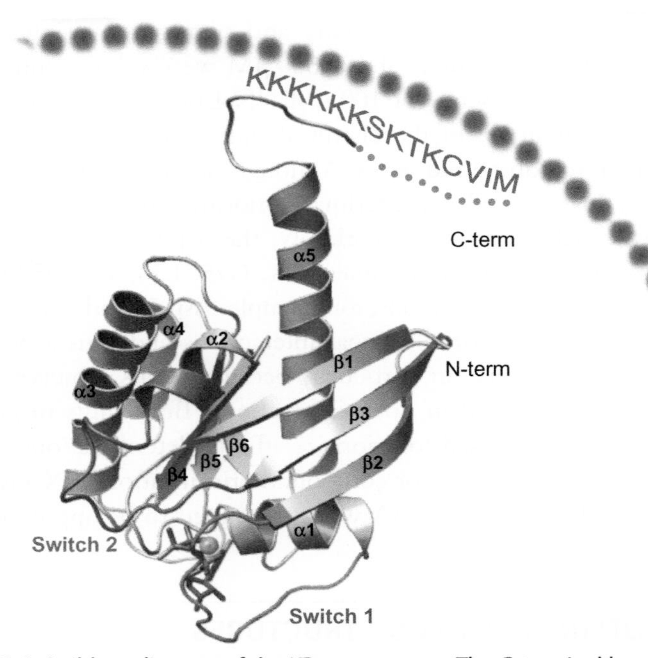

Figure 2.6 A ribbon diagram of the $KRas_m$ structure. The C-terminal loop is disordered and is shown in a dotted line. The blue dots represent plasma membrane where the poly-lysine and lipid-modified C-terminal loop interact with. *Adapted from Ref. [58].* (See color plate.)

referred to as switch 1 region (residues 25–40), form part of Ras effector interface. Another structural element termed switch 2 region (α2-helix and preceding loop, residues 57–70) is also critical for Ras signaling. Both switch regions respond to the presence of γ-phosphate by changing into a conformation that favors effector protein binding. These regions are also known to take part in GEF and GAP interactions. Among various crystal structures, the switch regions exhibit conformational heterogeneity as they are controlled by the nucleotide state and, to some extent, also influenced by the crystal packing environment.

3.2. The high-resolution view of ligand and $KRas_m$ interaction

$KRas_m$-ligand cocrystals were obtained by soaking with individual compounds at a concentration of 100 mM. The high compound concentration was required to compensate for the weak binding affinities. The complexes of $KRas_m$ with benzamidine (BZDN), benzimidazole (BZIM), and 4,6-dichloro-2-methyl-3-aminoethyl-indole (DCAI) are shown in Fig. 2.7.

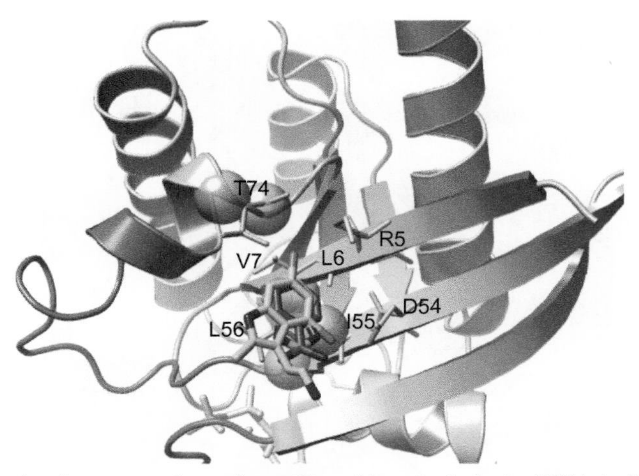

Figure 2.7 The fragments bound to KRas. Ligands include DCAI (yellow), BZDN (magenta), and BZIM (cyan). Residues directly interacting with the fragments are shown as sticks. The NMR-mapped residues are show in blue. The NMR amide cross peaks shifted in the NMR experiments are shown as spheres. (See color plate.)

Consistent with the NMR observation, all three compounds bind in a small pocket between helix α2 and β1–β3 strands. The interior of the pocket is hydrophobic and has a size equivalent to 10% of the volume enclosed in a typical ATP-binding site found in protein kinases.

Residues surrounding the binding pocket include K5, L6, V7, I55, L56, and T74 (Fig. 2.7). This is again consistent with the mapping based on the NMR chemical shift perturbations. Indeed, Fig. 2.7 shows that all of the NMR-perturbed residues are located in or around the binding pocket. The surface area outside the pocket is more hydrophilic and engages the polar portion of the compounds. D54 on the rim accepts hydrogen bonds from the NH group of BZDN or BZIM, therefore affecting the ligand orientation. H, N, and KRas share identical amino acid sequences in the binding site, suggesting a high likelihood of compounds binding to all three isoforms. We confirmed that DCAI and BZIM bind to HRas using STD NMR.

3.3. The ligand-binding pocket expands upon DCAI binding

By comparing the crystal structures on hand, we noted that the ligand-binding pocket is associated with a high degree of plasticity. The structure of DCAI-bound $KRas_m$ is most informative in this respect. Shown in Fig. 2.8, DCAI uses the chloro-substituent at the 6-position to anchor itself

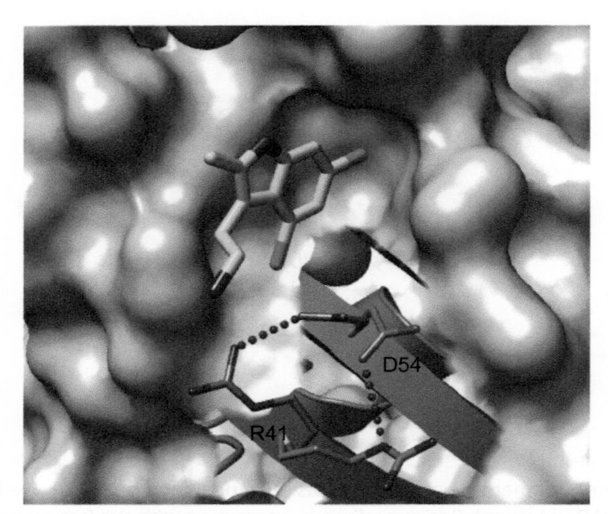

Figure 2.8 The ligand-binding pocket is expandable upon DCAI binding. *Adapted from Ref. [58].* (For color version of this figure, the reader is referred to the online version of this chapter.)

into the hydrophobic pocket. The 4-chloro group makes DCAI much larger than BZDN or BZIM (Fig. 2.7). This part is accommodated in the binding pocket first by rotating the indole core by 65° relative to BZIM, then by pushing the side chain of D54 away by 2.5 Å (Fig. 2.8). D54 normally forms a salt bridge with R41. This compound-induced rotamer conformation switch in D54 does not break the salt bridge but rather effects an outward flipping (7.5 Å) of R41, rendering the pocket to expand by about 3.5 Å (Fig. 2.8). These two residues normally adopt the closed position in most of crystal structures; however, their temperature factors are generally high indicating flexibility. DCAI binding increases the pocket opening by about 50%, although the extended region is relatively shallow.

4. INHIBITION OF SOS1-MEDIATED NUCLEOTIDE EXCHANGE FOR RAS

Structural comparison between $KRas_m$-DCAI and Ras–SOS1 complex [69] indicated that the ligand-binding pocket is adjacent to the Ras–SOS interface (Fig. 2.9), suggesting that ligand binding can affect SOS1-mediated nucleotide exchange for KRas. This hypothesis was validated using nucleotide exchange assays (see detailed description in [58]). DCAI was found to inhibit nucleotide exchange and nucleotide release with IC_{50}'s of 357 and 185 μM, respectively, while BZIM and indole (INDL) had

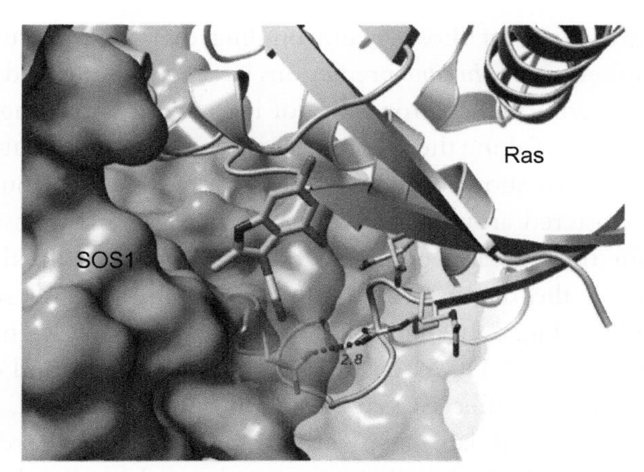

Figure 2.9 Structural basis of DCAI interfering with the SOS1/Ras interaction. The DCAI structure is modeled based on the crystal structure of SOS1/Ras (PDB code 1NVV). *Adapted from Ref. [58].* (See color plate.)

minimal effects. These data led to the conclusion that DCAI blocks the first phase of the exchange reaction by preventing the release of nucleotide from KRas. Further biochemical investigation demonstrated that DCAI inhibits nucleotide exchange by blocking the interactions between Ras and SOS1. This phenomenon can be rationalized using the crystal structures. DCAI binding interferes with the formation of the exchange reaction intermediates through two mechanisms. First, the 2-methyl and 3-ethylamine groups of DCAI sterically block the binding of SOScat to Ras. Second, D54 and R41 participate in two salt bridges to SOS1 H911 and SOS1 D910, respectively, that stabilize the Ras–SOS interaction (Fig. 2.9). DCAI binding forces D54and R41 to adopt alternative side chain rotamers, thereby interrupting this set of interactions. Consistent with this model, the KRas R41S mutant was shown to reduce the rate of SOS-mediated nucleotide release reaction by 2.5-fold, highlighting the important role of R41 in the exchange reaction. In contrast, structural modeling suggested that the trimmed down analogue of DCAI, BZIM would not sterically hinder SOS binding and would not alter the ability of D54 and R41 to participate in the interactions (Fig. 2.9) at the apoRas–SOS interface.

5. LIGAND-BINDING SITES ON Ras

A number of Ras binders have been reported in the literature. They target a variety of sites in the Ras structure mostly on the surface.

Encouragingly, many of those ligand binding exhibited certain levels of functional consequence. *In silico* predictions have also pinpointed some areas on Ras surface with higher propensity of interacting with drug-like small molecules [70–73]. Taking the theoretical and experimental results together, a systematic analysis showed that significant consensus was found at only several sites clustered around the switches [74]. These consensus sites are of interest for their greater potential for drug discovery. In this section, we will expand the discussion by highlighting these consensus sites (A through site D in Fig. 2.10) and describe evidence of true ligand binding.

As described above, DCAI occupies a pocket at site A, and compound binding to this site can affect SOS mediate nucleotide exchange. In an independent study, the Fesik lab at Vanderbilt University discovered a series of indole analogues with similar effects. Sun et al. conducted an NMR-based fragment screen with a different strategy. Here, a library of 11,000 compounds was screened against isotopically labeled KRas as the primary screen. No ligand-based detection methods or unlabeled protein was employed. They discovered primary hits that bind to site A with relatively low affinity of 1.3–2 mM [75]. Further chemical modification extended the compounds to site B and improved binding affinities to KRas^{GDP} with a K_d of 190 μM in the best case. This represents the first reported example of successful affinity improvement of Ras-binding molecules using

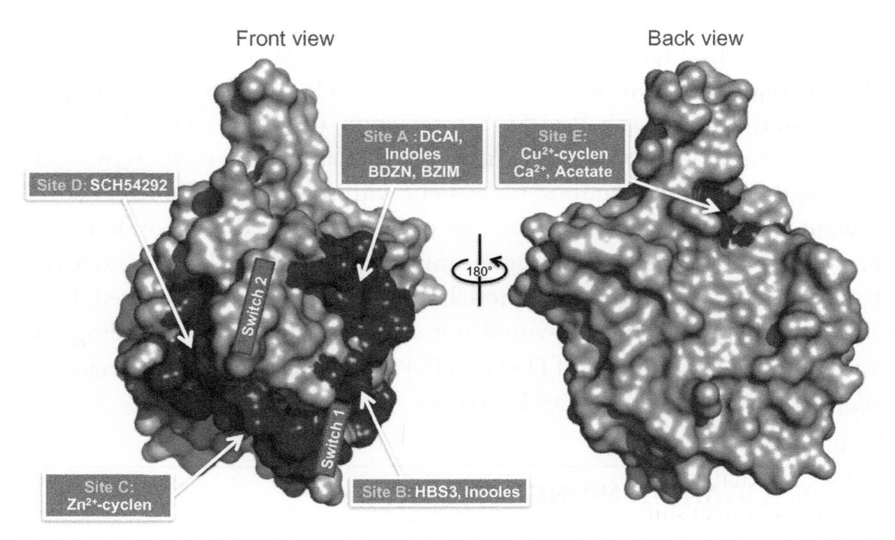

Figure 2.10 A global mapping of ligand-binding site on the Ras surface. *Adapted from Ref. [73].* (See color plate.)

structure-guided design. Similar to DCAI, these compounds inhibit SOS-mediated nucleotide exchange [75]. Intriguingly, DCAI does not reach into site B, indicating that site B occupancy is not a prerequisite for disrupting the Ras–SOS interaction. However, expanding into the direction of site B does seem to increase the chance of orthosterically blocking the effector RBD binding. This series of compounds represents an interesting starting point for further medicinal chemistry efforts aimed at enhancing affinity and activity.

Site B is on the outer surface of switch 1 and is directly involved in PPI. This site undergoes considerable conformational changes throughout the on–off cycles of Ras, due to the intrinsic flexibility of switch 1 and its role in sensing the nucleotide binding [30,76]. Directly targeting site B is undoubtedly desirable. Some solvent molecules were found to bind this site in multiple solvent crystal soaking studies [72,73]. There are also reports of site B-specific small-molecule binders that cause functional effects. For instance, the above-mentioned indole analogues have come very close, albeit they are still dependent on site A binding [75]. A more clear-cut example is the stapled helix. The αH helix of SOS1 is known to be the key structural element directly contacting Ras during the nucleotide exchange reaction. The Bar-Sagi lab pioneered the use of peptidyl mimics as an orthosteric blocking strategy [77]. HBS3 is a stapled peptide mimicking the αH helix in SOS1. They showed that HBS3 binds to nucleotide-free Ras with a K_d of ~28 μM and to GDP-bound Ras with a K_d of ~158 μM. Their $^1H^{15}N$ HSQC NMR titration experiments indicated that HBS3 binding involved the nucleotide-binding pocket and the area between site A and site B, supporting the prediction that HBS3 can act as a direct mimic of the αH helix in SOS1 [77].

Site C is located on the rim of the GTP-binding site. It involves a part of switch 1 which is most sensitive to the presence of the γ-phosphate. Therefore, ligands binding at this site can potentially counteract the GTP-induced effects. Using ^{31}P NMR, Spoerner et al. demonstrated that in the solution state this part of the Ras structure exists in multiple conformations in equilibrium [78]. They discovered that Zn^{2+} cyclen shifted the Ras conformational equilibrium toward an inactive conformation. It was later confirmed that both Zn^{2+}- and Cu^{2+}-cyclen [79] stabilize a conformational state of RasGTP that displays a low affinity to RBD [79]. Protein-observed NMR chemical shifts indicated that Zn^{2+} cyclen affects both site C and site E (Fig. 2.10) on Ras, and the effects can be saturated at compound concentrations of 2 and 6 mM, respectively.

Site D is a shallow grove adjacent to switch 2 and α3 helix. As switch 2 undergoes conformational changes upon GTP/GDP exchange, a cleft can transiently form underneath switch 2, which can be targeted by certain compounds. SCH-54292 [80] was found to bind to Ras and inhibit the intrinsic nucleotide exchange with IC_{50} in the range of 0.5–0.7 μM [81]. Guided by NMR chemical shift mapping, SCH-54292 was computationally modeled in the crystal structure of HRas [82]. In this docking model, SCH-54292 is placed at site D and is shown to chelate the Mg^{2+} ion. This binding mode is consistent with SCH-54292 being selective for the RasGDP state. Some of the SCH analogues displayed moderate inhibitory activity (EC_{50} ~10–20 μM) in a cellular assay, while other compounds were inactive in this assay, presumably due to poor cell permeability [80]. Guided by molecular modeling, a D-arabinose-derived novel bicyclic scaffold was used to replace the core motif of SCH-54292, while maintaining the benzyl and the phenylhydroxylamine moieties essential for Ras binding. These compounds displayed improved inhibition to GDP dissociation, and better cell permeability [83].

Site E is on the backside of the Ras molecule and is far removed from the protein–protein interface. For any ligand binding at this site and to cause a functional consequence, it will likely to employ an allosteric mechanism. Ligand binding to this site has been captured in crystal structures, for instance in the Cu^{2+}-cyclen/Ras cocrystal structure of HRas [79]. Calcium acetate was also found to bind at site D [72], although no functional activity was reported. Further investigation is necessary to assess the general utility of this site for the regulation of Ras activity.

6. SUMMARY AND CONCLUSIONS

Ras is a promising target for cancer therapy but is also a challenging protein for small-molecule drug discovery. NMR-based fragment screening provides a powerful tool of probing proteins lacking significant pockets for ligand binding. Using a carefully designed screening cascade, we discovered a series of ligands that bind to KRas at a common site. Further characterization of the binding site using solution state NMR, X-ray crystallography, and biochemical assays demonstrated the functional relevance of the binding site in SOS1-dependent nucleotide exchange. While Ras inhibitors are still scarce, extensive probing on the Ras protein through interdisciplinary efforts has yielded valuable knowledge of potential binders and binding sites.

These will form the basis facilitating future endeavors of pharmacological intervention against this important oncoprotein.

REFERENCES

[1] W.H. Kirsten, L.A. Mayer, Morphologic responses to a murine erythroblastosis virus, J. Natl. Cancer Inst. 39 (2) (1967) 311–335.

[2] J.J. Harvey, An unidentified virus which causes the rapid production of tumours in mice, Nature 204 (1964) 1104–1105.

[3] D. Stehelin, H.E. Varmus, J.M. Bishop, P.K. Vogt, DNA related to the transforming gene(s) of avian sarcoma viruses is present in normal avian DNA, Nature 260 (5547) (1976) 170–173.

[4] S. Pulciani, E. Santos, A.V. Lauver, L.K. Long, K.C. Robbins, M. Barbacid, Oncogenes in human tumor cell lines: molecular cloning of a transforming gene from human bladder carcinoma cells, Proc. Natl. Acad. Sci. U.S.A. 79 (9) (1982) 2845–2849.

[5] M. Perucho, M. Goldfarb, K. Shimizu, C. Lama, J. Fogh, M. Wigler, Human-tumor-derived cell lines contain common and different transforming genes, Cell 27 (3 Pt 2) (1981) 467–476.

[6] C. Shih, L.C. Padhy, M. Murray, R.A. Weinberg, Transforming genes of carcinomas and neuroblastomas introduced into mouse fibroblasts, Nature 290 (5803) (1981) 261–264.

[7] T.G. Krontiris, G.M. Cooper, Transforming activity of human tumor DNAs, Proc. Natl. Acad. Sci. U.S.A. 78 (2) (1981) 1181–1184.

[8] D.J. Capon, P.H. Seeburg, J.P. McGrath, J.S. Hayflick, U. Edman, A.D. Levinson, D.V. Goeddel, Activation of Ki-ras2 gene in human colon and lung carcinomas by two different point mutations, Nature 304 (5926) (1983) 507–513.

[9] M. Malumbres, M. Barbacid, RAS oncogenes: the first 30 years, Nat. Rev. Cancer 3 (6) (2003) 459–465.

[10] A.D. Cox, C.J. Der, Ras history: the saga continues, Small GTPases 1 (1) (2010) 2–27.

[11] A.T. Baines, D. Xu, C.J. Der, Inhibition of Ras for cancer treatment: the search continues, Future Med. Chem. 3 (14) (2011) 1787–1808.

[12] S. Schubbert, K. Shannon, G. Bollag, Hyperactive Ras in developmental disorders and cancer, Nat. Rev. Cancer 7 (4) (2007) 295–308.

[13] H. Ying, A.C. Kimmelman, C.A. Lyssiotis, S. Hua, G.C. Chu, E. Fletcher-Sananikone, J.W. Locasale, J. Son, H. Zhang, J.L. Coloff, H. Yan, W. Wang, S. Chen, A. Viale, H. Zheng, J.H. Paik, C. Lim, A.R. Guimaraes, E.S. Martin, J. Chang, A.F. Hezel, S.R. Perry, J. Hu, B. Gan, Y. Xiao, J.M. Asara, R. Weissleder, Y.A. Wang, L. Chin, L.C. Cantley, R.A. DePinho, Oncogenic Kras maintains pancreatic tumors through regulation of anabolic glucose metabolism, Cell 149 (3) (2012) 656–670.

[14] T. Tanaka, R.L. Williams, T.H. Rabbitts, Tumour prevention by a single antibody domain targeting the interaction of signal transduction proteins with RAS, EMBO J. 26 (13) (2007) 3250–3259.

[15] P.T. Tran, A.C. Fan, P.K. Bendapudi, S. Koh, K. Komatsubara, J. Chen, G. Horng, D.I. Bellovin, S. Giuriato, C.S. Wang, J.A. Whitsett, D.W. Felsher, Combined inactivation of MYC and K-Ras oncogenes reverses tumorigenesis in lung adenocarcinomas and lymphomas, PLoS One 3 (5) (2008) e2125.

[16] A.A. Adjei, Blocking oncogenic Ras signaling for cancer therapy, J. Natl. Cancer Inst. 93 (14) (2001) 1062–1074.

[17] B.B. Friday, A.A. Adjei, K-ras as a target for cancer therapy, Biochim. Biophys. Acta 1756 (2) (2005) 127–144.

[18] I.R. Vetter, A. Wittinghofer, The guanine nucleotide-binding switch in three dimensions, Science 294 (5545) (2001) 1299–1304.

[19] M.C. Willingham, I. Pastan, T.Y. Shih, E.M. Scolnick, Localization of the src gene product of the Harvey strain of MSV to plasma membrane of transformed cells by electron microscopic immunocytochemistry, Cell 19 (4) (1980) 1005–1014.
[20] A. Papageorge, D. Lowy, E.M. Scolnick, Comparative biochemical properties of p21 ras molecules coded for by viral and cellular ras genes, J. Virol. 44 (2) (1982) 509–519.
[21] P.H. Warne, P.R. Viciana, J. Downward, Direct interaction of Ras and the amino-terminal region of Raf-1 in vitro, Nature 364 (6435) (1993) 352–355.
[22] X.F. Zhang, J. Settleman, J.M. Kyriakis, E. Takeuchi-Suzuki, S.J. Elledge, M.S. Marshall, J.T. Bruder, U.R. Rapp, J. Avruch, Normal and oncogenic p21ras proteins bind to the amino-terminal regulatory domain of c-Raf-1, Nature 364 (6435) (1993) 308–313.
[23] A.B. Vojtek, S.M. Hollenberg, J.A. Cooper, Mammalian Ras interacts directly with the serine/threonine kinase Raf, Cell 74 (1) (1993) 205–214.
[24] S.A. Moodie, B.M. Willumsen, M.J. Weber, A. Wolfman, Complexes of Ras.GTP with Raf-1 and mitogen-activated protein kinase kinase, Science 260 (5114) (1993) 1658–1661.
[25] P. Rodriguez-Viciana, P.H. Warne, R. Dhand, B. Vanhaesebroeck, I. Gout, M.J. Fry, M.D. Waterfield, J. Downward, Phosphatidylinositol-3-OH kinase as a direct target of Ras, Nature 370 (6490) (1994) 527–532.
[26] J. Colicelli, Human RAS superfamily proteins and related GTPases, Sci. STKE 2004 (250) (2004) RE13.
[27] A.E. Karnoub, R.A. Weinberg, Ras oncogenes: split personalities, Nat. Rev. Mol. Cell Biol. 9 (7) (2008) 517–531.
[28] A.M. Rojas, G. Fuentes, A. Rausell, A. Valencia, The Ras protein superfamily: evolutionary tree and role of conserved amino acids, J. Cell Biol. 196 (2) (2012) 189–201.
[29] J.L. Bos, H. Rehmann, A. Wittinghofer, GEFs and GAPs: critical elements in the control of small G proteins, Cell 129 (5) (2007) 865–877.
[30] P.A. Boriack-Sjodin, S.M. Margarit, D. Bar-Sagi, J. Kuriyan, The structural basis of the activation of Ras by Sos, Nature 394 (6691) (1998) 337–343.
[31] T.Y. Shih, A.G. Papageorge, P.E. Stokes, M.O. Weeks, E.M. Scolnick, Guanine nucleotide-binding and autophosphorylating activities associated with the p21src protein of Harvey murine sarcoma virus, Nature 287 (5784) (1980) 686–691.
[32] K. Scheffzek, A. Lautwein, W. Kabsch, M.R. Ahmadian, A. Wittinghofer, Crystal structure of the GTPase-activating domain of human p120GAP and implications for the interaction with Ras, Nature 384 (6609) (1996) 591–596.
[33] J.B. Gibbs, I.S. Sigal, M. Poe, E.M. Scolnick, Intrinsic GTPase activity distinguishes normal and oncogenic ras p21 molecules, Proc. Natl. Acad. Sci. U.S.A. 81 (18) (1984) 5704–5708.
[34] J.P. McGrath, D.J. Capon, D.V. Goeddel, A.D. Levinson, Comparative biochemical properties of normal and activated human ras p21 protein, Nature 310 (5979) (1984) 644–649.
[35] V. Manne, E. Bekesi, H.F. Kung, Ha-ras proteins exhibit GTPase activity: point mutations that activate Ha-ras gene products result in decreased GTPase activity, Proc. Natl. Acad. Sci. U.S.A. 82 (2) (1985) 376–380.
[36] R.W. Sweet, S. Yokoyama, T. Kamata, J.R. Feramisco, M. Rosenberg, M. Gross, The product of ras is a GTPase and the T24 oncogenic mutant is deficient in this activity, Nature 311 (5983) (1984) 273–275.
[37] C.J. Der, T. Finkel, G.M. Cooper, Biological and biochemical properties of human rasH genes mutated at codon 61, Cell 44 (1) (1986) 167–176.
[38] L. Pauling, Nature of forces between large molecules of biological interest, Nature 161 (4097) (1948) 707–709.

[39] T.L. Amyes, J.P. Richard, Specificity in transition state binding: The Pauling Model Revisited, Biochemistry 52 (12) (2013) 2021–2035.
[40] G. Chessari, A.J. Woodhead, From fragment to clinical candidate–a historical perspective, Drug Discov. Today 14 (13–14) (2009) 668–675.
[41] M. Baker, Fragment-based lead discovery grows up, Nat. Rev. Drug Discov. 12 (1) (2013) 5–7.
[42] R. Macarron, Critical review of the role of HTS in drug discovery, Drug Discov. Today 11 (7–8) (2006) 277–279.
[43] R.A. Bauer, J.M. Wurst, D.S. Tan, Expanding the range of 'druggable' targets with natural product-based libraries: an academic perspective, Curr. Opin. Chem. Biol. 14 (3) (2010) 308–314.
[44] D.C. Rees, M. Congreve, C.W. Murray, R. Carr, Fragment-based lead discovery, Nat. Rev. Drug Discov. 3 (8) (2004) 660–672.
[45] P.J. Hajduk, J. Greer, A decade of fragment-based drug design: strategic advances and lessons learned, Nat. Rev. Drug Discov. 6 (3) (2007) 211–219.
[46] M.N. Schulz, R.E. Hubbard, Recent progress in fragment-based lead discovery, Curr. Opin. Pharmacol. 9 (5) (2009) 615–621.
[47] T. Maurer, Advancing fragment binders to lead-like compounds using ligand and protein-based NMR spectroscopy, Methods Enzymol. 493 (2011) 469–485.
[48] A.M. Giannetti, From experimental design to validated hits a comprehensive walk-through of fragment lead identification using surface plasmon resonance, Methods Enzymol. 493 (2011) 169–218.
[49] A.M. Petros, J.R. Huth, T. Oost, C.M. Park, H. Ding, X. Wang, H. Zhang, P. Nimmer, R. Mendoza, C. Sun, J. Mack, K. Walter, S. Dorwin, E. Gramling, U. Ladror, S.H. Rosenberg, S.W. Elmore, S.W. Fesik, P.J. Hajduk, Discovery of a potent and selective Bcl-2 inhibitor using SAR by NMR, Bioorg. Med. Chem. Lett. 20 (22) (2010) 6587–6591.
[50] F. Jeppsson, S. Eketjall, J. Janson, S. Karlstrom, S. Gustavsson, L.L. Olsson, A.C. Radesater, B. Ploeger, G. Cebers, K. Kolmodin, B.M. Swahn, S. von Berg, T. Bueters, J. Falting, Discovery of AZD3839, a potent and selective BACE1 inhibitor clinical candidate for the treatment of Alzheimer disease, J. Biol. Chem. 287 (49) (2012) 41245–41257.
[51] B. Meyer, T. Peters, NMR spectroscopy techniques for screening and identifying ligand binding to protein receptors, Angew. Chem. Int. Ed Engl. 42 (8) (2003) 864–890.
[52] D.L. Andrews, A unified theory of radiative and radiationless molecular-energy transfer, Chem. Phys. 135 (2) (1989) 195–201.
[53] C. Dalvit, G. Fogliatto, A. Stewart, M. Veronesi, B. Stockman, WaterLOGSY as a method for primary NMR screening: practical aspects and range of applicability, J. Biomol. NMR 21 (4) (2001) 349–359.
[54] G. Bodenhausen, D.J. Ruben, Natural abundance nitrogen-15 NMR by enhanced heteronuclear spectroscopy, Chem. Phys. Lett. 69 (1) (1980) 5.
[55] E.L. Ulrich, H. Akutsu, J.F. Doreleijers, Y. Harano, Y.E. Ioannidis, J. Lin, M. Livny, S. Mading, D. Maziuk, Z. Miller, E. Nakatani, C.F. Schulte, D.E. Tolmie, R. Kent Wenger, H. Yao, and J.L. Markley, BioMagResBank. Nucleic Acids Res. 36 (Database) (2007) D402-D408.
[56] M. Congreve, R. Carr, C. Murray, H. Jhoti, A 'rule of three' for fragment-based lead discovery? Drug Discov. Today 8 (19) (2003) 876–877.
[57] M. Mayer, B. Meyer, Group epitope mapping by saturation transfer difference NMR to identify segments of a ligand in direct contact with a protein receptor, J. Am. Chem. Soc. 123 (25) (2001) 6108–6117.

[58] T. Maurer, L.S. Garrenton, A. Oh, K. Pitts, D.J. Anderson, N.J. Skelton, B.P. Fauber, B. Pan, S. Malek, D. Stokoe, M.J. Ludlam, K.K. Bowman, J. Wu, A.M. Giannetti, M.A. Starovasnik, I. Mellman, P.K. Jackson, J. Rudolph, W. Wang, G. Fang, Small--molecule ligands bind to a distinct pocket in Ras and inhibit SOS-mediated nucleotide exchange activity, Proc. Natl. Acad. Sci. U.S.A. 109 (14) (2012) 5299–5304.

[59] A.G. Palmer, J. Cavanagh, P.E. Wright, and M. Rance, Sensitivity improvement in proton-detected two-dimensional heteronuclear correlation NMR spectroscopy. J. Magn. Reson. 93 (1) (1991) 151-170.

[60] P.J. Kraulis, P.J. Domaille, S.L. Campbell-Burk, T. Van Aken, E.D. Laue, Solution structure and dynamics of ras p21.GDP determined by heteronuclear three- and four-dimensional NMR spectroscopy, Biochemistry 33 (12) (1994) 3515–3531.

[61] C. O'Connor, E.L. Kovrigin, Assignments of backbone (1)H, (1)(3)C and (1)(5)N resonances in H-Ras (1–166) complexed with GppNHp at physiological pH, Biomol. NMR Assign. 6 (1) (2012) 91–93.

[62] F.H. Schumann, H. Riepl, T. Maurer, W. Gronwald, K.P. Neidig, H.R. Kalbitzer, Combined chemical shift changes and amino acid specific chemical shift mapping of protein-protein interactions, J. Biomol. NMR 39 (4) (2007) 275–289.

[63] L.A. Tong, A.M. de Vos, M.V. Milburn, J. Jancarik, S. Noguchi, S. Nishimura, K. Miura, E. Ohtsuka, S.H. Kim, Structural differences between a ras oncogene protein and the normal protein, Nature 337 (6202) (1989) 90–93.

[64] M.V. Milburn, L. Tong, A.M. deVos, A. Brunger, Z. Yamaizumi, S. Nishimura, S.H. Kim, Molecular switch for signal transduction: structural differences between active and inactive forms of protooncogenic ras proteins, Science 247 (4945) (1990) 939–945.

[65] E.F. Pai, U. Krengel, G.A. Petsko, R.S. Goody, W. Kabsch, A. Wittinghofer, Refined crystal structure of the triphosphate conformation of H-ras p21 at 1.35 A resolution: implications for the mechanism of GTP hydrolysis, EMBO J. 9 (8) (1990) 2351–2359.

[66] F. Shima, Y. Ijiri, S. Muraoka, J. Liao, M. Ye, M. Araki, K. Matsumoto, N. Yamamoto, T. Sugimoto, Y. Yoshikawa, T. Kumasaka, M. Yamamoto, A. Tamura, T. Kataoka, Structural basis for conformational dynamics of GTP-bound Ras protein, J. Biol. Chem. 285 (29) (2010) 22696–22705.

[67] Tong Y, Tempel W, Shen L, Arrowsmith CH, Edwards AM, Sundstrom M, et al. Human K-Ras in complex with a GTP analogue, 3GFT: Protein Data Bank 2012.

[68] Nedyalkova L, Tong Y, Tempel W, Shen L, Loppnau P, Arrowsmith CH, et al. Crystal structure of the human NRAS GTPase bound with GDP, 3CON: Protein Data Bank 2009.

[69] S.M. Margarit, H. Sondermann, B.E. Hall, B. Nagar, A. Hoelz, M. Pirruccello, D. Bar-Sagi, J. Kuriyan, Structural evidence for feedback activation by Ras.GTP of the Ras-specific nucleotide exchange factor SOS, Cell 112 (5) (2003) 685–695.

[70] H. te Heesen, A.M. Schlitter, J. Schlitter, Empirical rules facilitate the search for binding sites on protein surfaces, J. Mol. Graph. Model. 25 (5) (2007) 671–679.

[71] R. Brenke, D. Kozakov, G.Y. Chuang, D. Beglov, D. Hall, M.R. Landon, C. Mattos, S. Vajda, Fragment-based identification of druggable 'hot spots' of proteins using Fourier domain correlation techniques, Bioinformatics 25 (5) (2009) 621–627.

[72] G. Buhrman, C. O'Connor, B. Zerbe, B.M. Kearney, R. Napoleon, E.A. Kovrigina, S. Vajda, D. Kozakov, E.L. Kovrigin, C. Mattos, Analysis of binding site hot spots on the surface of Ras GTPase, J. Mol. Biol. 413 (4) (2011) 773–789.

[73] B.J. Grant, S. Lukman, H.J. Hocker, J. Sayyah, J.H. Brown, J.A. McCammon, A.A. Gorfe, Novel allosteric sites on Ras for lead generation, PLoS One 6 (10) (2011) e25711.

[74] W. Wang, G. Fang, J. Rudolph, Ras inhibition via direct Ras binding—is there a path forward? Bioorg. Med. Chem. Lett. 22 (18) (2012) 5766–5776.

[75] Q. Sun, J.P. Burke, J. Phan, M.C. Burns, E.T. Olejniczak, A.G. Waterson, T. Lee, O.W. Rossanese, S.W. Fesik, Discovery of small molecules that bind to K-Ras and inhibit Sos-mediated activation, Angew. Chem. Int. Ed Engl. 51 (25) (2012) 6140–6143.
[76] M. Geyer, T. Schweins, C. Herrmann, T. Prisner, A. Wittinghofer, H.R. Kalbitzer, Conformational transitions in p21ras and in its complexes with the effector protein Raf-RBD and the GTPase activating protein GAP, Biochemistry 35 (32) (1996) 10308–10320.
[77] A. Patgiri, K.K. Yadav, P.S. Arora, D. Bar-Sagi, An orthosteric inhibitor of the Ras–Sos interaction, Nat. Chem. Biol. 7 (9) (2011) 585–587.
[78] M. Spoerner, T. Graf, B. Konig, H.R. Kalbitzer, A novel mechanism for the modulation of the Ras-effector interaction by small molecules, Biochem. Biophys. Res. Commun. 334 (2) (2005) 709–713.
[79] I.C. Rosnizeck, T. Graf, M. Spoerner, J. Trankle, D. Filchtinski, C. Herrmann, L. Gremer, I.R. Vetter, A. Wittinghofer, B. Konig, H.R. Kalbitzer, Stabilizing a weak binding state for effectors in the human ras protein by cyclen complexes, Angew. Chem. Int. Ed Engl. 49 (22) (2010) 3830–3833.
[80] A.G. Taveras, S.W. Remiszewski, R.J. Doll, D. Cesarz, E.C. Huang, P. Kirschmeier, B.N. Pramanik, M.E. Snow, Y.S. Wang, J.D. del Rosario, B. Vibulbhan, B.B. Bauer, J.E. Brown, D. Carr, J. Catino, C.A. Evans, V. Girijavallabhan, L. Heimark, L. James, S. Liberles, C. Nash, L. Perkins, M.M. Senior, A. Tsarbopoulos, S.E. Webber, et al., Ras oncoprotein inhibitors: the discovery of potent, ras nucleotide exchange inhibitors and the structural determination of a drug-protein complex, Bioorg. Med. Chem. 5 (1) (1997) 125–133.
[81] A.K. Ganguly, B.N. Pramanik, E.C. Huang, S. Liberles, L. Heimark, Y.H. Liu, A. Tsarbopoulos, R.J. Doll, A.G. Taveras, S. Remiszewski, M.E. Snow, Y.S. Wang, B. Vibulbhan, D. Cesarz, J.E. Brown, J. del Rosario, L. James, P. Kirschmeier, V. Girijavallabhan, Detection and structural characterization of ras oncoprotein-inhibitors complexes by electrospray mass spectrometry, Bioorg. Med. Chem. 5 (5) (1997) 817–820.
[82] A.K. Ganguly, Y.S. Wang, B.N. Pramanik, R.J. Doll, M.E. Snow, A.G. Taveras, S. Remiszewski, D. Cesarz, J. del Rosario, B. Vibulbhan, J.E. Brown, P. Kirschmeier, E.C. Huang, L. Heimark, A. Tsarbopoulos, V.M. Girijavallabhan, R.M. Aust, E.L. Brown, D.M. DeLisle, S.A. Fuhrman, T.F. Hendrickson, C.R. Kissinger, R.A. Love, W.A. Sisson, S.E. Webber, et al., Interaction of a novel GDP exchange inhibitor with the Ras protein, Biochemistry 37 (45) (1998) 15631–15637.
[83] F. Peri, C. Airoldi, S. Colombo, E. Martegani, A.S. van Neuren, M. Stein, C. Marinzi, F. Nicotra, Design, synthesis and biological evaluation of sugar-derived Ras inhibitors, Chembiochem 6 (10) (2005) 1839–1848.

CHAPTER THREE

The Allosteric Switch and Conformational States in Ras GTPase Affected by Small Molecules

Christian W. Johnson, Carla Mattos[1]
Department of Chemistry and Chemical Biology, Northeastern University, 360 Huntington Ave, Boston, Massachusetts, USA
[1]Corresponding author: e-mail address: c.mattos@neu.edu

Contents

Abstract

Ras is a hub protein in signal transduction pathways leading to the control of cell proliferation, migration, and survival and a major target for drug discovery due to the presence of its mutants in about 20% of human cancers. Yet, the discovery of small molecules that can directly interfere with its function has been elusive in spite of intense efforts. This is most likely due to its highly flexible nature and the lack of a well-ordered active site. This chapter contains a discussion of our current understanding of conformational states in Ras-GTP, with focus on a recently discovered allosteric switch mechanism that may promote intrinsic hydrolysis of GTP in the presence of Raf. We discuss the manner in which small molecules are known to affect the equilibrium of states in Ras-GTP and suggest novel strategies to go forward in the search for inhibitors of this master signaling protein.

The Enzymes, Volume 33
ISSN 1874-6047
http://dx.doi.org/10.1016/B978-0-12-416749-0.00003-8

1. INTRODUCTION

In the early 1980s, Ras was discovered by its ability to transform mammalian cells to a neoplastic state and has since been a rich system of study in the field of cancer research [1,2]. Soon after its discovery, it was determined that the oncogenic potential of Ras is related to its ability to act as a binary switch [3,4], where specific mutations result in the inability of this enzyme to catalyze the hydrolysis of bound GTP to GDP, disturbing the balance that keeps neoplastic growth in check [5,6]. Ras now is known to be involved not only in pathways controlling cell proliferation, but also apoptosis, invasion, survival, and angiogenesis [7]. It is therefore a central participant in many of the processes characterized as the hallmarks of cancer [8]. In addition to its oncogenic capacity, Ras also plays a significant role in various developmental disorders, in concert with an emerging understanding of its role in organism development [2,9]. Despite the considerable need for therapeutic engagement focused on Ras, this protein has been largely elusive as a target in drug discovery [2,10].

Oncogenic Ras is found in about 20% of all cancers, with mutations in the K-Ras isoform being particularly prominent in human tumors [11]. The most common mutations found in Ras-driven tumors involve point mutations at residue G12, G13, or Q61. These mutations result in proteins that favor the GTP-bound state, producing enzymes that are constitutively active due to impaired intrinsic hydrolysis rates [6,12] and insensitivity to GAPs [13–15]. Research in this field has placed GAP as the primary and universal mode of attenuation of pathways that depend on Ras activation, and it provides a strong model for understanding the oncogenic potential of Ras [16]. However, research performed in the mid-1990s suggested that the Ras/Raf/MEK/ERK pathway, a pathway most often associated with cellular proliferation, may be regulated by a different set of interactions. Competition experiments showed that active Ras strongly and exclusively binds Raf-1 and B-Raf when also in the presence of GAP [14]. The affinity of Ras-GTP for Raf is 3.4 nM [17], three orders of magnitude stronger than for GAP [18]. Considering the measured concentrations of Ras (0.1–1.6 μM) and Raf (3–500 nM) in cells, it is unlikely that GAP by itself would be an effective downregulator of the Ras/Raf/MEK/ERK pathway [19,20].

While the classic view of the Ras-GTP/Ras-GDP cycle regulated by guanine nucleotide exchange factors (GEFs) and GAPs portrays a common regulatory mechanism for GTPases in general, it misses the nuances that may be particular to each one. In the case of Ras, recent structural work has elucidated two kinds of dynamical properties that may lead to novel opportunities for targeting it directly. One is associated with a recently discovered allosteric switch that appears to function to increase intrinsic hydrolysis of GTP in the presence of Raf [20,21]. The regulation through an allosteric switch mechanism is probably not general to all Ras pathways, but may be of critical importance in attenuation of the Ras/Raf/MEK/ERK pathway [21], a major regulator of cell proliferation [7]. This would abrogate the need for GAP to displace Raf in order to turn off the signal, although GAP may still contribute by quickly depleting the pool of available Ras-GTP. The other recent finding has to do with nucleotide-specific orientations of the catalytic domain with respect to the membrane and binding site hot spots that may be targeted to interfere with these critical interactions [22–24]. Together, these two recent paradigm-shifting discoveries—allosteric switch and nucleotide-specific orientation with respect to the membrane—open novel venues for drug discovery that have not yet been explored. Here, we describe in detail the structural features associated with the allosteric switch in Ras and its connection with the membrane in Ras-GTP. Drawing on our experience with Ras, we suggest that some Ras oncogenic mutants have the potential to shift the equilibrium between natural conformational states, disfavoring intrinsic hydrolysis of GTP by stabilizing a catalytically incompetent state in the presence of Raf. We discuss the effect of small molecule interactions on the allosteric state of Ras-GTP, with an overview of allosteric states found in Ras–ligand complexes that have been deposited in the PDB.

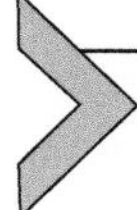

2. Ras ARCHITECTURE: THE G DOMAIN AND ITS FUNCTION

Ras GTPase is composed of a catalytic G domain and a C-terminal hypervariable region (HVR) that is posttranslationally modified for insertion in the membrane. The HVR (residues 171–189 in H-Ras numbering) is highly divergent in the H-, K-, and N-Ras isoforms and is responsible for the membrane localization associated with specific function [25]. The extent of posttranslational modification on the HVR and its effect on Ras has been reviewed elsewhere [26]. In regards to our current review, the importance of the HVR is that it places the Ras G domain at the cytoplasmic membrane surface. The interaction of Ras with the membrane is a nucleotide-dependent process [22], mediated by posttranslational modifications at the HVR [26,27], as well as by residues in helices α4 and α5 [23,24] that face the membrane in Ras-GTP. The nucleotide-dependent orientations of the G domain with respect to the membrane have been shown to regulate the specificity and strength of effector interactions with Ras [22,24,28].

The G domain is the defining structural feature of the Ras superfamily, of which Ras is the canonical member [4,29]. The architecture of the G domain consists of six β-sheets and five α-helices, forming a modified Rossman fold [30] that produces an overall globular structure with α-helices 2, 3, and 4 on one side of the β-sheet, and α-helices 1 and 5 on the other (Fig. 3.1a). This fold consists of a three-layered αβα sandwich (β1–α1–β2–β3–α2–β4–α3–β5–α4–β6–α5) with intervening loops (L1–L10) between secondary structural motifs. Unlike the classical Rossman fold motif, which consists of each β-sheet in parallel alignment in the general form β3–β2–β1–β4–β5–β6, the G domain of the Ras superfamily forms a parallel/antiparallel mixed β-sheet [31]. The structural orientation of β2 and β3 is flipped and

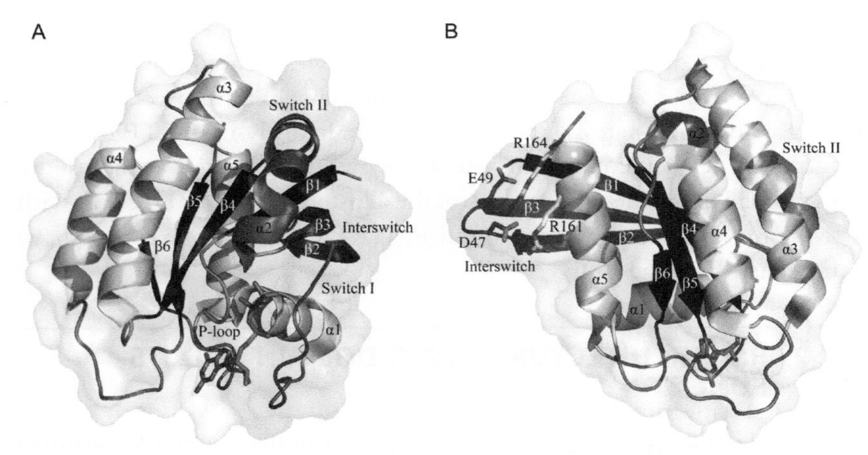

Figure 3.1 Ribbon diagram of Ras-GppNHp showing the secondary structural elements that make up its architecture. The β-sheets are in black and the α-helices are in light gray, except for the P-loop consisting of loop L1 (yellow), switch I consisting of L2 (brown), the interswitch region consisting of β2 and β3 (ruby), and the switch II composed of L4 and α2 (pink). The nucleotide is shown with carbon and phosphorous atoms in orange. The N-terminal effector lobe of the G domain is indicated with the surface in green and the C-terminal allosteric lobe is indicated with the surface in white. Views in (a) and (b) are rotated 90° relative to one another. In (b), the nucleotide sensor residues D47, E49, R161, and R164 are shown in stick and are found at the membrane-interacting surface of Ras. (See color plate.)

forms a β-hairpin at the start of the β-sheet (β2–β3–β1–β4–β5–β6) [30]. The β-hairpin produces an important structural element in the G domain, referred to as the interswitch region, leading away from switch I and back into switch II in the protein sequence, with the turn containing residues that together with residues on α5 are nucleotide-sensing residues near the membrane-interacting region of Ras [32] (Fig. 3.1b). The interswitch, along with switch I and switch II, has been shown to be integral in the evolution and diversification of the Ras superfamily, providing sites of deformation that act to both conserve function across the superfamily and at the same time allow specialization [30].

The N-terminal lobe consisting of β1–α1–β2–β3–α2–β4 represents the catalytic portion of the G domain, including the P-loop (residues 10–14), switch I (30–40), and switch II (60–76). This lobe exhibits 100% conservation between the three isoforms of Ras (H-, K-, and N-Ras) [33]. The P-loop, switch I, and switch II regions converge over the β-γ phosphate end of GTP, providing the catalytic elements for hydrolysis of GTP to GDP. At the same time, switches I and II also act as the binding regions for Ras effectors and regulators, providing a logical scheme by which the nucleotide bound state of Ras can produce signaling [34,35]. We therefore refer to this lobe as the effector lobe (residues 1–86, green surface in Fig. 3.1) [23]. We refer to the C-terminal half of the protein, α3–β5–α4–β6–α5, which harbors the allosteric site as well as sites of membrane interaction, as the allosteric lobe (87–171, white surface in Fig. 3.1) [20,24]. The allosteric lobe exhibits 90% conservation between Ras isoforms and contains all of the isoform-specific differences outside of the C-terminal HVR [33].

Overlap of catalytic and effector regions produces an elegant solution to the problem of signal transduction, a solution that is likely responsible for the ubiquity of GTPases in almost every cellular process [36]. The general scheme of GTPase regulation starts with the inactive, GDP-bound form of Ras. Activation of Ras involves binding of a GEF which opens up the active site and destabilizes the interaction of Ras with GDP and Mg^{2+} [37]. Subsequent binding of GTP allows Ras to interact with various effectors, including RalGDS [38], PI3K [39], Raf [40], and NORE1A [41], each promoting the activation of one or more cellular responses [7]. Interaction with protein-binding partners results in the stabilization of switches I and II in a specific conformation to carry out the associated functional role [42,43].

Cessation of signaling then must occur through hydrolysis of GTP to GDP, resulting in the inactive GDP-bound form of Ras. Until recently, this catalytic event was thought to be solely dependent on GAPs [16]. Ras is capable of intrinsic hydrolysis of GTP to GDP [12], but the possibility that its slow rate could be enhanced by factors in the cell other than GAPs has been largely ignored due to a lack of evidence for this possible mode of regulation. The discovery of the allosteric switch has provided a venue through which GAP-independent hydrolysis could be enhanced, suggesting the possibility that Ras, in some pathways, may be regulated by more subtle mechanisms involving the intrinsic hydrolysis reaction.

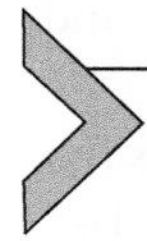

3. THE ALLOSTERIC SWITCH IN Ras-GTP AND ITS ROLE IN INTRINSIC HYDROLYSIS

The interplay of interactions between the two lobes of the catalytic domain of Ras in its GTP-bound form is a key feature of the allosteric switch mechanism that links the active site for GTP hydrolysis with membrane-interacting sites on Ras [20]. Active GTP-bound Ras can be in several conformational states identified by a combination of X-ray crystallography and NMR spectroscopy, obtained using GTP analogs such as GppNHp. These states are associated with Ras' ability to hydrolyze GTP and can be regulated by proteins or other factors that shift the population of states. In the absence of binding partners, both switch I and switch II are disordered [44], resulting in slow hydrolysis of GTP [45]. There are two main states thus far observed for switch I by ^{31}P NMR, based on the position of residue Y32 with respect to the nucleotide: one is associated with GAP-catalyzed hydrolysis and the other with intrinsic hydrolysis, termed state 1 and state 2, respectively [46,47]. In state 1, the switch I residue Y32 is away from the nucleotide, making room for the GAP arginine finger to complement the active site as seen in the Ras/RasGAP complex (PDB code 1WQ1) [16]. In state 2, Y32 is intimately placed in the active site as seen in the Raps/Raf-RBD complex (PDB code 1GUA), where Raps is the homologue Rap with switch I residues mutated to those in Ras and RBD is the Ras-binding domain of Raf that interacts with the effector region [40,48]. In the canonical crystal form of Ras-GppNHp, with symmetry of the space group $P3_221$, switch I is stabilized by crystal contacts in state 1 and the resulting model has been the standard for Ras since the early 1990s [49]. In our more recently discovered crystal form with symmetry R32, switch I is stabilized in state 2 and switch II is free of crystal contacts [50]. With respect to switch I, our crystals represent a situation that we expect favors intrinsic hydrolysis and has a structure identical to that of the Raps/Raf-RBD complex [40]. Given the switch I similarity in our crystals to switch I in the Raps/Raf-RBD complex and that Raf interacts with Ras through switch I but not switch II [51], we use our crystals as a model for studying the effects of Raf on the Ras allosteric switch.

With switch I stabilized in the state 2 conformation, switch II is able to adopt a range of conformations affected by the position of helix 3 and loop 7 (α3/L7) associated with the two states of the allosteric switch mechanism [20]. In one of the states, which we obtained by using calcium chloride in our crystallization conditions, switch II is disordered and α3/L7 (residues 87–110) is positioned toward the effector lobe (PDB code 2RGE) (Fig. 3.2) [50]. This corresponds to the T state in which hydrolysis is slow due to an incomplete active site with catalytic residue Q61 disordered (T for tardy in this case). In the other state, obtained in our crystals in the presence of calcium acetate, α3/L7 is shifted by nearly 3 Å toward α4, making room for an ordered switch II helix with R68 at the center of an H-bonding network that completely orders the active site and places Q61 for catalysis (PDB code 3K8Y) (Fig. 3.2) [20]. This is the R state (R for reactive in this case).

We referred to the R state in our previous publications as the "on" state of the allosteric switch, since we associate this state with a catalytically "on" conformation. This promotes turning the Ras signal "off." The T state, which is catalytically "off," and thus originally referred to as the "off" state of the allosteric switch, keeps Ras in the GTP-bound state where signaling is "on." This is a confusing nomenclature that we wish to avoid in the future.

The R state is stabilized by binding of calcium and acetate ions at an allosteric site remote from the active site. The acetate ion makes a direct interaction with R97 positioned toward the C-terminal end of α3. The Ca^{2+} coordinates with one of the acetate oxygen atoms; with the carbonyl groups of D107 in L7 and Y137 at the C-terminal end of α4; and with three water molecules (Fig. 3.2). We propose that Ca^{2+} binds at the allosteric site along with a negatively charged membrane component mimicked by the acetate ion [20], as this site nestles up against the membrane in Ras-GTP [22,24]. R97 is at the center of a network of ionic and H-bonding interactions that stabilizes α3 in the R state. The shift of α3 is further stabilized by hydrogen bonds between the side chains of Y137 in α4 and H94 in α3 (Fig. 3.2). This positioning of α3 facilitates a second network of H-bonding interactions centered on switch II residue R68 and extensively mediated by water molecules. This network involves several residues in α3, which forms an anchor for the catalytic conformation of switch II that positions Q61 in the active site. The network involves no less than six water molecules that facilitate

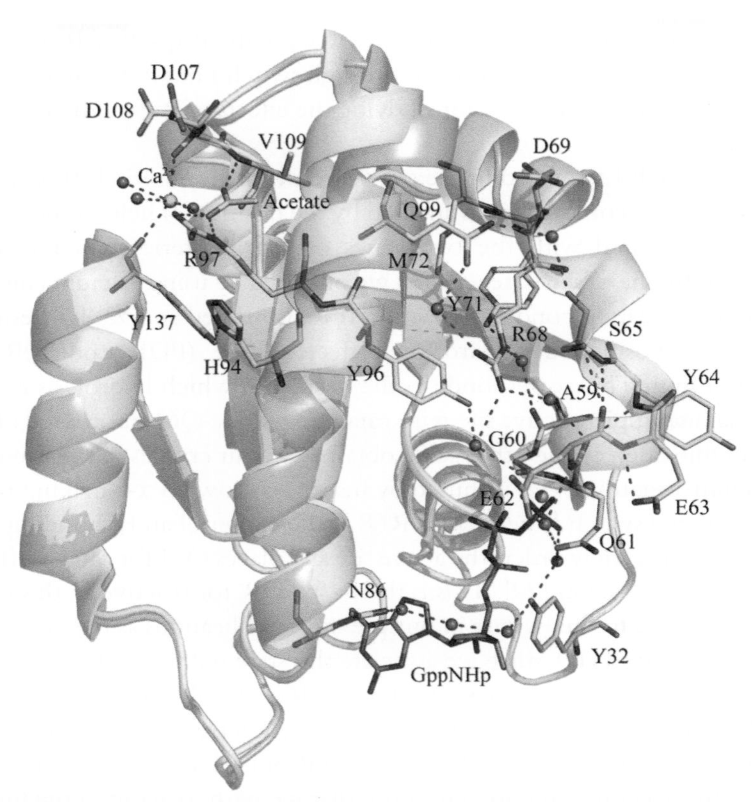

Figure 3.2 Ribbon diagram of the T and R states of the allosteric switch in Ras-GppNHp. The T state is shown in gray (PDB code 2RGE) and the R state in green with water molecules in red spheres and the calcium ion in green (PDB code 3K8Y). The nucleotide is shown in stick with carbon and phosphorous atoms in orange. The Ras residues in the R state that contribute to binding of calcium and acetate at the allosteric site and to the water-mediated H-bonding networks associated with ordering of switch II are shown in sticks, as is the acetate molecule. (See color plate.)

H-bonding interactions throughout the entire length of switch II, starting at the C-terminal end of α2 and ending at the N-terminus of switch II that lies over switch I and the γ-phosphate of GppNHp (Fig. 3.2). The placement and conformation of R68 is stabilized by van der Waals interactions with Y71 and M72. The guanidinium group of R68 then provides a locus around which the hydrogen-bonding network can form, with all three side chain nitrogen atoms making water-mediated interactions to various parts of α3 and switch II. Through these water-mediated H-bonds, R68 links the side chains of Y96 and Q99 in α3 to the backbone carbonyl groups of A59, G60, Q61, and S65. To complete the ordering of switch II side chains at the active site, the backbone amide and side chain hydroxyl group of S65 directly hydrogen bond with the backbone carbonyl group of E62. The E62 side chain and backbone amide then hydrogen bond to a water molecule that in turn stabilizes the side chain of Q61 [20]. This network extends to the N-terminal end of α3 through another series of water molecules. In addition to the nucleophilic water molecule present in all high-resolution structures of Ras in the GTP-bound form, the active site with switch I in state 2 and the allosteric switch in the R state has a bridging water molecule that H-bonds simultaneously to the side chains of Y32 in switch I and Q61 in switch II, as well as to the γ-phosphate of GppNHp. The network continues on the other side of Y32 through three water molecules that connect it to the side chain of N86, at the N-terminus of α3 (Fig. 3.2) [20]. Thus, in the R state, the effector lobe in Ras is anchored to the allosteric lobe through water-mediated H-bonding interactions at both ends of α3. In the T state, these water-mediated networks are impaired and there is a much looser connection at the interface between the two lobes. The allosteric site is unoccupied and the aliphatic portion of the R97 side chain participates in a hydrophobic core between α3 and α4 formed by residues C80, I93, M111, Y137, and I139, while its guanidinium group is exposed to the protein surface.

The conformational changes observed in going from the T to R state of the allosteric switch in our crystals are consistent with dynamics in the millisecond time scale obtained by NMR [52] and by molecular dynamics simulations [53]. In both cases, these motions involve global features of Ras-GppNHp that are contributed in great part by α3/L7. NMR has also been used to study the binding of ions at the allosteric site in solution [54]. The study shows that there is weak binding of either Ca^{2+} or Mg^{2+} in the allosteric site in Ras-GppNHp, with K_d in the low millimolar range and that this is not affected by the presence of acetate. It also shows that the allosteric site in solution has a significant degree of flexibility [54]. The observed disorder of switch II and a sluggish reaction rate for GTP hydrolysis measured *in vitro* [12,20] are consistent with the T state being the predominant state of the allosteric switch in Ras-GTP (or GppNHP) in solution. The NMR result is different from what we observe in the crystal where the allosteric site shows a clear preference for Ca^{2+} over Mg^{2+}, and Ca^{2+} binds in the presence of $Ca(OAc)_2$, but not $CaCl_2$ [20]. Although the transition from the T to the R state in the crystals is clearly dependent on both calcium and acetate and results in highly ordered allosteric and active sites, this is not the case in solution, where weak binding of either Ca^{2+} or Mg^{2+} is not reported to alleviate the broadening of the signal associated with disorder in switch II [54]. It may be that the interaction between Ras and the membrane at several sites, including sites containing residues R128 and R135 shown to make salt bridges with membrane phospholipids [24,32], helps stabilize the allosteric site, making it specific for Ca^{2+} and a negatively charged component. In our crystals, at least one of these arginine residues makes a salt bridge with negatively charged residues in a symmetry-related molecule of Ras, perhaps mimicking this stabilization effect. In any case, in both the crystals and in the NMR experiments, the concentration of Ca^{2+} used is much higher (100 and 20 mM, respectively) than the physiological concentrations of calcium in the cell (0.01–1 μM), even accounting for local transient concentrations produced through stimulation ($<$10 μM) [55]. However, the two-dimensionality of the membrane may result in an increase in effective concentrations between binding partners up to 1000-fold [54]. Given the K_d of 5.9 mM measured by NMR for the allosteric site affinity for Ca^{2+} in the T state, and that an ordered allosteric site in the R state most likely has a greater affinity and increased specificity for calcium, the cellular concentration Ca^{2+} could very well be in the range for binding to have a regulatory role in intrinsic hydrolysis of GTP by Ras.

The conformation of the active site in the R state (PDB code 3K8Y) is consistent with a substrate-assisted mechanism [56] involving two catalytic water molecules: the nucleophilic water positioned for an in-line attack on the γ-phosphate of GTP and the bridging water molecule that makes three excellent H-bonding interactions: one with the γ-phosphate of GTP (2.6 Å), one with Y32 in switch I (2.6 Å), and one with Q61 in switch II (2.8 Å) (Fig. 3.3). This arrangement led to our proposed mechanism for intrinsic hydrolysis where a proton abstracted from the nucleophilic water molecule is shuttled through the γ-phosphate to the bridging water molecule, which in turn develops a partial positive charge [20]. This positively charged bridging water molecule is near the oxygen atom that links the β and γ phosphates of GTP, where negative charge accumulates in the dissociative-like transition state of the hydrolysis reaction [57,58]. Thus the bridging water molecule in the transition state for intrinsic hydrolysis could play a similar role to the Arg finger in the GAP-catalyzed reaction. Toward the end of the reaction, the proton is delivered from the bridging water molecule to the GDP leaving group and the side chain of Q61 flips outward, facilitating the exit of the newly formed inorganic phosphate (P_i) [20]. Vibrational spectroscopy experiments support the catalytic role of Q61 as stabilizing water in the active site, consistent with our proposed

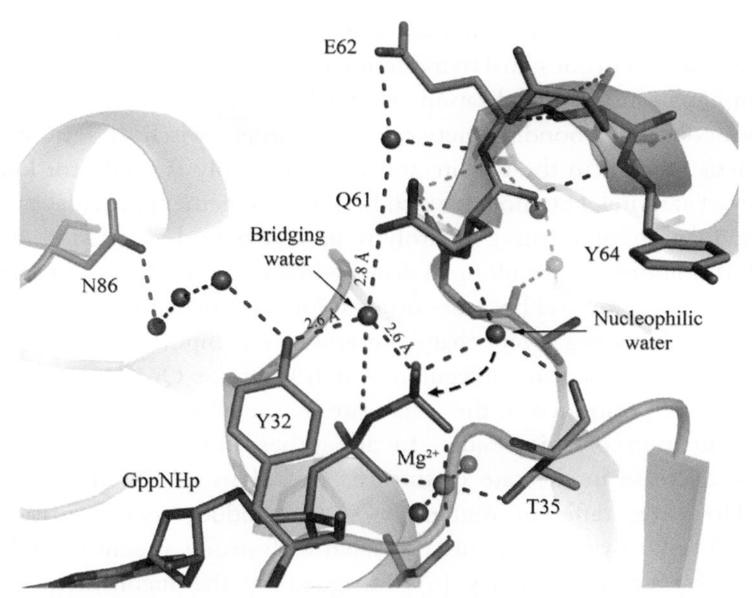

Figure 3.3 The active site in Ras-GppNHp with switch I in state 2 and switch II ordered with the allosteric switch in the R state. The P-loop, switch I, and switch II are shown centered on the active site and the water network extending to the N86 at the N-terminal end of α3 is included. The model with PDB code 3K8Y was used with the protein carbon atoms shown in green, oxygen atoms in red, and nitrogen atoms in blue. Water molecules are shown in red spheres and the magnesium ion is represented as a green sphere. The nucleotide is shown with carbon and phosphorous atoms in orange. The structure is of the ground state. A black dashed arrow indicates the attack of the nucleophilic water molecule on the γ-phosphate of GTP. The orange dashed line indicates the connection between the bridging water molecule and the oxygen atom between the β- and γ-phosphates, which are expected to become closer during the transition state of the reaction in our proposed mechanism for intrinsic hydrolysis of GTP by Ras. (See color plate.)

mechanism [59]. This mechanism is fundamentally different from the two-water model proposed in the early days for intrinsic hydrolysis based on the canonical crystal form with symmetry $P3_221$ [60] where switch I is in state 1, a situation that today we know disfavors the intrinsic hydrolysis reaction [47].

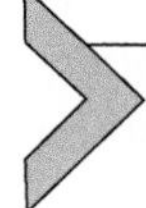

4. THE ALLOSTERIC SWITCH LINKED TO THE Ras/Raf/MEK/ERK PATHWAY

The fact that switch I is stabilized in state 2 in crystals with symmetry R32 with the conformation seen in the Ras/Raf complex led to the idea that the allosteric switch mechanism may be associated with the Ras/Raf/MEK/ERK pathway, a major signal transduction cascade involved in the control of cell proliferation [7]. Stabilization of switch I in state 2 is critical for the water-mediated H-bonding networks that order switch II for intrinsic hydrolysis of GTP in the R state and Raf is the only Ras effector known to interact at switch I but not at switch II. This arrangement provides a venue for modulation of intrinsic hydrolysis in Raf-bound Ras-GTP, where switch II is disordered, resulting in slow hydrolysis in the absence of calcium in the allosteric site, and becomes ordered for catalysis with bound calcium and presumably a negatively charged membrane component (mimicked in our crystals by the acetate molecule). Switch II residue Q61 is a critical catalytic residue positioned in the active site by the allosteric switch mechanism. Mutation of this residue to any other amino acid results in a 10-fold decrease in the rate of intrinsic hydrolysis measured *in vitro* for Ras. However, different mutants of Q61 produce foci formation in NIH3T3 cells to various extents associated with strong, medium, and poor cell transformation efficiency [6]. Intrigued by the inconsistent results obtained *in vitro* and in cells for the various RasQ61 mutants, our group determined the structure of three strongly transforming mutants, Q61L (PDB: 2RGD), Q61V (PDB: 2RGC), and Q61K (PDB: 2RGB); a moderately transforming mutant Q61I (PDB: 2RGA); and compared them to a previously published Q61G structure (PDB: 1ZW6) [61], all from crystals with symmetry of the R32 space group [50]. Recall that in this crystal form, switch I is stabilized in the conformation seen in the Ras/Raf complex, state 2, associated with intrinsic hydrolysis. However, in all of the Q61 mutants, Y32 moves closer to the nucleotide, resulting in a direct H-bond between its hydroxyl group and the γ-phosphate of GTP, excluding the bridging water molecule that is normally found in the wild-type structure (Fig. 3.4). The distinguishing feature between the highly, moderately, and nontransforming mutants is the conformation of switch II [50]. The highly transforming mutants all have an aliphatic portion of the residue 61 side chain in a hydrophobic cluster that also involves Y32, P34, I36, and Y64 (Fig. 3.4). This is identical to the noncatalytic conformation seen in the Ran/Importin-β complex (PDB: 1IBR) where hydrolysis of GTP is prevented during trafficking of cargo between the nucleoplasm and cytoplasm [62]. The moderately transforming mutant Q61I can attain this conformation with some steric hindrance and the nontransforming mutant Q61G has an open active site exposed to bulk solvent [50]. Since the switch I conformation in the structures of highly transforming mutants is critical for positioning Y32 in

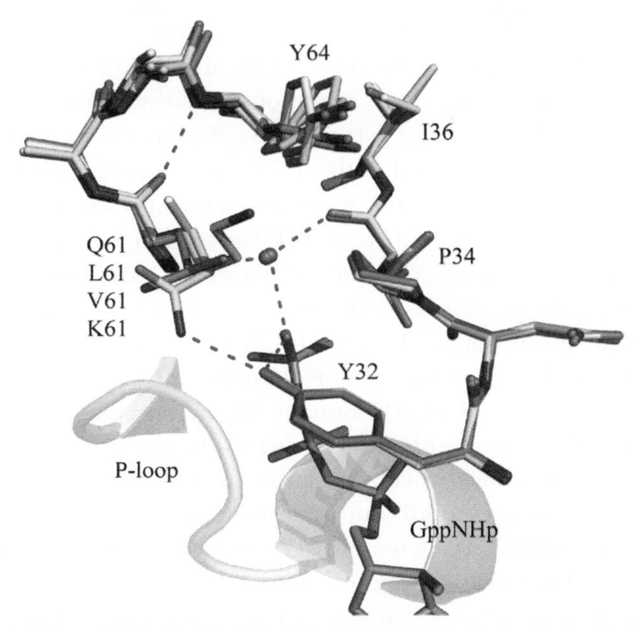

Figure 3.4 Ras-GppNHp in the T state with the anticatalytic conformation of switch II. This conformation is seen in the highly transforming mutants of Ras at residue 61: RasQ61L is in hot pink, RasQ61V is in yellow, and RasQ61K is in wheat. This conformation is also observed for wild-type Ras, shown in cyan, when a small molecule such as DTE or DTT is bound in a pocket between switch II and α3 (pocket not shown in this figure). The nucleotide is shown with carbon and phosphorous atoms in orange. The nucleophilic water molecule is represented as a red sphere. Dashed red lines indicate hydrogen bonds. (See color plate.)

the hydrophobic cluster that stabilizes the anticatalytic conformation of switch II, we reasoned that the potency of these mutants in forming foci in NIH3T3 cells could be associated with the Ras/Raf complex. We thus expected to show with *in vitro* experiments that a highly transforming mutant such as RasQ61L does not hydrolyze GTP in the presence of Raf, while in the absence of Raf it is only 10-fold slower than the wild type as previously observed. We were indeed able to show this both qualitatively with size exclusion chromatography experiments over many hours [50] and quantitatively by measuring the single turnover rate for intrinsic hydrolysis of GTP catalyzed by Ras in the absence and presence of Raf [21]. The anticatalytic conformation that we associate with the transforming RasQ61 mutants in the presence of Raf was obtained from crystals grown in the presence of $CaCl_2$ with the allosteric switch in the T state, under similar conditions to those where the wild type shows a disordered switch II (PDB code 2RGE). Mutants of Ras at residue 12 are also highly transforming and RasG12V, like RasQ61L, is a common mutation found in human tumors [11]. Unlike the RasQ61 mutants, the switch II in RasG12V (PDB code 3OIV) is disordered in the T state as it is in the wild type [21]. This structure, however, shows a direct interaction between the hydroxyl group of Y32 and the γ-phosphate of GTP as seen in the Ras61 mutants. We were able to obtain the structures of both RasG12V and RasQ61L in the R state by growing crystals in the presence of $Ca(OAc)_2$. Interestingly, while the R state was obtained for RasG12V in 100 mM $Ca(OAc)_2$ as for the wild type, a higher concentration of 345 mM $Ca(OAc)_2$ was necessary to obtain the R state for the RasQ61L mutant. This is most likely due to greater stabilization of the T state in RasQ61L due to the hydrophobic core associated with the anticatalytic conformation in this mutant. Cell biology experiments are consistent with our interpretation of the structural results. By transforming NIH3T3 cells with RasG12V and RasQ61L and measuring the output of MEK and ERK phosphorylation exclusively due to signaling through Ras/Raf, we were able to detect a large difference between the two mutants in the activation of the Ras/Raf/MEK/ERK pathway, with RasQ61L resulting in saturating basal levels of MEK and ERK phosphorylation not observed for RasG12V [21]. Taken together, the structural, biochemical, and cell biology experiments support a model in which the duration of the Ras/Raf interaction is modulated by intrinsic hydrolysis promoted through the allosteric switch mechanism. The equilibrium between the T and R states is shifted toward the T state in the RasQ61 mutants by promotion of an anticatalytic conformation in the presence of Raf, resulting in constitutive activation of the Ras/Raf/MEK/ERK pathway for these mutants [21].

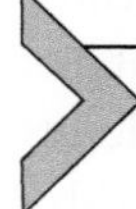

5. MOST Ras STRUCTURES IN THE PROTEIN DATA BANK ARE IN THE R STATE

A survey of Ras structures found in the protein data bank and solved at 3.0 Å resolution or better shows that the majority have α3/L7 in the overall position associated with the R state of the allosteric switch, with α3 shifted toward α4. However, a detailed look at the conformations of switches I and II shows large variation in structure. In all structures of Ras in complex with effector proteins other than Raf, switch I is in state 2 and switch II is ordered. This includes complexes with RalGDS (PDB code ILFD), PLC-ε (PDB code 2C5L), PI3K (PDB code 1HE8), Byr2 (PDB code 1K8R), as well as an intracellular antibody Anti-Ras FV (PDB codes 2V5H, 2UZI). Since most of these complexes were obtained with RasG12V, the active site in these structures resembles that seen in our crystals of RasG12V-GppNHp in the R state (PDB code 3OIW), with Q61 interacting with the nucleophilic water molecule and a direct H-bond between the hydroxyl group of Y32 and the γ-phosphate of the GTP. In these complexes, stabilization of switch II by the effector proteins is independent of ligand binding at the allosteric site. Here, the R state is a consequence of protein binding, resulting in the shift of α3 to avoid steric hindrance with the ordered conformation of switch II promoted by complex formation. This is also the case in the complex with GAP (PDB code 1WQ1), where the binding of GAP stabilizes switch II with R68 at the center of H-bonding interactions in the active site [16]. Interestingly, the overall switch II conformation observed in the Ras/GAP complex is somewhat different from the conformation seen in our R32 crystal form that mimics the Ras/Raf interaction (PDB code 3K8Y). Although the effector proteins that bind both switch I and switch II have a switch I in state 2 similar to that observed for Ras/Raf, the switch II near the active site is closer in detail to the conformation seen in the Ras/RasGAP complex [43]. Without the bridging water molecule in place, this conformation is unlikely to promote intrinsic hydrolysis and termination of the signal is likely to require displacement of the effectors for GAP-catalyzed hydrolysis. The one exception is the structure of the Ras/NORE1A complex (PDB code 3DDC) that has an active site identical to that observed in our structure of wild-type Ras with calcium and acetate bound at the allosteric site (PDB code 3K8Y). This is the conformation that we associate with intrinsic hydrolysis, with the active site bridging water molecule making H-bonds to Y32, Q61, and the γ-phosphate of the GTP analog. Interestingly, NORE1A is a tumor suppressor protein and we have proposed that its stabilization of the catalytic conformation may be associated with turning off signaling through Ras [63]. The Ras/Sos complex (PDB: 1NVW), where Ras is in a GTP-bound state and bound to the allosteric site of Sos [64], deviates from the other complexes in that α3 is in a intermediate position between that of the T and R states, while the N-terminal end of switch II is shifted toward the N-terminal end of α3, so that switch II residue E62 makes a direct hydrogen bond with K88 in α3. This distortion of the active site may contribute to prolonged duration of Ras in its GTP-bound state when bound to the Sos allosteric site.

An inspection of Ras crystallized with symmetries of different space groups also shows Ras mostly in the R state, with switch II either ordered or partially ordered due to crystal contacts. In general, the T state has only been observed in Ras structures obtained from crystals with symmetry of the R32 space group, which is why the allosteric switch mechanism was only discovered so recently. In the canonical space group with symmetry $P3_221$, structures of Ras and its mutants are always found in the R state, with α3 shifted toward α4. Due to the lack of crystal contacts near switch II, the R32 crystal form perhaps provides the closest mimic of the allosteric state of Ras in solution, with a highly disordered switch II associated with the T state in the wild-type structure. Furthermore, the anticatalytic conformation observed in the RasQ61 mutants is only seen associated with the T state and again can be accessed due to a lack of crystal contacts in the region. Transition to the R state in this crystal form is dependent on the presence of Ca^{2+} and acetate at the allosteric site and our serendipitous discovery of this transition has opened up new questions regarding the regulation of Ras discussed in this chapter. Among the several mutants of Ras whose structures have been obtained in this crystal form [21,50,61,65,66], the only structure that has an R state conformation in the absence of $Ca(OAc)_2$ is that of the RasT35S mutant (PDB code 3KKM). This structure is unusual in that it has a disordered switch II, commonly associated with the T state. This atypical situation may be explained by the appearance of two $(PO_4)^{3-}$ molecules near α3 that interact on either side of the aromatic rings of Y137 and H94, which are H-bonded to each other as in our canonical R state. The $(PO_4)^{3-}$ molecule closest to the allosteric site makes H-bonding interactions with R97 and K101, as well as with numerous water molecules, and R97 forms a hydrogen bond with the carbonyl group of D107. The second $(PO_4)^{3-}$ molecule H-bonds with the backbone carbonyl group of E91 and with a water molecule that forms H-bonds with Y137 and H94, but it also interacts with backbone carbonyl groups of a symmetry-related molecule. The effect of these interactions is to stabilize α3 toward α4 in the R state. In this structure, switch I is pulled away from the nucleotide in state 1 [66] and in the absence of crystal contacts switch II is disordered. This is an interesting situation showing that a shift of α3 to make room for ordering switch II, although necessary, is not sufficient to promote order. In fact, an ordered switch II usually does not occur without an ordered switch I, suggesting that switch I conformations may be required for the formation of an ordered switch II. A notable exception to this occurs when Ras is bound to the diaminobenzophenonephosphoramidate-GTP (DABP-GTP) analog [67]

(PDB codes 1RVD, 1CLU). This analog is a derivative of GTP with a large hydrophobic group attached at the γ-phosphate. It is possible that this analog provides its own packing surface for switch II.

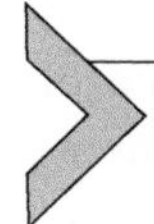

6. MODULATION OF THE ALLOSTERIC STATES IN Ras-GTP BY SMALL MOLECULES

The allosteric switch mechanism through which intrinsic hydrolysis may be modulated opens new opportunities for targeting Ras, although this is not without significant challenges. The Ras/Raf complex appears to be unaffected by whether Ras is in the T or R state, as long as Ras is bound to GTP. Our hypothesis is that in the R state, intrinsic hydrolysis is promoted and, once Ras is bound to GDP, dissociation of the complex takes place and the signal through the Ras/Raf interaction is turned off. Thus promotion of the R state by small molecule binding could be a new venue for targeting Ras. While we had already shown that this is possible through ligand binding at the allosteric site, another serendipitous discovery brought to our attention the fact that the equilibrium between allosteric states can also be shifted toward the T state through binding of small molecules at a site between switch II and α3 near the active site [63]. This site happens to be a hot spot for ligand binding [23,68], but one that must be avoided in targeting Ras-GTP unless it leads to the disruption of Ras/effector interactions.

In the search for an optimal stabilization solution for transferring crystals out of the mother liquor, we discovered that in the presence of 60% PEG (30% PEG3350 and 30% PEG400) a molecule of the reducing agent DTE or DTT present at the concentration of 1 mM was found to bind at the site between switch II and α3, stabilizing the anticatalytic conformation that we previously observed for the highly transforming RasQ61 mutants and for wild-type Ran bound to Importin-β [63]. In wild-type Ras, Q61 in the anticatalytic conformation forms a direct H-bond to the hydroxyl group of Y32, which also interacts directly with the γ-phosphate group of the nucleotide, contributing to the ordered conformation of switch II associated with the T state (Fig. 3.4). The bridging water molecule, present in the wild-type structure with a disordered switch II but absent in all of the mutants, is also absent in the wild type where switch II adopts the anticatalytic conformation in the presence of DTE/DTT.

It is clear that the binding of $Ca(OAc)_2$ at the allosteric site and either DTE or DTT near the active site between switch II and α3 are mutually exclusive. The conformation of the allosteric site in the T state is not conducive to coordinating the calcium and acetate ions, whereas the R state conformation of the active site would place R68 in a position that overlaps with the DTE/DTT-binding pocket. Interestingly, we were not able to obtain the R state in the high PEG conditions, even in the presence of 200 mM $Ca(OAc)_2$. We interpreted this as being due to preferential interaction of water with PEG leading to protein dehydration and maximizing interactions between protein atoms associated with the anticatalytic conformation. We thus used lower PEG conditions (30% PEG 3350) in a series of soaking experiments where we varied the $Ca(OAc)_2$ and DTE/DTT contents of the solutions into which crystals were transferred. In the absence of either of the ligands, Ras molecules in the crystal sample both the T and R states, with clear electron density to support both conformations of the α3/L7 structural element associated with the allosteric switch. Starting from these crystals, we could then stabilize either the R state or T state at will, with 100 mM $Ca(OAc)_2$ or 100 mM DTE/DTT, respectively. Competition between the two ligands was also explored and showed that at 100 mM $Ca(OAc)_2$, the R state is stabilized in the presence of 100 mM DTT, but that the T state predominates in the presence of 100 mM DTE [63].

In addition to showing the reversibility between the T and R states in the crystals and the sensitivity to small molecules, this study also brought insight into the effect of the bulk solvent composition in facilitating ligand binding at the site between switch II and α3. There is a clear effect of the PEG concentration on the ability of DTE or DTT to bind at that site. Measurement of hydrolysis rates in the presence of 300 mM DTE in solution where PEG is not present did not show the expected decrease in rate associated with the anticatalytic conformation, leading to the conclusion that DTE does not bind significantly under these conditions. In the presence of 30% PEG, DTE or DTT binds in the crystal but can be displaced upon competition with calcium and acetate binding at the allosteric site. In the 60% PEG conditions, the anticatalytic conformation of switch II is stabilized, further facilitating binding of DTE/DTT, even in the presence of high concentration of $Ca(OAc)_2$. These experiments indicate that without the bulk solvent effect due to the presence of PEG, it is unlikely that small molecules such as DTE or DTT can by themselves promote the ordering of switch II that accompanies binding. This brings up the interesting point that screening for small molecules that bind Ras *in vitro* under dilute solution conditions could lead to hits that may behave very differently in the crowded environment of the cell, as the environment clearly influences the conformational behavior of Ras.

The direct targeting of Ras with small molecules has been elusive, in spite of significant recourses devoted to this effort by both pharmaceutical companies and academic laboratories [68]. There are a few compounds that have been shown to block the Ras/Raf interaction [69–72], but there are no structures of Ras in complex with these inhibitors and it is unclear where the compounds bind on the surface of Ras. Another set of compounds target the Ras-GDP/Sos interaction with inhibition of nucleotide exchange [73,74]. These studies include high-resolution crystal structures of several molecules bound to Ras-GDP in a pocket near Y71 on the exposed surface of switch II. Interestingly, this pocket is very similar in some structures of Ras bound to GTP analogs and there are three crystal structures in the PDB (4DSO, 4DSN, and 4DST) with the Sos-inhibiting compounds bound to Ras in complex with GTP analogs (GTPγS or GppNHP) [73]. Binding of small molecules at this site was also observed by a fragment-based computational method, FTMap [75], using a template representing Ras-GppNHp in the R state and switch I in state 2, with Y32 closed over the nucleotide [23]. The three Ras-GTP structures with compounds bound at the Y71 site also have switch I in state 2 and α3/L7 in the conformation associated with the R state. In addition to inhibiting the Ras-GDP/Sos complex associated with nucleotide exchange, binding at this site would be expected to interfere with effectors such as PI3K and RalGDS, but not with Raf, since this effector does not interact at switch II. The functional assays, however, did not focus on the Ras/effector interactions [73]. Another set of compounds, the Zn^{2+}-cyclens, have been shown by NMR to bind selectively to the Ras state 1 conformation of switch I in solution, thus disfavoring the state 2 effector-interacting conformation [76]. These compounds bind at two distinct sites shown by NMR to be near the active site and on the opposite side of loop 7 relative to the allosteric site (Loop7 site). A crystal structure of Ras-GppNHp obtained with crystals of R32 symmetry in the presence of Zn^{2+}-cyclen (PDB code 3L8Y) shows the compound binding at the Loop7 site, but the switch I is in state 2 and the structure is found in the T state with a partially disordered switch II. The next generation of compounds that more effectively interact with Ras-GppNHp to stabilize the effector binding incompetent state 1 is the M^{2+}-Bis(2-picolyl)amine (M^{2+}-BPA) series of molecules [77]. These compounds are known by NMR to bind at the Loop7 site and at a site that partially overlaps with the Y71 site that binds the Sos-inhibiting molecules. Although these recent compounds have been shown by NMR to effectively stabilize the state 1 conformation of switch I and to inhibit the Ras/Raf interaction, there are no crystal structures of Ras-GppNHp/M^{2+}-BPA complexes. NMR experiments indicate that these inhibitors bind in a pocket that is seen in the structures of Ras-GppNHp crystallized with symmetry of the *I*222 space group in which switch I is clearly in state 1 [66,78]. The pocket is present in wild-type Ras-GppNHp (PDB code 4EFL), RasG12V-GppNHp (PDB code 4EFM), and RasQ61L-GppNHp (PDB code 4EFN). These structures are all in the R state with respect to the position of α3/L7.

The idea of modulating the conformation of switch I with small molecules that inhibit the Ras/effector interactions has emerged in the past few years and the first attempts have been made to develop molecules to this effect. Modulating the conformation of switch II to promote hydrolysis in the Ras/Raf complex is now emerging as a new concept that needs to be further validated and explored. The approach would be expected to be specific to the Ras/Raf/MEK/ERK pathway. We have shown that both RasG12V and RasQ61L bound to GppNHp can access the T and R states but that, in terms of abrogating hydrolysis *in vitro* and promotion of signal transduction in cells, Raf has a much more significant effect on RasQ61L than on RasG12V [21]. The Q61 mutations are particularly prominent in melanomas associated with N-Ras and this is also the only isoform in which a human tumor has been found to have a mutation at the allosteric site [11]. It is becoming increasingly clear that therapies must be developed for specific cancers. Targeting the allosteric switch mechanism could provide an effective venue to develop compounds specific for a particular type of mutation in Ras.

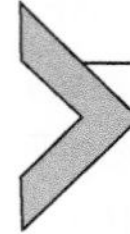

7. MAPPING THE BINDING SITE HOT SPOTS IN Ras: TARGETING THE PROTEIN/MEMBRANE INTERFACE

The catalytic or G domain of Ras is tethered to the cytoplasmic side of the cell membrane through the posttranslationally modified C-terminal HVR [26]. This sets the stage for complex dynamics between the allosteric lobe of Ras and the membrane, with residues R161 and R164 in α5 along with effector lobe residues D47 and E49 in the interswitch region near the membrane sensing the state of the nucleotide and somehow leading to a switch in orientation of the G domain with respect to the membrane (Fig. 3.1b). In the GDP-bound form, Ras interacts with the membrane primarily through the HVR and residues R169 and K170 immediately preceding it, located at the very end of the catalytic G domain. Upon binding to GTP, Ras undergoes an orientational switch that results in much more extensive interaction of the G domain with the membrane, involving a large surface of the allosteric lobe, including the allosteric site region [24,33]. In this orientation, residues R128 and R135 on α4 make important interactions with membrane phospholipids, confirmed by mutation studies, FRET experiments, and functional assays [22,28,32]. These studies show that this Ras-GTP orientation is clearly coupled to signaling output through the effectors Raf, PI3K, and galectin-1. The importance of residues R128 and R135 in mediating a signaling-competent orientation of the Ras G domain with respect to the plasma membrane suggests yet another venue for direct targeting of Ras at binding sites on the Ras-GTP surface that has to date been unexplored.

Our group recently used a combination of two complementary approaches to determine binding site hot spots on the surface of Ras-GppNHp [23]. The Multiple Solvent Crystal Structures method uses organic solvents as small molecule probes to experimentally locate binding site hot spots on proteins [79,80] and FTMap has been developed as a computational counterpart of this approach and shown to be effective in determining druggable sites on protein surfaces [75,81,82]. The MSCS experiments included 10 crystal structures of Ras-GppNHp solved in different solvent environments. Superposition of these structures identified eight clusters on the surface of Ras-GppNHp, most of which were found in the interlobal region between the effector and allosteric lobes [23]. Cluster 1 with the largest number of probe molecules is found in the pocket between α3 and switch II where DTE or DTT is observed to bind. Interestingly, the organic solvents appear to have a similar effect in stabilizing the T state of the allosteric switch as we observed with PEG, with poor hydration promoting interactions between protein atoms associated with the anticatalytic conformation of switch II [23,43,63]. Cluster 2, the second largest cluster found by MSCS, was located between helices α3 and α4 about 15 Å from the allosteric site. The FTMap calculations were performed in three sets of structures: Ras-GppNHp in the T state, Ras-GppNHp in the R state with switch I in state 1, and Ras-GppNHp in the R state with switch I in state 2 (Fig. 3.5). The calculations identified the two experimentally determined clusters as strong druggable sites (Clusters 1 and 2). In addition, they picked up the allosteric site, which did not bind organic solvents experimentally, probably due to a combination of the polar nature of the site and the presence of a highly conserved water molecule not present in the protein models used in FTMap.

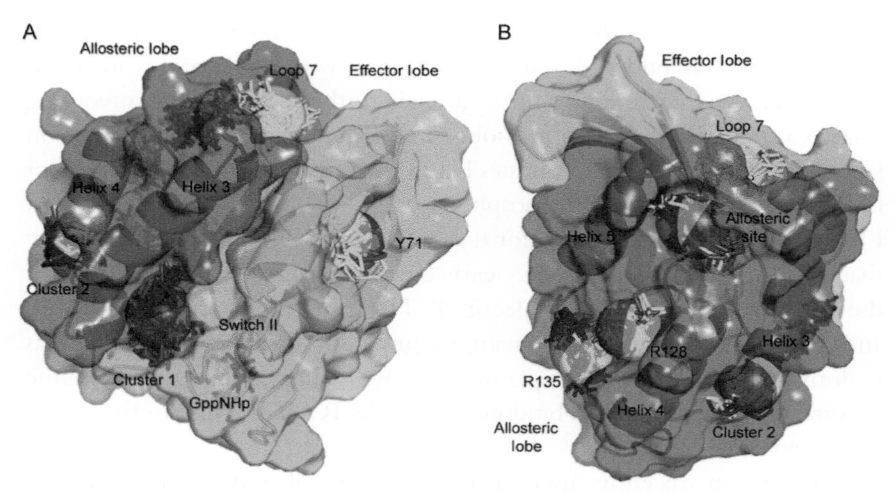

Figure 3.5 FTMap results for Ras in three conformational states associated with the GTP-bound form. The protein surface is shown in green for the effector lobe and in gray for the allosteric lobe. Hot spots found for Ras in the T state are shown in purple, those found for Ras in the R state with switch I in state 1 are in blue, and those for Ras in the R state with switch I in state 2 are in cyan. All sites where small molecules have been observed to bind experimentally are identified as druggable. In addition, sites R128 and R135 on the allosteric lobe are identified as druggable, but have not yet been explored. *[Reprinted with permission from Elsevier from Buhrman et al. [23].]* (See color plate.)

Site Y71, observed to bind the Sos inhibitors, was located by FTMap but not by MSCS, most likely due to the dynamic nature of switch II in the R32 crystal form. Most importantly, FTMap identified a site between α4 and α5 containing residue R135 in all three conformational states of Ras-GTP and another one containing residue R128, also between α4 and α5, in Ras-GTP in the T state and in the R state with switch I in state 2. These sites are in areas of extensive crystal contacts in the crystals used for the MSCS experiments and therefore were not accessible to bind small molecules. Nevertheless, the FTMap results suggest that the binding pockets containing R128 and R135, both allosteric lobe residues determined to be important in the Ras-GTP/membrane interaction, are druggable sites that could be targeted directly to interfere with Ras signaling.

8. CONCLUSIONS

Ras has been a great challenge as a direct target to drug discovery. It is a highly dynamic protein with a G domain that contains an extensive binding surface that interacts with different effector proteins as well as with regulators such as GAPs and GEFs. In addition, it interacts with the membrane in a nucleotide-specific manner. The protein-binding partners are able to select biologically relevant conformations that are sampled in uncomplexed Ras by the intrinsically disordered switch I and switch II regions. In addition to this relatively local disorder associated with the effector lobe near the β- and γ-phosphates of the bound nucleotide, correlated motions involving α3/L7 and switch II span across the effector and allosteric lobes, affecting the molecule in a more global scale. Given such high degree of flexibility, it is hardly surprising that it has been so difficult to modulate Ras with small molecules. Even so, there has been significant progress in recent years, with a few compounds now known to bind directly to Ras and interfere with its function.

In the future, the search for direct binders to Ras has the potential to progress in a more focused manner in three new directions: selective modulation of switch I to state 1 with decreased affinity to effectors; selective stabilization of the R state that allows access to the catalytic conformation of switch II, promoting intrinsic hydrolysis; and targeting of the membrane-interacting regions of Ras-GTP. Selective binding to state 1 in Ras-GTP has been shown to be possible *in vitro* and validated to a limited extent by functional assays. Modulation of the allosteric switch with small molecules has been shown to be possible *in vitro*, but the effectiveness of this approach still needs to be validated by functional assays. It has the potential of selectively targeting mutants that preferentially activate the Ras/Raf/MEK/ERK pathway. Finally, interfering with the Ras/membrane interaction in Ras-GTP has emerged as an idea supported by identification of hot spots through computational solvent mapping (FTMap) and by elegant molecular dynamics and experimental approaches showing the functional relevance of key arginine residues at those sites. However, no small molecules have yet been used to test the viability of this approach either *in vitro* or *in vivo*. As the search for Ras inhibitors continues, these venues need to be further explored and new functional assays developed to test their efficacies in targeting Ras-mediated signaling in the cell.

REFERENCES

[1] E.P. Reddy, et al., A point mutation is responsible for the acquisition of transforming properties by the T24 human bladder carcinoma oncogene, Nature 300 (1982) 149–152.

[2] A.D. Cox, C.J. Der, Ras history: the saga continues, Small GTPases 1 (1) (2010) 2–27.

[3] H.R. Bourne, D.A. Sanders, F. McCormick, The GTPase superfamily: conserved structure and molecular mechanism, Nature 349 (1991) 117–127.
[4] A. Wittinghofer, I.R. Vetter, Structure-function relationships of the G domain, a canonical switch motif, Annu. Rev. Biochem. 80 (2011) 943–971.
[5] O. Fasano, et al., Analysis of the transforming potential of the human H-ras by random mutagenesis, Proc. Natl. Acad. Sci. U.S.A. 81 (1984) 4008–4012.
[6] C.J. Der, T. Finkel, G.M. Cooper, Biological and biochemical properties of human rasH genes mutated at codon 61, Cell 44 (1) (1986) 167–176.
[7] Y. Pylayeva-Gupta, E. Grabocka, D. Bar-Sagi, RAS oncogenes: weaving a tumorigenic web, Nat. Rev. Cancer 11 (11) (2011) 761–774.
[8] D. Hanahan, R.A. Weinberg, Hallmarks of cancer: the next generation, Cell 144 (5) (2011) 646–674.
[9] E. Castellano, E. Santos, Functional specificity of ras isoforms: so similar but so different, Genes Cancer 2 (3) (2011) 216–231.
[10] R. Blum, A.D. Cox, Y. Kloog, Inhibitors of chronically active Ras: potential for treatment of human malignancies, Recent Pat. Anticancer Drug Discov. 3 (2008) 31–47.
[11] I.A. Prior, P.D. Lewis, C. Mattos, A comprehensive survey of Ras mutations in cancer, Cancer Res. 72 (10) (2012) 2457–2467.
[12] J. John, M. Frech, A. Wittinghofer, Biochemical properties of Ha-ras encoded p21 mutants and mechanism of the autophosphorylation reaction, J. Biol. Chem. 263 (24) (1988) 11792–11799.
[13] U. Krengel, et al., Three-dimensional structures of H-ras p21 mutants: molecular basis for their inability to function as signal switch molecules, Cell 62 (3) (1990) 539–548.
[14] S.A. Moodie, et al., Different structural requirements within the switch II region of the Ras protein for interactions with specific downstream targets, Oncogene 11 (3) (1995) 447–454.
[15] M. Trahey, F. McCormick, A cytoplasmic protein stimulates normal N-Ras p21 GTPase, but does not affect oncogenic mutants, Science 238 (4826) (1987) 542–545.
[16] K. Scheffzek, et al., The Ras-RasGAP complex: structural basis for GTPase activation and its loss in oncogenic Ras mutants, Science 277 (5324) (1997) 333–338.
[17] T. Minato, et al., Quantitative analysis of mutually competitive binding of human Raf-1 and Yeast Adenylyl Cyclase to Ras proteins, J. Biol. Chem. 269 (33) (1994) 20845–20851.
[18] U.S. Vogel, et al., Cloning of bovine GAP and its interaction with oncogenic ras p21, Nature 335 (1988) 90–93.
[19] A. Fujioka, et al., Dynamics of the Ras/ERK MAPK cascade as monitored by fluorescent probes, J. Biol. Chem. 281 (13) (2006) 8917–8926.
[20] G. Buhrman, et al., Allosteric modulation of Ras positions Q61 for a direct role in catalysis, Proc. Natl. Acad. Sci. U.S.A. 107 (11) (2010) 4931–4936.
[21] G. Buhrman, et al., Allosteric modulation of Ras-GTP is linked to signal transduction through RAF kinase, J. Biol. Chem. 286 (5) (2011) 3323–3331.
[22] D. Abankwa, A.A. Gorfe, J.F. Hancock, Mechanisms of Ras membrane organization and signalling: Ras on a rocker, Cell Cycle 7 (17) (2008) 2667–2673.
[23] G. Buhrman, et al., Analysis of binding site hot spots on the surface of Ras GTPase, J. Mol. Biol. 413 (4) (2011) 773–789.
[24] A.A. Gorfe, et al., Structure and dynamics of the full-length lipid-modified H-Ras protein in a 1,2-Dimyristoylglycero-3-phosphocholine bilayer, J. Med. Chem. 50 (2007) 674–684.
[25] Y.I. Henis, J.F. Hancock, I.A. Prior, Ras acylation, compartmentalization and signaling nanoclusters (Review), Mol. Membr. Biol. 26 (1) (2009) 80–92.
[26] H.B. Hodges-Loaiza, L.E. Parker, A.D. Cox, Prenylation and phosphorylation of Ras superfamily small GTPases, in: C.A. Hrycyna, M.O. Bergo, F. Tamanoi

(Eds.), The Enzymes: Protein Prenylation Part B, Academic Press, London, 2011, pp. 43–69.
[27] O. Rocks, et al., An acylation cycle regulates localization and activity of palmitoylated ras isofroms, Science 307 (5716) (2005) 1746–1752.
[28] D. Abankwa, et al., Ras membrane orientation and nanodomain localization generate isoform diversity, Proc. Natl. Acad. Sci. U.S.A. 107 (3) (2010) 1130–1135.
[29] S.R. Sprang, G protein mechanisms: insights from structural analysis, Annu. Rev. Biochem. 66 (1997) 639–678.
[30] F. Raimondi, M. Orozco, F. Fanelli, Deciphering the deformation modes associated with function retention and specialization in members of the Ras superfamily, Structure 18 (3) (2010) 402–414.
[31] E.F. Pai, et al., Refined crystal structure of the triphosphate conformation of H-ras p21 at 1.35 Å resolution: implications for the mechanism of GTP hydrolysis, EMBO J. 9 (8) (1990) 2351–2359.
[32] D. Abankwa, et al., A novel switch region regulates H-ras membrane orientation and signal output, EMBO J. 27 (5) (2008) 727–735.
[33] A.A. Gorfe, B.J. Grant, J.A. McCammon, Mapping the nucleotide and isoform-dependent structural and dynamical features of Ras proteins, Structure 16 (6) (2008) 885–896.
[34] I.R. Vetter, A. Wittinghofer, The guanine nucleotide-binding switch in three dimensions, Science 294 (5545) (2001) 1299–1304.
[35] I. Schlichting, et al., Time-resolved x-ray crystallographic study of the conformational change in Ha-Ras p21 protein on GTP hydrolysis, Nature 345 (1990) 309–315.
[36] A.M. Rojas, et al., The Ras protein superfamily: evolutionary tree and role of conserved amino acids, J. Cell Biol. 196 (2) (2012) 189–201.
[37] P.A. Boriack-Sjodin, et al., The structural basis of the activation of Ras by Sos, Nature 394 (1998) 337–343.
[38] L. Huang, et al., Structural basis for the interaction of Ras with RalGDS, Nat. Struct. Biol. 5 (1998) 422–426.
[39] M.E. Pacold, et al., Crystal structure and functional analysis of Ras binding to its effector phosphoinositide 3-kinase gamma, Cell 103 (2000) 931–943.
[40] N. Nassar, et al., Ras/Rap effector specificity determined by charge reversal, Nat. Struct. Biol. 3 (8) (1996) 723–729.
[41] B. Stieglitz, C. Bee, D. Schwarz, O. Yildiz, A. Moshnikova, A. Khokhlatchev, et al., Novel type of Ras effector interaction established between tumour suppressor NORE1A and Ras switch II, EMBO J. 27 (2008) 1995–2005.
[42] Y. Ito, K. Yamasaki, J. Iwahara, T. Terada, A. Kamiya, M. Shirouza, et al., Regional polysterism in the GTP-bound form of the human c-Ha-Ras protein, Biochemistry 36 (1997) 9109–9119.
[43] G. Buhrman, V. de Serrano, C. Mattos, Organic solvents order the dynamic switch II in Ras crystals, Structure 11 (7) (2003) 747–751.
[44] Y. Ito, et al., Regional polysterism in the GTP-bound form of the human c-Ha-Ras protein, Biochemistry 36 (30) (1997) 9109–9119.
[45] J. John, et al., C-Terminal truncation of p21H preserves crucial kinetic and structural properties, J. Biol. Chem. 264 (22) (1989) 13086–13092.
[46] M. Spoerner, et al., Dynamic properties of the Ras switch I region and its importance for binding to effectors, Proc. Natl. Acad. Sci. U.S.A. 98 (9) (2001) 4944–4949.
[47] M. Spoerner, et al., Conformational states of human rat sarcoma (Ras) protein complexed with its natural ligand GTP and their role for effector interaction and GTP hydrolysis, J. Biol. Chem. 285 (51) (2010) 39768–39778.
[48] N. Nassar, et al., The 2.2 A crystal structure of the Ras-binding domain of the serine/threonine kinase c-Raf1 in complex with Rap1A and a GTP analogue, Nature 375 (6532) (1995) 554–560.

[49] E.F. Pai, et al., Refined crystal structure of the triphosphate conformation of H-ras p21 at 1.35 A resolution: implications for the mechanism of GTP hydrolysis, EMBO J. 9 (8) (1990) 2351–2359.
[50] G. Buhrman, G. Wink, C. Mattos, Transformation efficiency of RasQ61 mutants linked to structural features of the switch regions in the presence of Raf, Structure 15 (12) (2007) 1618–1629.
[51] R. Thapar, J.G. Williams, S.L. Campbell, NMR characterization of full-length farnesylated and non-farnesylated H-Ras and its implications for Raf activation, J. Mol. Biol. 343 (5) (2004) 1391–1408.
[52] C. O'Connor, E.L. Kovrigin, Global conformational dynamics in ras, Biochem. J. 47 (2008) 10244–10246.
[53] B.J. Grant, A.A. Gorfe, J.A. McCammon, Ras conformational switching: simulating nucleotide-dependent conformational transitions with accelerated molecular dynamics, PLoS Comput. Biol. 5 (3) (2009) e1000325.
[54] C. O'Connor, E.L. Kovrigin, Characterization of the second ion-binding site in the G domain of H-Ras, Biochemistry 51 (48) (2012) 9638–9646.
[55] R.H. Kretsinger, Calcium-binding proteins, Annu. Rev. Biochem. 45 (1976) 239–266.
[56] T. Schweins, M. Geyer, K. Scheffzek, A. Warshel, H.R. Kalbitzer, A. Wittinghofer, Substrate-assisted catalysis as a mechanism for GTP hydrolysis of p21ras and other GTP-binding proteins, Nat Struct Biol. 2 (1) (1995) 36–44.
[57] X. Du, et al., Kinetic isotope effects in Ras-catalyzed GTP hydrolysis: evidence for a loose transition state, Proc. Natl. Acad. Sci. U.S.A. 101 (24) (2004) 8858–8863.
[58] K.A. Maegley, S.J. Admiraal, D. Herschlag, Ras-catalyzed hydrolysis of GTP: a new perspective from model studies, Proc. Natl. Acad. Sci. U.S.A. 93 (1996) 8160–8166.
[59] A.J. Stafford, D.M. Walker, L.J. Webb, Electrostatic effects of mutations of Ras glutamine 61 measured using vibrational spectroscopy of a thiocyanate probe, Biochemistry 51 (13) (2012) 2757–2767.
[60] A.J. Scheidig, C. Burmester, R.S. Goody, The pre-hydrolysis state of p21ras in complex with GTP: new insights into the role of water molecules in the GTP hydrolysis reaction of ras-like, Structure 7 (11) (1999) 1311–1324.
[61] B. Ford, et al., Structure of a transient intermediate for GTP hydrolysis by ras, Structure 14 (3) (2006) 427–436.
[62] I.R. Vetter, et al., Structural view of the Ran-Importin beta interaction at 2.3 Å resolution, Cell 97 (1999) 635–646.
[63] G. Holzapfel, G. Buhrman, C. Mattos, Shift in the equilibrium between on and off states of the allosteric switch in Ras-GppNHp affected by small molecules and bulk solvent composition, Biochemistry 51 (31) (2012) 6114–6126.
[64] S.M. Margarit, et al., Structural evidence for feedback activation by Ras-GTP of the Ras-specific nucleotide exchange factor SOS, Cell 112 (2003) 685–695.
[65] B.E. Hall, D. Bar-Sagi, N. Nassar, The structural basis for the transition from Ras-GTP to Ras-GDP, Proc. Natl. Acad. Sci. U.S.A. 99 (19) (2002) 12138–12142.
[66] F. Shima, et al., Structural basis for conformational dynamics of GTP-bound Ras protein, J. Biol. Chem. 285 (29) (2010) 22696–22705.
[67] M.R. Ahmadian, et al., Guanosine triphosphatase stimulation of oncogenic Ras mutants, Proc. Natl. Acad. Sci. U.S.A. 96 (1999) 7065–7070.
[68] W. Wang, G. Fang, J. Rudolph, Ras inhibition via direct Ras binding—is there a path forward? Bioorg. Med. Chem. Lett. 22 (18) (2012) 5766–5776.
[69] V. Gonzalez-Perez, et al., Genetic and functional characterization of putative Ras/Raf interaction inhibitors in C. elegans and mammalian cells, J. Mol. Signal. 5 (2010) 2.
[70] C. Herrmann, et al., Sulindac sulfide inhibits Ras signaling, Oncogene 17 (14) (1998) 1769–1776.

[71] J. Kato-Stankiewicz, et al., Inhibitors of Ras/Raf-1 interaction identified by two-hybrid screening revert Ras-dependent transformation phenotypes in human cancer cells, Proc. Natl. Acad. Sci. U.S.A. 99 (22) (2002) 14398–14403.
[72] H. Waldmann, et al., Sulindac-derived Ras pathway inhibitors target the Ras-Raf interaction and downstream effectors in the Ras pathway, Angew. Chem. Int. Ed Engl. 43 (4) (2004) 454–458.
[73] T. Maurer, et al., Small-molecule ligands bind to a distinct pocket in Ras and inhibit SOS-mediated nucleotide exchange activity, Proc. Natl. Acad. Sci. U.S.A. 109 (14) (2012) 5299–5304.
[74] Q. Sun, et al., Discovery of small molecules that bind to K-Ras and inhibit Sos-mediated activation, Angew. Chem. Int. Ed Engl. 51 (25) (2012) 6140–6143.
[75] R. Brenke, et al., Fragment-based identification of druggable 'hot spots' of proteins using Fourier domain correlation techniques, Bioinformatics 25 (5) (2009) 621–627.
[76] I.C. Rosnizeck, et al., Stabilizing a weak binding state for effectors in the human ras protein by cyclen complexes, Angew. Chem. Int. Ed Engl. 49 (22) (2010) 3830–3833.
[77] I.C. Rosnizeck, et al., Metal-bis(2-picolyl)amine complexes as state 1(T) inhibitors of activated Ras protein, Angew. Chem. Int. Ed Engl. 51 (42) (2012) 10647–10651.
[78] S. Muraoka, et al., Crystal structures of the state 1 conformations of the GTP-bound H-Ras protein and its oncogenic G12V and Q61L mutants, FEBS Lett. 586 (12) (2012) 1715–1718.
[79] M. Dechene, et al., Multiple solvent crystal structures of ribonuclease A: an assessment of the method, Proteins 76 (4) (2009) 861–881.
[80] C. Mattos, et al., Multiple solvent crystal structures: probing binding sites, plasticity and hydration, J. Mol. Biol. 357 (5) (2006) 1471–1482.
[81] D.R. Hall, D. Kozakov, S. Vajda, Analysis of protein binding sites by computational solvent mapping, Methods Mol. Biol. 819 (2012) 13–27.
[82] C.H. Ngan, et al., FTMAP: extended protein mapping with user-selected probe molecules, Nucleic Acids Res. 40 (Web Server issue) (2012) W271–W275.

CHAPTER FOUR

State 1(T) Inhibitors of Activated Ras

Hans Robert Kalbitzer[1], Michael Spoerner
Institute for Biophysics and Physical Biochemistry and Centre of Magnetic Resonance in Chemistry and Biomedicine, University of Regensburg, Regensburg, Germany
[1]Corresponding author: e-mail address: hans-robert.kalbitzer@biologie.uni-regensburg.de

Contents

Abstract

Oncogenic mutations in the Ras (rat sarcoma) protein lead to a permanent activation of the Ras pathway and are found in approximately 30% of all human tumors. During signal transduction, Ras is transiently activated by GTP binding and interacts with effector proteins such as Raf kinase. Ras complexed with GTP (T) occurs in at least two conformational states, states 1(T) and 2(T), where state 2(T) represents the true effector-interaction state and state 1(T) has only a low affinity for effectors. Stabilization of state 1(T) by small molecules such as metal-cyclens can reduce the affinity for effectors and thus it can lead to an interruption of the signal transduction chain. Metal-cyclens bind inside the nucleotide-binding pocket to GTP, shifting the conformational equilibrium of Ras toward state 1(T). In contrast, Zn^{2+}-BPA (bis(2-picolyl)amine) binds outside the nucleotide-binding pocket but nevertheless allosterically stabilizes state 1(T) and thus inhibits Raf interaction. It shows a higher affinity for the oncogenic mutant Ras(G12V) than for wild type in contrast to other compounds such as Zn^{2+}-cyclen.

1. INTRODUCTION

The small guanine nucleotide-binding (GNB) protein Ras is involved in cellular signal transduction that controls proliferation, differentiation, and

The Enzymes, Volume 33
ISSN 1874-6047
http://dx.doi.org/10.1016/B978-0-12-416749-0.00004-X

apoptosis. It cycles between two main states, the inactive GDP-bound state and the active GTP-bound state. Only in complex with GTP, it can bind effector proteins with high affinity (see, e.g., [1–4]). Mutations in amino acid positions 12, 13, or 61 of Ras are found in more than 30% of all human malignancies [5–7]. Those mutants stay permanently activated since the intrinsic as well as the GAP (GTPase-activating protein) catalyzed hydrolysis of GTP is very inefficient.

Therefore, Ras is a target for cancer therapy (for reviews see, e.g., Refs. [7–9]). To date, there were also many efforts to find small compounds which directly interact with Ras with the aim to interrupt the Ras-mediated signal transduction [10]. Different strategies were followed: (1) targeting Ras activation by inhibition of nucleotide exchange [11–16] or (2) increasing the GTPase activity of oncogenic Ras mutants thus deactivating Ras [17–20]. However, suitable compounds could not be found yet. (3) A further strategy is the inhibition of Ras–effector interaction by small compounds [21–24] or peptides [25,26]. For the latter-mentioned approach, selective stabilization of the weak effector-binding state 1(T) by small compounds represents a promising novel strategy for the inhibition of oncogenic Ras signaling [27–31], which will be the topic of this chapter.

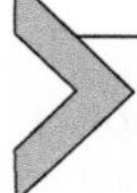

2. ALLOSTERIC REGULATION OF THE Ras–EFFECTOR INTERACTION

Detailed theoretical models for allosteric regulation were first developed for oligomeric proteins using hemoglobin as an example [32,33]. Here, the binding of the allosteric inhibitor mainly determines the quaternary structure of the different subunits. However, on the general idea, it is not restricted to oligomeric proteins, but as minimum condition it only needs two locally distinct binding sites on a protein where binding of a ligand to one site influences the structure at a second site and thus the affinity of a second ligand. If the two binding sites are selective for different ligands (heterotropic allostery) and binding of the first ligand changes the affinity of the second ligand, then a situation typical for allosteric regulation is given.

However, the Monod–Wyman–Changeux model (MWC model) can be generalized further for a multiple state system [34]: if one assumes that the interaction of Ras·Mg^{2+}·GTP (Ras(T)) with effectors, GTPase-activating proteins (GAPs), and guanine nucleotide exchange factors (GEFs) requires different local (or global) conformations of Ras and binding can be described by conformational selection, then these states should also

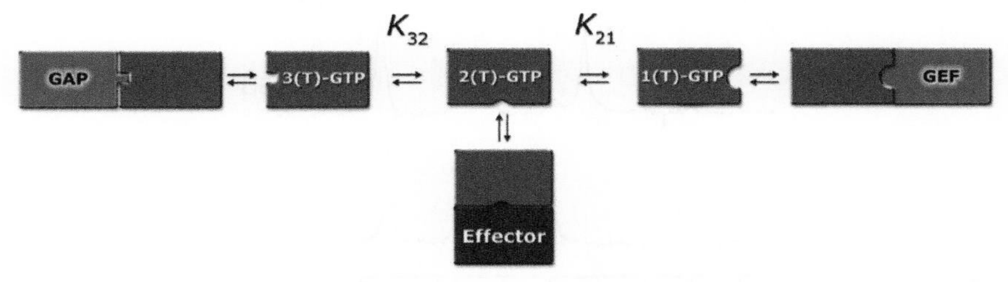

Figure 4.1 Interaction of activated Ras in the conformational selection model. Ras·Mg^{2+}·GTP (Ras(T)) should exist in different conformational states: state 1(T) interacts preferentially with guanine nucleotide exchange factors (GEFs), state 2(T) with effectors, and state 3(T) with GTPase-activating proteins (GAPs).

exist in the absence of the interaction (Fig. 4.1). In activated Ras, one would expect a GEF-interaction state 1(T), an effector-interaction state 2(T), and a GAP-interaction state 3(T). It is very likely that these conformational states will have different affinities for the different types of interacting proteins; that is, state 1(T) should have a higher affinity for GEFs than for effectors and GAPs. With this idea, one creates a new type of allosteric regulation if one finds a compound that selectively binds to a given state and thus enhances its relative population. This in turn will influence the occupancy of all other states connected in a regulatory cycle and therefore modulate their activity. We have demonstrated earlier that this mechanism is really working [27,28].

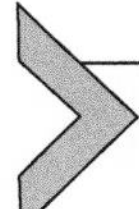

3. IDENTIFICATION OF STATE 1(T) IN ACTIVATED Ras BY NMR SPECTROSCOPY

We could show by ^{31}P NMR spectroscopy that wild-type Ras complexed with a nucleoside triphosphate (T) occurs in at least two conformational states in dynamic equilibrium which interconvert on the millisecond time scale at room temperature [35–38]. This observation fits well to the MWC model presented above. Initially, these states were defined by their spectroscopic properties in the complex of Ras with the GTP analog GppNHp: at low temperatures, two sets of ^{31}P NMR resonance lines were observed; here in state 1(T), the resonance lines of the α- and γ-phosphate groups were downfield shifted relative to those corresponding to state 2(T) (Fig. 4.2). The relative populations of the two states that are directly obtained by the integrals of the resonance lines are dependent on different factors such as temperature, pH, and ionic strength. Surprisingly, the relative population is also dependent on the nucleotide analog used (Fig. 4.2). For wild-type H-Ras(1–166), the

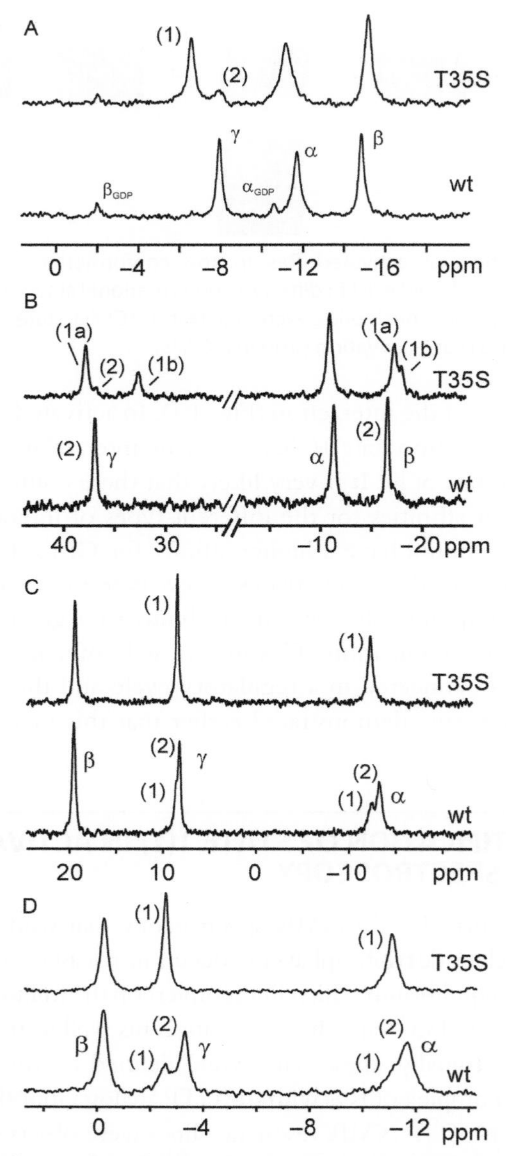

Figure 4.2 Dependence of conformational equilibria on the nucleotide analog used. ^{31}P NMR spectra of human *H*-Ras(1–189) and *H*-Ras(1–189) (T35S) complexes at 202.4 MHz and 278 K. Complexes with Mg^{2+}·GTP (A), Mg^{2+}·GTPγS (B), Mg^{2+}·GppCH$_2$p (C), and Mg^{2+}·GppNHp (D). In the complex of Mg^{2+}·GTPγS with Ras(T35S), two substates 1a and 1b of state 1(T) are observed. Buffer: 40 mM Hepes/NaOH pH 7.5, 10 mM $MgCl_2$, 2 mM dithioerythritol, 10% D_2O, 0.2 mM DSS.

equilibrium constant K_{12} = [state 2(T)]/[state 1(T)] for the two states is 11.3 in the presence of its natural ligand GTP [38] (Table 4.1). In the complex of Ras(wt) with Mg^{2+}·GTPγS state 2(T) also predominates [39]. In contrast, for Ras(wt) complexes with the nucleotide analogs GppNHp and $GppCH_2p$,

Table 4.1 Influence of the nucleotide bound and specific mutations on the populations of state 1(T) and state 2(T)[a]

Protein complex	K_{12}	ΔG_{12} [kJ mol^{-1}]
Ras1-166(wt)·Mg^{2+}·GTP[b]	11.3	−5.7
Ras(wt)·Mg^{2+}·GTP[b]	>20	< −6.9
Ras(wt)·Mg^{2+}·GTPγS[c]	> 10	< −5.3
Ras1-166(wt)·Mg^{2+}·GppNHp[d]	1.7	−1.2
Ras(wt)·Mg^{2+}·GppNHp[d]	1.9	−1.5
Ras(wt)·Mg^{2+}·$GppCH_2p$[e]	2.0	−1.6
Ras(G12V)·Mg^{2+}·GTP[b]	> 10	< −5.3
Ras(G12V)·Mg^{2+}·GppNHp[d]	0.9	0.2
Ras(G13R)·Mg^{2+}·GTP[b]	> 8	< −4.8
Ras(G13R)·Mg^{2+}·GppNHp	1.4	−0.8
Ras(T35S)·Mg^{2+}·GTP[b]	0.29	2.1
Ras(T35S)·Mg^{2+}·GTPγS[c]	0.06	6.5
Ras(T35S)·Mg^{2+}·GppNHp[d]	< 0.1	> 5.3
Ras(T35S)·Mg^{2+}·$GppCH_2p$[e]	< 0.05	> 6.9
Ras(T35A)·Mg^{2+}·GTP[b]	< 0.04	> 7.4
Ras(T35A)·Mg^{2+}·GTPγS[c]	< 0.05	> 6.9
Ras(T35A)·Mg^{2+}·GppNHp[d]	< 0.05	> 6.9
Ras(Q61A)·Mg^{2+}·GTP[b]	> 20	< − 6.9
Ras(Q61A)·Mg^{2+}·GppNHp	2.6	−2.2

[a]If not stated otherwise, measurements were performed with full-length human *H*-Ras(1–189) at 278 K in 40 mM HEPES/NaOH, pH 7.4, 10 mM $MgCl_2$, 2 mM 1,4-dithioerythritol, 0.1 mM 2,2-dimethyl-2-silapentane-5-sulfonate in 5% D_2O, 95% H_2O. Truncated variants are labeled Ras(1–166). The equilibrium constant K_{12} is defined by K_{12} = [Ras2(T)]/[Ras1(T)], and the free energy ΔG_{12} by $\Delta G_{12} = -RT \ln K_{12}$.
[b]Data taken from ref. [38].
[c]Data taken from ref. [39].
[d]Data taken from ref. [45].
[e]Data taken from ref. [37].

the equilibrium constant is reduced by a factor of 5 and is strongly shifted toward state 1(T) [37]. These results also stress to be very careful when interpreting data obtained with different nucleotide analogs in a quantitative way. That also holds, in principle, for fluorescent GTP analogs where it first has to be established that they have identical thermodynamical and kinetic properties. The crystal structure of Ras in complex with the commonly used fluorescent nucleotide analog mantGppNHp looks quite similar to the complex obtained with unmodified GppNHp [40]. NMR experiments, which allow investigation of functional properties of the protein in complex with both nucleotides, showed that the intrinsic GTPase activity was rather similar for both complexes in Ras protein, but reactions including regulatory-binding proteins such as the GAP-catalyzed GTP hydrolysis or the GEF-accelerated nucleotide exchange of GDP/mantGDP loaded proteins significantly differed. Especially in the Ras-related proteins RhoA and Rheb, even the detected intrinsic GTPase activity was reported to be significantly different in an unpredictable way [41].

Mutations of the protein can also influence the observed equilibria. Here, a typical mutation used in protein biochemistry is an N- or C-terminal truncation to improve the physico-chemical properties of a protein. In Ras, the C-terminal truncated form Ras(1–166) is often used for crystallization and structure determination. As shown by ^{31}P NMR spectroscopy, the truncation leads to a significant shift toward state 1(T) compared to full-length Ras(1–189) (Table 4.1).

Partial loss-of-function mutants, such as Ras(T35A) or Ras(T35S), which only activate a subset of the known Ras effectors, predominantly exist in state 1(T) (Table 4.1, Fig. 4.2) [36–38]. Here, primarily the specific interaction between the amide and hydroxyl groups of T35 with the magnesium ion and the γ-phosphate of the nucleotide is perturbed. In crystal structures of Ras(T35S), different structures can be identified; in one structure, the region around T35 is not visible (disordered) [36], and in other published structures [42] the switch I (residues 32–38) and switch II (residues 60–75) regions are visible. In crystal form 1 the hydrogen bond of the amide group of S35 and in a different crystal form 2 also the hydrogen bond of G60 with the γ-phosphate of GppNHp are broken. The solution NMR structure of Ras(T35S) [43] shows similar structural features: again the stabilization of switch I through the interaction of residue 35 with the magnesium ion and the γ-phosphate group and the hydrogen bond of G60 with the γ-phosphate group is abolished. However, a reanalysis of the solid-state ^{31}P NMR data [44] suggests that in solution only a minor proportion of the mutant occurs in form 2 (details will be published elsewhere).

A strong shift toward state 1(T) can also be induced by mutations in other residues of the switch regions, for example, Ras(V29G), Ras(V29G/I36G), Ras(Y32C/C118S), Ras(Y32W), Ras(I36G), Ras(Y40C), and Ras(G60A) [45,46]. In Fig. 4.3C, the spectra of selected partial loss-of-function mutants are shown together with the spectrum of *H*- and *K*-Ras wild type (Fig. 4.3A). One can expect that these mutations induce indirectly the same structural

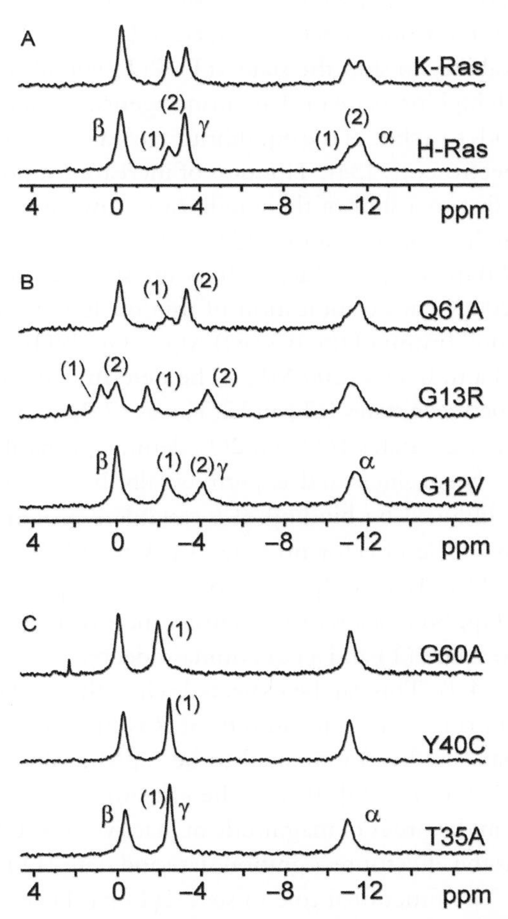

Figure 4.3 Influence of point mutation on the conformational equilibria of activated Ras. ^{31}P NMR spectra of human Ras complexes with Mg^{2+}·GppNHp at 202.4 MHz and 278 K. (A) Wild-type *H*-Ras(1–189) and wild-type *K*-Ras(1–189), (B) oncogenic mutants of *H*-Ras(1–189), and (C) state 1(T) mutants of *H*-Ras(1–189). Buffer: 40 mM Hepes/NaOH pH 7.5, 10 mM $MgCl_2$, 2 mM dithioerythritol, 10% D_2O, 0.2 mM DSS. The Ras variants are indicated at each spectrum.

changes around the γ-phosphate group that lead to a release of T35 from its interactions with the metal ion and the γ-phosphate group. An example is the X-ray structure of Ras(G60A) complexed with GppNHp [46]: again the hydrogen bonds of the amide protons of T35 and G60 with the γ-phosphate group are broken and the interaction of the side chain hydroxyl group of T35 with the magnesium ion is abolished. In contrast, several oncogenic variants of Ras (Fig. 4.3B) also in the GppNHp-bound state present a significant population of state 2(T) which dominates in the GTP-bound complex and is necessary for a strong effector interaction [38].

Another possibility to shift the state 1(T)–2(T) equilibrium is the treatment of Ras with high pressure or denaturing agents. High pressure of some hundred MPa leads to a shift of the equilibrium toward state 1(T) as observed by ^{31}P NMR spectroscopy [34]. Titration of increasing amounts of GdmCl to Ras(wt) lead first to a shift of the equilibrium toward state 1(T) conformation, and at higher concentration additional resonances corresponding to free Mg^{2+}·GppNHp complexes arose. In contrast, when using hydrostatic pressure up to 200 MPa, no indication of free nucleotide signals could be found, but the equilibrium of the Ras(wt)·Mg^{2+}·GppNHp complex shifted from 1.9 at 3 MPa to 0.44 at 200 MPa. The determined change in molar-specific volume between states 1(T) and 2(T) was 17.2 ± 0.5 mL mol^{-1} [34].

The function of the states 1(T) and 2(T) defined primarily by ^{31}P NMR spectroscopy has been elucidated experimentally by the observation of a population shift induced by binding of a candidate protein. Adding Ras-binding domains of the effector proteins Raf kinase [35–39,45], RalGDS [35–39,47], AF6 [48], Byr2 [49], and Nore1 to a sample containing wild-type Ras·Mg^{2+}·GppNHp leads to an intensity increase of the resonance line corresponding to state 2(T) and a concomitant decrease of the line assigned to state 1(T) (Fig. 4.4). This can be expected when the RBDs preferentially bind to state 2(T) (Fig. 4.1). The affinity of Ras in state 1(T) to the main effector Raf kinase has been estimated to be 7 μM at 283 K compared to 11.7 nM for state 2(T) [28,36], that is, the effector affinity of Ras-GTP in state 1(T) is within the order of magnitude of inactive Ras–GDP complexes. Thus state 2(T) is the effector recognition state and state 1(T) has a low affinity for effectors. The functional role of state 1(T) has been elucidated with analogous experiments by adding the exchange factor Sos [34]. However, from the experimental setup, these studies are more complicated since Sos contains not only the catalytic site responsible for the nucleotide exchange but also a second regulatory site that recognizes activated Ras in state 2(T). In the ^{31}P NMR spectra with wild-type Sos, the interaction

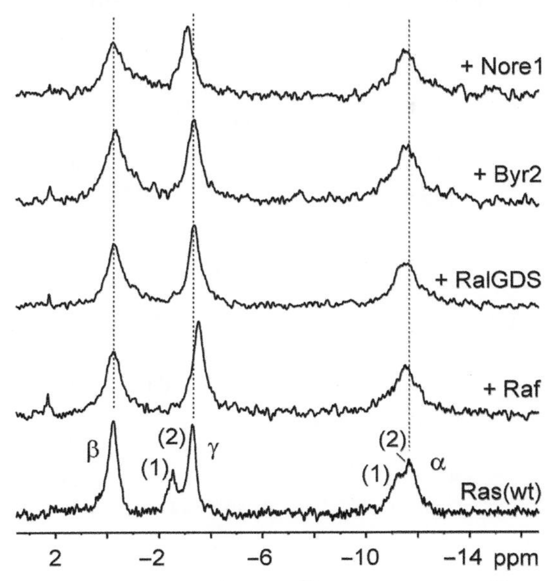

Figure 4.4 Interaction of Ras with effectors. ^{31}P NMR spectra of human *H*-Ras(1–189) complexes with Mg^{2+}·GppNHp at 202.4 MHz and 278 K. To a sample of wild-type Ras (bottom), RBDs of different effectors were added in saturating amounts. The corresponding Ras effectors are indicated. Buffer: 40 mM Hepes/NaOH pH 7.5, 150 mM NaCl, 10 mM $MgCl_2$, 2 mM dithioerythritol, 10% D_2O, 0.2 mM DSS.

with both states can be observed. In the mutant W729E of Sos, the affinity of the regulatory side is substantially reduced, allowing a specific interaction of Ras with the catalytic side only [34].

Since for state 1(T) mutants the typical hydrogen bond pattern around the γ-phosphate is destroyed and the backbone structure in the P-loop, switch I, and switch II is changed, state 1(T) and state 2(T) conformers can also be distinguished in the [^{1}H,^{15}N] HSQC spectra of isotope-enriched Ras. As to be expected, the resonance position of G60 changes significantly in state 1(T) compared to state 2(T) (Fig. 4.5). The differences in the HSQC spectra of Ras(wt)·Mg^{2+}·GppNHp and Ras(T35S)·Mg^{2+}·GppNHp were analyzed in detail in Ref. [43] and show that for some residues such as G12, G13, and G60 in the wild-type protein two cross peaks can be observed with relative intensities corresponding to the same populations that we observed earlier in the ^{31}P NMR spectra. In the T35S mutant, only the peaks assignable to state 1(T) are visible.

In most cases, it can be directly decided from the ^{31}P NMR spectra if a protein is in state 1(T) or 2(T) using the nucleotide as probe. However,

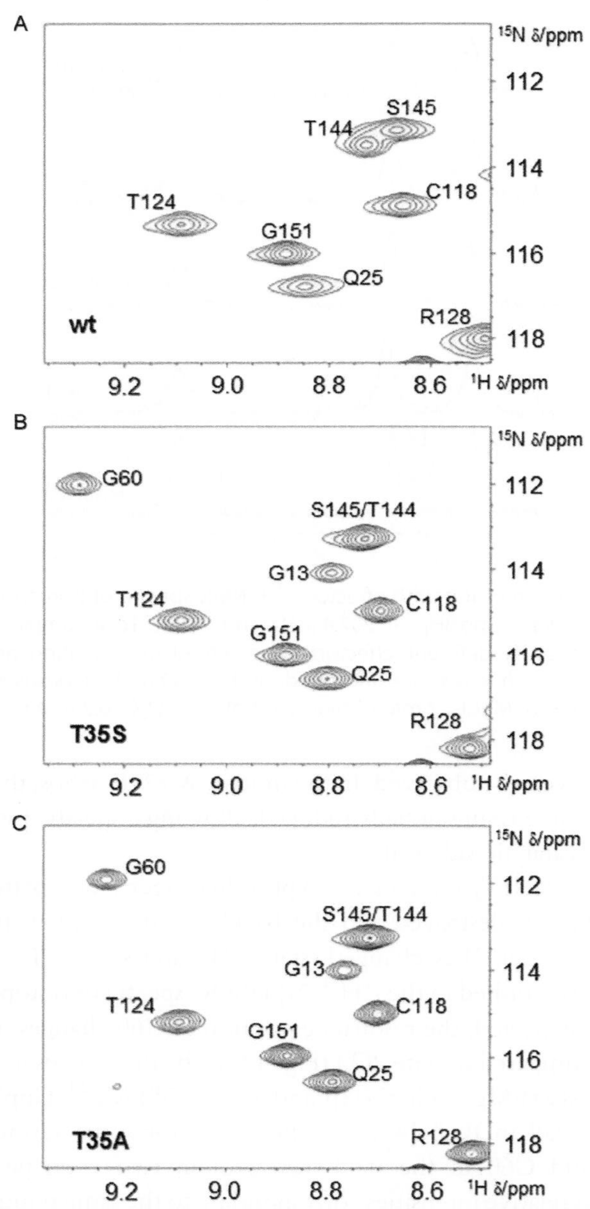

Figure 4.5 Recognition of state 1(T) and state 2(T) conformers in [^1H,^{15}N]-HSQC spectra of activated Ras. (A) Part of the [^1H,^{15}N]-HSQC spectrum of wild-type Ras(1–166)·Mg^{2+}·GppNHp (50 mM Tris/HCl pH 7.5, 10 mM $MgCl_2$, 2 mM dithioerythritol, 10% D_2O, 0.2 mM DSS, T = 278 K) showing the signal of G60, same part of the spectra of (B) Ras(T35S)·Mg^{2+}·GppNHp and (C) Ras(T35A)·Mg^{2+}·GppNHp.

mutations have also direct local effects and cause additional shifts. Therefore, sometimes it is difficult to obtain an unambiguous assignment to a given state from the chemical shifts alone if only one state can be observed. In this case the structural definition of state 2(T) as effector-binding state can help: if binding of an effector leads to a significant highfield shift of the γ-phosphate resonance, the protein originally existed in state 1(T) (see also Fig. 4.4).

To summarize, state 1(T) is the Sos-interacting state characterized spectroscopically by a typical downfield shift of ^{31}P NMR lines of the α- and γ-phosphate resonances, structurally by the missing hydrogen bond between T35 and the γ-phosphate group, and a weak affinity to effectors. Correspondingly, state 2(T) is the effector-interacting state characterized spectroscopically by a typical upfield shift of ^{31}P NMR lines of the α- and γ-phosphate resonances, structurally by a hydrogen bond between T35 and the γ-phosphate group, and a weak affinity to the catalytic site of Sos. From Fig. 4.1, additional properties can be predicted: (1) an increased nucleotide exchange rate, (2) a reduced intrinsic GTPase activity, (3) an increased affinity to GEFs, and (4) a decreased GEF-catalyzed exchange activity. Point (2) has been verified experimentally [38]: state 1(T) mutants have a lower intrinsic GTPase than state 2(T) mutants; a shift of the equilibrium by effector binding to state 2(T) leads to a GTPase activity typical for wild-type Ras existing predominantly in state 2(T). An increased affinity for GEFs and a reduction in the GEF-activity have been observed for the 1(T)-mutant Ras(G60A) [46,50].

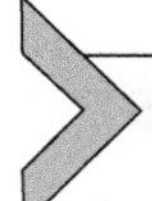

4. IDENTIFICATION OF 1(T) INHIBITORS BY NMR SPECTROSCOPY

Principally, state 1(T) inhibitors are defined by their property that they have a higher affinity to state 1(T) than to state 2(T). Therefore, they can be identified by assays where their affinity to a typical state 1(T) mutant is compared with the affinity to a typical 2(T) mutant. In an NMR-based assay, the use of mutants is not obligatory since the two states can be directly distinguished in one spectrum. Usually, the identification of 1(T) inhibitors of Ras is a two-step procedure. In the first step, compounds with appropriate properties of "drug likeness" (e.g., low molecular mass, water solubility, number of donor/acceptor to hydrogen bonds) [51] are screened for interaction by NMR methods such as saturation transfer difference (STD) spectroscopy [52] or WaterLOGSY [53] using the state 1(T) mutant Ras(T35A)·

Mg^{2+}·GppNHp as target. From these experiments, the binding constant as well as information on the binding epitope of the ligand is obtained.

Titration of active Ras with the ligand monitored by ^{31}P NMR spectroscopy allows us in the second step to identify state 1(T) stabilizers. In a simple 1D spectrum of wild-type Ras·Mg^{2+}·GppNHp, we can directly observe the selective 1(T) binding of a ligand by an intensity increase in the resonances corresponding to state 1(T) and the slow disappearance of the resonances of state 2(T) [28]. The partial loss-of-function mutant Ras (T35S)·Mg^{2+}·GppNHp is very suitable candidate for demonstrating the power of a ligand to perturb Ras–effector interaction. The free protein exists predominantly in state 1(T), but after effector binding the equilibrium is shifted toward state 2(T), the effector interaction state. If a compound is capable to interfere with effector binding to Ras, the resonances of free Ras are restored [31].

Using 2D [^{1}H,^{15}N] Sofast HMQC experiments [54] and ^{15}N isotope-labeled Ras(T35A), in principle the typical spectral differences between the two states (Fig. 4.5) can be used to observe a shift from state 2(T) to state 1(T) after ligand binding. However, since ligand binding itself will also induce shift changes, the data evaluation can become rather complicated. However, if only the effect that concerns is the removal of an effector by binding of ligands, a highly informative NMR assay uses the complex of the 1(T) mutant in a 1:1 complex with the Raf-binding domain: in the presence of Raf, a spectrum typical for state 2(T) is obtained; after the release of Raf by an inhibitor, a state 1(T) spectrum is obtained.

5. METAL-CYCLEN DERIVATIVES AS 1(T) INHIBITORS

The first 1(T) inhibitors found were metal complexes of 1,4,7,10-tetraazacyclododecane (M^{2+}-cyclens) (Fig. 4.6A) and their derivatives [28]. ^{31}P NMR spectroscopy on the titration of Ras(wt)·Mg^{2+}·GppNHp with Zn^{2+}-cyclen shows the expected decrease in the resonance signals of Ras corresponding to conformational state 2(T) and the signals corresponding to state 1(T) increases at the same time (Fig. 4.7A). With higher concentrations of Zn^{2+}-cyclen, we can shift the equilibrium in Ras(wt)·Mg^{2+}·GppNHp completely toward state 1(T) conformation. Metal free or trivalent Co^{3+}-cyclens did not show any effect in the ^{31}P NMR spectrum at the low millimolar concentrations tested [28].

Effects due to Zn^{2+}-cyclen binding are best seen for β- and γ-phosphate resonances. Whereas the γ-phosphate signal shows a strong downfield shift, the β-resonance signal shifts upfield. The resonances corresponding to state

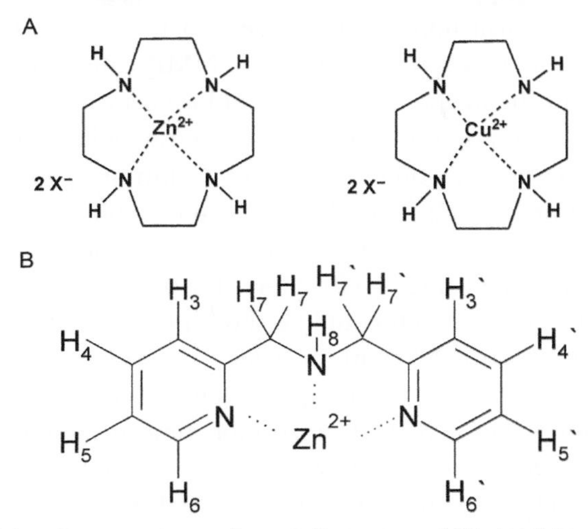

Figure 4.6 Molecular structure of two Ras state 1(T) inhibitors. (A) 1,4,7,10-tetraazacyclododecane (cyclen) zinc(II) (1) and copper(II) (2) complexes. (B) zinc(II) bis(2-picolyl)amine. Counter ion X used here is chloride.

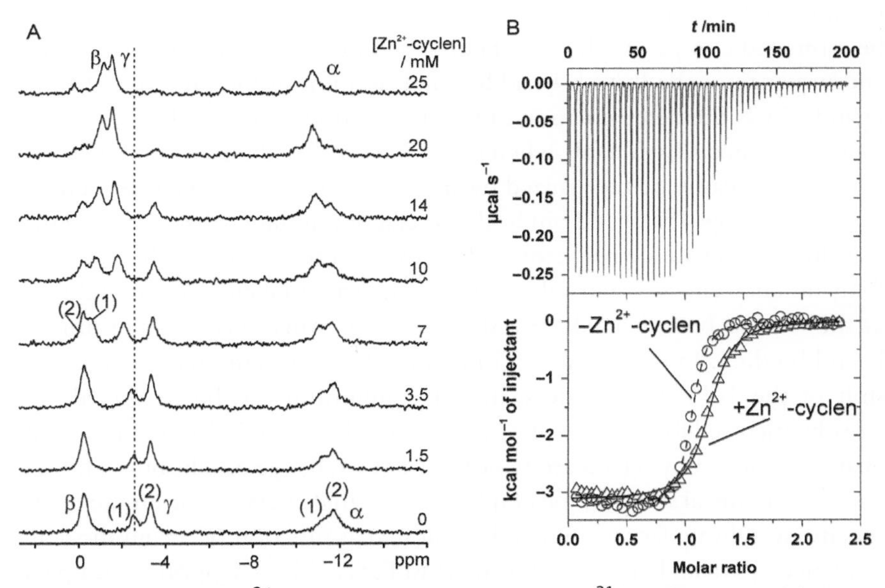

Figure 4.7 Impact of Zn^{2+}-cyclen binding to active Ras. (A) ^{31}P NMR spectra on the titration of Zn^{2+}-cyclen to initially 1.5 mM wild-type Ras(1–189) complexed with Mg^{2+}·GppNHp (40 mM Hepes/NaOH pH 7.4, 10 mM $MgCl_2$, 2 mM dithioerythritol, 10% D_2O, 0.2 mM DSS, T = 278 K). The corresponding concentrations of Zn^{2+}-cyclen are indicated. (B) Isothermal titration calorimetry on the Ras–effector interaction in the absence and presence of 4 mM Zn^{2+}-cyclen. A sample containing 40 μM *H*-Ras(wt)·Mg^{2+}·GppNHp in 50 mM Tris/HCl pH 7.5, 100 mM NaCl, 5 mM $MgCl_2$, 2 mM DTE. The sample was titrated with a 400 μM solution of Raf-RBD in the same buffer. Temperature was 298 K. *For further details see Ref. [29].*

2(T) only show a small upfield shift. The association and dissociation of M^{2+}-cyclen to/from Ras-GppNHp is fast on NMR time scale (no line splitting by binding of the ligand is observed) and is characterized by a continuous shift of the corresponding resonance lines with higher concentration of the ligand (Fig. 4.7A). By isothermal titration calorimetry experiments, we could show the impact of Zn^{2+}-cyclen on the interaction between Ras(wt)·Mg^{2+}·GppNHp and its main effector Raf kinase. The interaction with Raf is significantly perturbed (Fig. 4.7B). Here, for the first time, the proof of principle of our new approach could be shown [29].

The titration of Ras(T35A) with the paramagnetic Cu^{2+}-cyclen lead to a strong line broadening and to an upfield shift of the ^{31}P signal of the γ-phosphate group (Fig. 4.8A). T_1 relaxation time measurements allowed the determination of the distances between the metal ion and the phosphorous nuclei [28,29]. The obtained distances showed that the metal ion of M^{2+}-cyclens is directly coordinated to the γ-phosphorus of bound nucleotide triphosphates of Ras selectively in state 1(T) conformation of the protein.

[^1H,^{15}N] HSQC NMR experiments [55] can give more detailed information on the ligand interaction: the ligand-binding site of the protein can be estimated comparing the spectra obtained in experiments in the absence and presence of the ligand. In addition, a titration of the protein with the ligand allows the determination of the dissociation constant. These experiments are especially valuable when more than one binding site for the ligand exists. It can easily be recognized by an analysis of the sequence-dependent spectral changes. If a ligand binds, the interaction induces changes in chemical shifts of the amide resonances close to the interaction site. In addition, long-distance effects occur when large-scale structural changes are induced as they are to be expected for a shift of a conformational equilibrium by ligand binding. Using a state 1(T) mutant, chemical shift changes due to a shift of the 1(T)–2(T) conformational equilibrium can be neglected. As already discussed for the phosphorus data, also the ^1H and ^{15}N chemical shift changes induced by ligand are typical for fast exchange conditions. The ^1H and ^{15}N chemical shifts can be combined to a single value using amino acid-specific weighting factors [56] and be plotted as function of the amino acid sequence. This has been done for the Ras(T35A)·Mg^{2+}·GppNHp complex with Zn^{2+}-cyclen (Fig. 4.8B).

Since the product of the exchange correlation time τ_e and the chemical shift difference $\Delta\omega$ decides on the NMR time scale, residues characterized by large chemical shift changes can show slow exchange features even when for most of the other residues fast exchange conditions apply. Instead of one peak, two cross peaks appear, and the volume V_i of a cross peak i decreases.

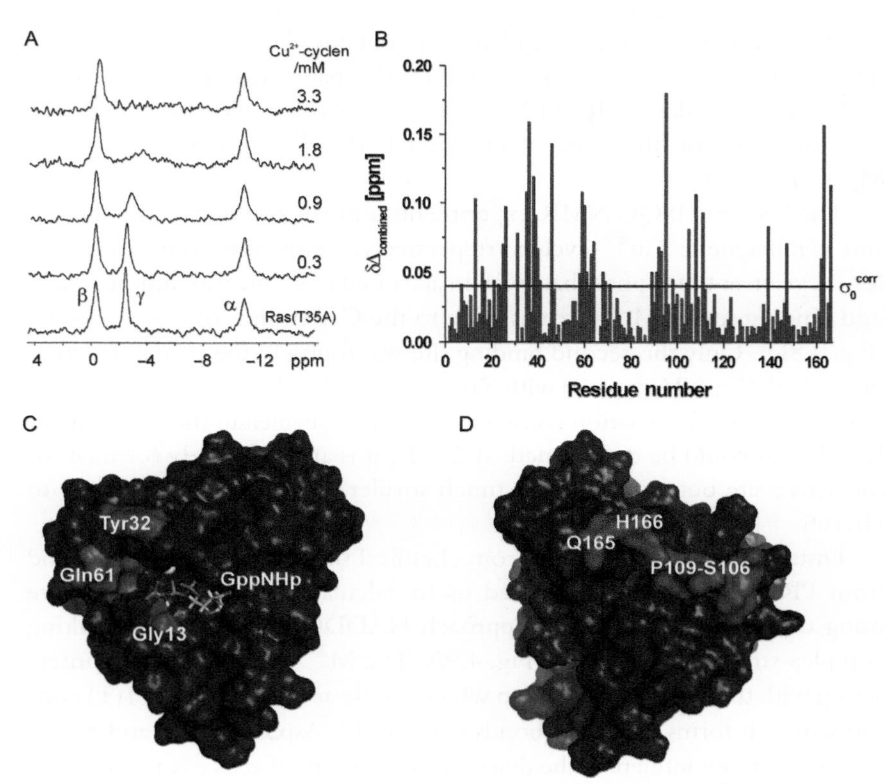

Figure 4.8 Identification of M^{2+}-cyclen binding site on active Ras: (A) ^{31}P NMR spectra of Ras(T35A)·Mg^{2+}·GppNHp in an absence and presence of Cu^{2+}-cyclen: the paramagnetic Cu^{2+}-ion induces strong line broadening of the γ-phosphate signal by PRE. (B) Results of chemical shift change during the titration of Ras(T35A)·Mg^{2+}·GppNHp with Zn^{2+}-cyclen detected by [^{1}H,^{15}N]-HSQC experiments. The combined chemical shift changes $\Delta\delta_{comb}$ are plotted as a function of the amino acid sequence. (C, D) Surface plots of the Ras(T35S)·Mg^{2+}·GppNHp solution structure. (D) structure (C) rotated by 180 degrees. Residues exhibiting strong PRE effects in the [^{1}H,^{15}N] HSQC spectra of Ras(T35A)·Mg^{2+}·GppNHp in the presence of Cu^{2+}-cyclen are given in red. *For further details see Refs. [28,29].* (See color plate.)

A plot of the relative cross peak volume changes gives also information about the ligand-binding site [29].

In M^{2+}-containing ligands, the diamagnetic metal ion can be replaced by a paramagnetic Cu^{2+}-ion. In that case, one gets further information using paramagnetic relaxation enhancement (PRE). That means resonances of nuclei which are close to the paramagnetic center are strongly broadened. In the 2D spectrum, the cross peaks become weaker or completely disappear in the spectrum in the presence of the paramagnetic ligand. Because this effect is strongly distance dependent, one obtains clear information of the direct binding site. In Fig. 4.8C and D, the paramagnetic effect due to Cu^{2+}-cyclen binding to Ras(T35A)·Mg^{2+}·GppNHp is plotted on the solution structure of the state 1(T) mutant Ras(T35S) complexed with Mg^{2+}·GppNHp.

The $^{1}H,^{15}N$ HSQC NMR experiments using diamagnetic Zn^{2+}-cyclen and paramagnetic Cu^{2+}-cyclen, respectively, verify the coordination of M^{2+}-cyclen at the γ-phosphate within activated Ras, but also indicate a second binding site of M^{2+}-cyclens close to the C-terminus of Ras at His166 (Fig. 4.8D). Only this second binding site was found in the crystal structure of Ras(wt)·GppNHp soaked with Zn^{2+}-cyclen [29]. From an analysis of the NMR data, the dissociation constant K_D of Zn^{2+}-cyclen to the 1(T) mutant Ras(T35A) could be determined: at 293 K, it is approximately 6.7 mM for the active site but with 0.6 mM much smaller for the second site close to His166.

Distance restraints obtained from chemical shift perturbation (CSP) and from PRE measurements allowed us to calculate the complex structure using the molecular dynamics approach HADDOCK [57]. The resulting complex structure is shown in Fig. 4.9A. The M^{2+}-cyclen is directly interacting with the γ-phosphate group which is only accessible in state 1(T) conformation. It forms hydrogen bonds with Gly12, Asp33, Thr35, and Ala59.

Zn^{2+}-cyclen influences the dynamic equilibrium of active Ras selectively stabilizing conformational state (1) and is able to perturb the Ras–effector

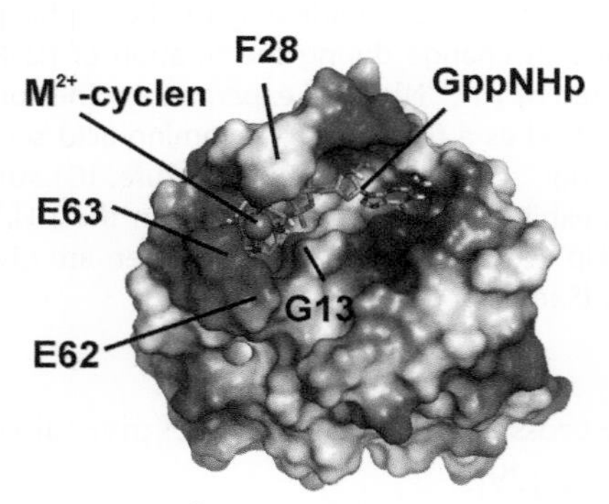

Figure 4.9 Complex structure of M^{2+}-cyclen and active Ras. Surface plot (electrostatic potential) of the structure of Ras(T35A)·Mg^{2+}·GppNHp complexed with M^{2+}-cyclen. The structure was determined by HADDOCK [57] considering restraints derived from CSP and PRE effects obtained in [$^{1}H,^{15}N$] HSQC experiments as well as distant restraints obtained by ^{31}P NMR relaxation PRE measurements. *For further details see Ref. [29].* (See color plate.)

interaction. But its application is limited by its low affinity, which has been determined to be in the millimolar range and the fact that more than one binding position is present in Ras for this compound. In order to increase both the affinity and the selectivity of Zn^{2+}-cyclen at the binding site in the active center of Ras, bivalent Ras ligands have been designed [30].

State (1) represents the Sos-recognition state; therefore, it can be assumed that the structure has similarities with the conformation of Ras found in complex with its exchange factor Sos [34,46]. Consequently, the crystal structure of the Ras–Sos complex [58] was used as a basis for the design of a peptide possibly interacting with Ras and bivalent Ras ligands have been designed. The peptide consists of amino acids from the sequence of the exchange factor Sos directly interacting with Ras close to the active center. Single amino acids in between, not contributing to the interaction with Ras, have been displaced by glycines in order to allow for more flexibility. The resulting peptide consisted of the amino acid sequence LGGIR. The peptide itself interacts with Ras(T35A)·GppNHp with 4 mM binding affinity as obtained from CSP in HSQC NMR experiments [30]. Strongest effects were obtained for the switch II region, which is in accordance with its interaction in the Ras–Sos complex. Unfortunately, the PEG-linked new bivalent cyclen-peptide compound showed no increase in binding affinity as obtained from STD-titration NMR experiment. The optimal choice of the linker length seems to be rather challenging. In contrast, PEG-linked bis(Zn^{2+}/Cu^{2+}-cyclen) ligands strongly interact with Ras-GppNHp but lead to a rapid precipitation of the protein, thus preventing any spectroscopic studies. This effect can be due to a crosslinking of Ras proteins by the bivalent ligand or the dissociation of the bound nucleotide which may lead to destabilization of the protein in solution. In both cases, Ras–effector interaction should strongly be perturbed [30].

6. METAL-BPA DERIVATIVES AS 1(T) INHIBITORS

Receptors based on metal(II)-bis(2-picolyl)amine (subsequently referred to as M^{2+}-BPA, Fig. 4.6B) are used in molecular recognition for phosphate sensing [59]. Consequently, the compound class was initially tested for its binding activity for Ras(T35A)·Mg^{2+}·GppNHp using STD NMR spectroscopy. As supposed, Zn^{2+}-BPA interacts with Ras(T35A)·Mg^{2+}·GppNHp. The affinity of 2 mM at 278 K was estimated from titration experiments. Further, the binding epitope of the compound could be elucidated [31].

It was of interest whether the compound binds to Ras in a similar position as Zn^{2+}-cyclen close to the active center and might influence the dynamic equilibrium in active Ras. Zn^{2+}-BPA was titrated to wild-type Ras complexed with Mg^{2+}·GppNHp up to a molar ratio of 1:28. During the titration, precipitation was observed and resonances representing free Mg^{2+}·GppNHp showed up and shifted upfield in the corresponding ^{31}P NMR spectra. SDS–Page analysis of the precipitant after the last titration step clearly revealed that Ras itself precipitated. The ^{31}P NMR spectra show that Zn^{2+}-BPA is able to shift the equilibrium toward state 1(T) conformation, and at higher concentration the protein precipitates most probably as a consequence of the release of the nucleotide (state 0(T)) [31]. In contrast to the results obtained for Zn^{2+}-cyclen, ^{31}P chemical shift changes of Ras-bound nucleotide induced by binding of this compound are in the limit of error. These results indicate that M^{2+}-BPA should interact in a different way with Ras compared to M^{2+}-cyclen. In Fig. 4.10A, the ^{31}P NMR spectrum of the Zn^{2+}-BPA titration to the oncogenic variant Ras(G12V) in complex with Mg^{2+}·GppNHp is shown. In this case, a complete shift of the equilibrium toward the weak effector-binding state 1(T) is reached at concentrations of the compound where most of the nucleotide is still bound.

The displacement of effectors by Zn^{2+}-BPA can be shown by a NMR assay (Fig. 4.10B) as well as by other spectroscopic methods. We know that the partial loss-of-function mutant Ras(T35S)·Mg^{2+}·GppNHp predominately exists in the weak effector-binding state 1(T). Upon addition of effector, the signals corresponding to conformational state 2(T) appear in the ^{31}P NMR spectrum and those corresponding to state 1(T) disappear (Fig. 4.10B). Stepwise addition of Zn^{2+}-BPA to the complex leads to an increase in the population of conformational state 1(T) and at the same time to a decrease in state 2(T). These results demonstrate that indeed the Zn^{2+}-BPA is able to inhibit the interaction of activated Ras with its main effector Raf kinase. One has to note that also an additional direct effect of Zn^{2+}-BPA on the effector binding should be caused since its binding site is part of the Raf–Ras interaction site, thus enhancing the inhibitory effect of the ligand. By stopped flow fluorescence assay using wild-type Ras in complex with Mg^{2+}·mantGppNHp, it was clearly shown that association rate of Raf drops significantly in the presence of M^{2+}-BPA (Fig. 4.10C) [31].

As in the case of M^{2+}-cyclens, the paramagnetic derivative Cu^{2+}-BPA can be used to determine its binding site on the protein. Cu^{2+}-BPA binding to Ras(T35A)·Mg^{2+}·GppNHp leads to a strong line broadening not only of the γ-phosphate resonance, but also of the β-phosphate resonance.

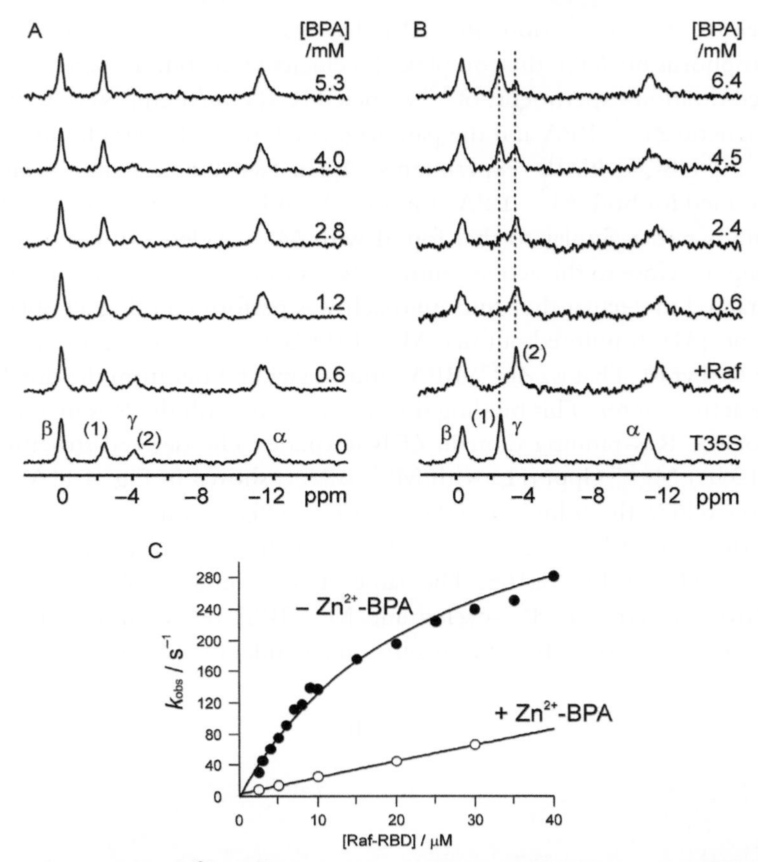

Figure 4.10 Impact of M^{2+}-BPA binding on active Ras. (A) Stabilization of state 1(T) conformation in Ras. ^{31}P NMR spectra on the titration of Zn^{2+}-BPA to initially 0.6 mM Ras (G12V) complexed with Mg^{2+}·GppNHp (40 mM Tris/HCl pH 7.5, 10 mM $MgCl_2$, 2 mM dithioerythritol, 10% D_2O, 0.2 mM DSS, T = 278 K). The corresponding concentrations of Zn^{2+}-cyclen are indicated. (B) Release of the effector by binding of Zn^{2+}-BPA. Initially, 0.8 mM Ras(T35S)·Mg^{2+}·GppNHp (bottom spectrum) Raf-RBD was added ([Ras] = 0.6 mM and [RBD] = 0.9 mM). To this sample, increasing amounts of highly concentrated Zn^{2+}-BPA were added up to a final concentration of 6.4 mM (top spectrum). (C) Fluorescence spectroscopy to determine the kinetics of the association between wild-type Ras·Mg^{2+}·mantGppNHp and Raf-RBD in the absence and presence of Zn^{2+}-BPA. Observed rate constants k_{obs} were obtained by fitting single-exponential function to the time-dependent fluorescence traces. The observed apparent rate constants k_{obs} are plotted against the Raf-RBD concentration in the absence and the presence of 10 mM Zn^{2+}-BPA. The measurements were performed in 15 mM Hepes/NaOH pH 7.5, 125 mM NaCl, 5 mM $MgCl_2$ at 283 K. *For further details see Ref. [31].*

Therefore, the Cu^{2+}-ion in BPA has similar distances to γ- and β-phosphorus nuclei in the complex. To elucidate the binding site, we performed titration experiments of ^{15}N labeled Ras(T35A)·GppNHp with the diamagnetic Zn^{2+}-BPA and the paramagnetic Cu^{2+}-BPA and followed by [^{1}H,^{15}N] HSQC NMR experiments. Again two binding sites could be determined for both M^{2+}-BPA compounds on Ras. The C-terminal binding site is rather similar to that found with M^{2+}-cyclens. In contrast, the binding site close to the active centre of Ras showed different results. Again a restrained molecular dynamic approach was performed using all obtained restraints [31]. It pointed out that M^{2+}-BPA binds close to the amino acids Ser38 to Tyr40. That is, Zn^{2+}-BPA binds at switch I but from the distal site of the active center. This binding site also overlaps with the Ras interaction site for the Ras-binding domain of Raf kinase. The derived structure of Ras(T35A)·Mg^{2+}·GppNHp with M^{2+}-BPA is shown in Fig. 4.11A.

To exclude the inhibitory effects on Ras–effector interaction resulting from the second binding site, we eliminated the second binding site by replacing His166 by Ala166. The same effects could be observed in the ^{31}P NMR spectra. If M^{2+}-cyclen and M^{2+}-BPA really bind at different positions, they should be able to bind independently. We performed ^{31}P

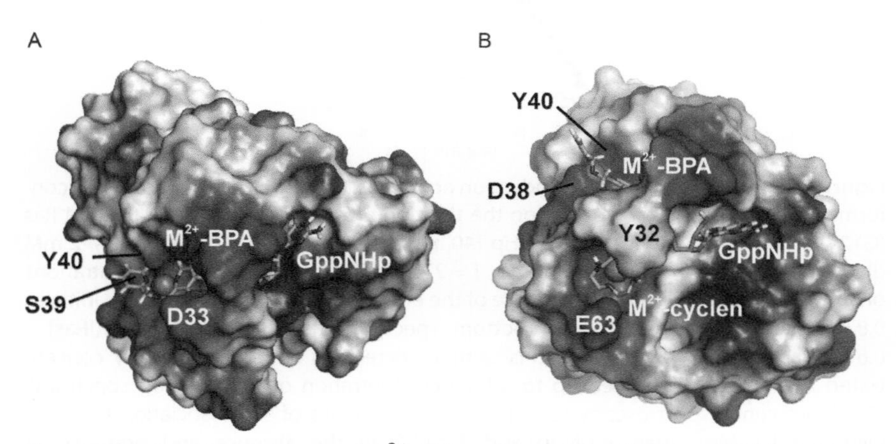

Figure 4.11 Complex structure of M^{2+}-BPA and active Ras. Surface plot (electrostatic potential) of the structure of Ras(T35A)·Mg^{2+}·GppNHp complexed with M^{2+}-BPA (A) and M^{2+}-BPA and M^{2+}-cyclen (B). Structures were determined by HADDOCK [57] considering restraints derived from CSP and PRE effect obtained in [^{1}H,^{15}N] HSQC experiments as well as distant restraints obtained by ^{31}P NMR relaxation experiments. Calculations for the structure presented in (B) started with the obtained structure shown in Fig. 4.9, where M^{2+}-cyclen was already bound. *For further details see Ref. [31].* (See color plate.)

NMR experiments on the Ras(H166A) variant by adding first Cu^{2+}-BPA, leading to the strong line broadening of the γ- and β-phosphorus resonances, and then Zn^{2+}-cyclen in high excess. If Zn^{2+}-cyclen would displace the Cu^{2+}-BPA, the resonances should become sharper and in addition they should shift with Zn^{2+}-cyclen binding. This was not the case, indicating that both compounds can interact with Ras simultaneously [31].

Based on this knowledge, we calculated the complex structure of Ras in complex with both compounds, starting from the Ras(T35A)·Zn^{2+}-cyclen complex [31]. The obtained structure is shown in Fig. 4.11B.

In activated Ras the metal ion of M^{2+}-cyclen interacts directly with the γ-phosphate of the bound nucleotide; the same interaction is not possible in inactive Ras since in GDP the γ-phosphate is missing. This is different for M^{2+}-BPA that does not require coordination with the γ-phosphate group. Correspondingly, changes in ^{31}P chemical shifts of the α-phosphate group in Ras–GDP are observable as to be expected for a binding of M^{2+}-BPA to Ras–GDP. In our nomenclature, metal-BPA complexes would also represent possible 1(D) inhibitors.

For using compounds in therapy of tumors, the inactivation of oncogenic variants of Ras is mandatory. Because Zn^{2+}-cyclen interacts with Ras directly in the active site close to where common mutations in oncogenic variants of Ras at amino acid positions 12, 13, and 61 occur, a direct contact of Zn^{2+}-cyclen and its derivatives with the side chains of the mutated residues could occur. This could lead to steric hindrance and thus to a decrease in the affinity. On the other hand, such a contact could help in a specific recognition of the mutant protein. In contrast, for M^{2+}-BPA, such a direct effect cannot occur. Indeed, the affinity of Zn^{2+}-BPA to Ras(G12V)·Mg^{2+}·GppNHp is somewhat higher compared to wild-type protein as estimated from ^{31}P NMR titration experiments.

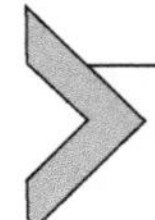

7. OUTLOOK: POSSIBLE APPLICATIONS TO OTHER SMALL GTPases

The existence of a conformational equilibrium between a GEF-interaction state 1(T) and a effector-interaction state 2(T) seems to be a common feature of Ras family GTP-binding proteins: it has also been reported for Ran [60], Rho [61], Cdc42 [62], Rap [63], Ral [64,65], and Arf1 [66] using ^{31}P NMR spectroscopy combined with the results of NMR or X-ray structure determination. The equilibrium constants between the two states are rather different for these proteins, indicating that they are adapted to specific requirements of these proteins. However, the existence of the two states opens a new avenue to address different Ras-like proteins with this novel approach to modulate their activity. The affinity of Cu^{2+}-cyclen to activated Ras is about 10 times weaker than that of Zn^{2+}-cyclen and thus too low to inhibit Ras–effector interaction. However, Cu^{2+}-cyclen presents a helpful tool to identify state 1(T) in Ras and Ras-like proteins. Due to its paramagnetic properties and low affinity, the resonance corresponding to state 1(T) selectively disappears in the ^{31}P NMR spectrum. Cu^{2+}-cyclen recognized state 1(T) in different tested Ras variants independent of the nucleotide bound [67]. This approach was so far also applied to the Ras-like proteins Ran and Arf1 [67].

From theory, the two (so far detected) states should exist in all Ras-like proteins (Fig. 4.1); the experimental proof of the existence of the two states has been provided by a remarkable number of small GTPase studied to date. It also means that their activity can be modulated in the same way as that of Ras, provided that drugs are found that selectively stabilize state 1(T).

REFERENCES

[1] A. Wittinghofer, H. Waldmann, Ras—a molecular Switch involved in tumor formation, Angew. Chem. Int. Ed. 39 (2000) 4192–4214. Angew. Chem. 112 (2000) 4360–4383.
[2] K. Wennerberg, K.L. Rossman, C.J. Der, The Ras superfamily at a glance, J. Cell Sci. 118 (2005) 843–846.
[3] C. Herrmann, Ras-effector interactions: after one decade, Curr. Opin. Struct. Biol. 13 (2003) 122–129.
[4] K. Rajalingam, R. Schreck, U.R. Rapp, S. Albert, Ras oncogenes and their downstream targets, Biochim. Biophys. Acta 1773 (2007) 1177–1195.
[5] J.L. Bos, Ras oncogenes in human cancer: a review, Cancer Res. 49 (1989) 4682–4689.
[6] A.E. Karnoub, R.A. Weinberg, Ras oncogenes: split personalities, Mol. Cell. Biol. 9 (2008) 517–531.
[7] A.T. Baines, D. Xu, C.J. Der, Inhibition of Ras for cancer treatment: the search continues, Future Med. Chem. 3 (2011) 1787–1808.
[8] A.A. Adjei, Blocking oncogenic Ras signaling for cancer therapy, J. Natl. Cancer Inst. 93 (2001) 1062–1074.
[9] B.B. Friday, A.A. Adjei, K-ras as a target for cancer therapy, Biochim. Biophys. Acta 1756 (2005) 127–144.
[10] W. Wang, G. Fang, J. Rudolph, Ras inhibition via direct Ras binding—is there a path forward? Bioorg. Med. Chem. Lett. 22 (2012) 5766–5776.
[11] A. Palmioli, E. Sacco, S. Abraham, C.J. Thomas, A. Di Domizio, L. De Gioia, V. Gaponenko, M. Vanoni, F. Peri, First experimental identification of Ras-inhibitor binding interface using a water-soluble Ras ligand, Bioorg. Med. Chem. Lett. 19 (2009) 4217–4222.
[12] S. Colombo, A. Palmioli, C. Airoldi, R. Tisi, S. Fantinato, S. Olivieri, L. De Gioia, E. Martegani, F. Peri, Structure-activity studies on arylamides and arysulfonamides Ras inhibitors, Curr. Cancer Drug Targets 10 (2010) 192–199.

[13] E. Sacco, D. Metalli, M. Spinelli, R. Manzoni, M. Samalikova, R. Grandori, A. Morrione, S. Traversa, L. Alberghina, M. Vanoni, Biotechnol. Adv. 30 (2012) 233–243.
[14] C. Müller, M.A. Gomez-Zurita Frau, D. Ballinari, S. Colombo, A. Bitto, E. Martegani, C. Airoldi, A.S. van Neuren, M. Stein, J. Weiser, C. Battistini, F. Peri, ChemMedChem 4 (2009) 524–528.
[15] T. Maurer, L.S. Garrenton, A. Oh, K. Pitts, D.J. Anderson, N.J. Skelton, B.P. Fauber, B. Pan, S. Malek, D. Stokoe, M.J. Ludlam, K.K. Bowman, J. Wu, A.M. Giannetti, M.A. Starovasnik, I. Mellman, P.K. Jackson, J. Rudolph, W. Wang, G. Fang, Small-molecule ligands bind to a distinct pocket in Ras and inhibit SOS-mediated nucleotide exchange activity, Proc. Natl. Acad. Sci. U.S.A. 109 (2012) 5299–5304.
[16] Q. Sun, J.P. Burke, J. Phan, M.C.J. Burns, E.T. Olejniczak, A.G. Waterson, T. Lee, O.W. Rossanese, S.W. Fesik, Discovery of small molecules that bind to K-Ras and inhibit Sos-mediated activation, Angew. Chem. Int. Ed. 51 (2012) 6140–6143. Angew. Chem. 124 (2012) 6244–6247.
[17] T. Zor, M. Bar-Yaacov, S. Elgavish, B. Shaanan, Z. Selinger, Rescue of a mutant G protein by substrate-assisted catalysis, Eur. J. Biochem. 249 (1997) 330–336.
[18] M.R. Ahmadian, T. Zor, D. Vogt, W. Kabsch, Z. Selinger, A. Wittinghofer, K. Scheffzek, Guanosine triphosphatase stimulation of oncogenic Ras mutants, Proc. Natl. Acad. Sci. U.S.A. 96 (1999) 7065–7070.
[19] R. Gail, B. Costisella, M.R. Ahmadian, A. Wittinghofer, Ras-mediated cleavage of a GTP analogue by a novel mechanism, Chembiochem 2 (2001) 570–575.
[20] L. Soulère, C. Aldrich, O. Daumke, R. Gail, L. Kissau, A. Wittinghofer, H. Waldmann, Synthesis of GTP-derived Ras ligands, Chembiochem 5 (2004) 1448–1453.
[21] O. Müller, E. Gourzoulidou, M. Carpintero, I.M. Karaguni, A. Langerak, C. Herrmann, T. Möröy, L. Klein-Hitpass, H. Waldmann, Identification of potent Ras signaling inhibitors by pathway-selective phenotype-based screening, Angew. Chem. Int. Ed Engl. 43 (2004) 450–454. Angew. Chem. 116 (2004) 456–460.
[22] H. Waldmann, I.M. Karaguni, M. Carpintero, E. Gourzoulidou, C. Herrmann, C. Brockmann, H. Oschkinat, O. Müller, Sulindac-derived Ras pathway inhibitors target the Ras-Raf interaction and downstream effectors in the Ras pathway, Angew. Chem. Int. Ed Engl. 43 (2004) 454–458. Angew. Chem. 116 (2004) 160–164.
[23] B.J. Grant, S. Lukman, H.J. Hocker, J. Sayyah, J.H. Brown, J.A. McCammon, A.A. Gorfe, Novel allosteric sites on Ras for lead generation, PLoS One 6 (2011) e25711.
[24] G. Buhrman, C. O'Connor, B. Zerbe, B.M. Kearney, R. Napoleon, E.A. Kovrigina, S. Vajda, D. Kozakov, E.L. Kovrigin, C. Mattos, Analysis of binding site hot spots on the surface of Ras GTPase, J. Mol. Biol. 413 (2011) 773–789.
[25] F. Barnard, H. Sun, L. Baker, M.S. Marshall, In vitro inhibition of Ras-Raf association by short peptides, Biochem. Biophys. Res. Commun. 247 (1998) 176–180.
[26] P.C. Gareiss, A.R. Schneekloth, M.J. Salcius, S.Y. Seo, C.M. Crews, Identification and characterization of a peptidic ligand for Ras, Chembiochem 11 (2010) 517–522.
[27] H.R. Kalbitzer, B. Koenig, Method and 1,4,7,10-tetraazacyclododecane metal complexes for influencing the spatial structure of Ras or other guanine nucleotide binding (GNB) proteins, and use as antitumor agents, PCT Int. Appl. 2004, 14 pp. WO 2004006934 A2 20040122 CAN 140:122769 AN 2004:60322.
[28] M. Spoerner, T. Graf, B. König, H.R. Kalbitzer, A novel mechanism for the modulation of the Ras-effector interaction by small molecules, Biochem. Biophys. Res. Commun. 334 (2005) 709–713.
[29] I.C. Rosnizeck, T. Graf, M. Spoerner, J. Tränkle, D. Filchtinski, C. Herrmann, L. Gremer, I.R. Vetter, A. Wittinghofer, B. König, H.R. Kalbitzer, Stabilizing a weak binding state for effectors in the human Ras-protein by small compounds, Angew. Chem. Int. Ed Engl. 49 (2010) 3830–3833. Angew. Chem. 122 (2010) 3918–3922.

[30] F. Schmidt, I.C. Rosnizeck, M. Spoerner, H.R. Kalbitzer, B. König, Zinc(II)cyclen-peptide conjugates interacting with the weak effector binding state of Ras, Inorg. Chim. Acta 365 (2011) 38–48.
[31] I.C. Rosnizeck, M. Spoerner, T. Harsch, D. Filchtinski, C. Herrmann, D. Engel, B. König, H.R. Kalbitzer, Metal BPA complexes as state 1(T) inhibitors of activated Ras protein, Angew. Chem. Int. Ed. 51 (2012) 10647–10651. Angew. Chem. 124 (2012) 10799–10804.
[32] J. Monod, J. Wyman, J.P. Changeux, On the nature of allosteric transitions: a plausible model, J. Mol. Biol. 12 (1965) 88–118.
[33] D.E. Jr Koshland, G. Némethy, D. Filmer, Comparison of experimental binding data and theoretical models in proteins containing subunits, Biochemistry 5 (1966) 365–368.
[34] H.R. Kalbitzer, M. Spoerner, P. Ganser, C. Hosza, W. Kremer, A fundamental link between folding states and functional states of proteins, J. Am. Chem. Soc. 131 (2009) 16714–16719.
[35] M. Geyer, T. Schweins, C. Herrmann, T. Prisner, A. Wittinghofer, H.R. Kalbitzer, Conformational transitions in p21ras and its complexes with the effector protein raf-RBD and the GTPase activating protein GAP, Biochemistry 35 (1996) 10308–10320.
[36] M. Spoerner, C. Herrmann, I.R. Vetter, H.R. Kalbitzer, A. Wittinghofer, Dynamic properties of the Ras switch I region and its importance for binding to effectors, Proc. Natl. Acad. Sci. U.S.A. 98 (2001) 4944–4949.
[37] M. Spoerner, A. Nuehs, P. Ganser, C. Herrmann, A. Wittinghofer, H.R. Kalbitzer, Conformational states of Ras complexed with the GTP analogue GppNHp or GppCH2p: implications for the interaction with effector proteins, Biochemistry 44 (2005) 2225–2236.
[38] M. Spoerner, C. Hozsa, J.A. Poetzl, K. Reiss, P. Ganser, M. Geyer, H.R. Kalbitzer, Conformational states of human rat sarcoma (Ras) protein complexed with its natural ligand GTP and their role for effector interaction and GTP hydrolysis, J. Biol. Chem. 285 (2010) 39768–39778.
[39] M. Spoerner, A. Nuehs, C. Herrmann, G. Steiner, H.R. Kalbitzer, Slow conformational dynamics of the guanine nucleotide-binding protein Ras complexed with the GTP analogue GTPγS, FEBS J. 274 (2007) 1419–1433.
[40] A.J. Scheidig, S.M. Franken, J.E. Corrie, G.P. Reid, A. Wittinghofer, E.F. Pai, R.S. Goody, X-ray crystal structure analysis of the catalytic domain of the oncogene product p21H-ras complexed with caged GTP and mant dGppNHp, J. Mol. Biol. 253 (1995) 132–150.
[41] M.T. Mazhab-Jafari, C.B. Marshall, M. Smith, G.M. Gasmi-Seabrook, V. Stambolic, R. Rottapel, B.G. Neel, M. Ikura, Real-time NMR study of three small GTPases reveals that fluorescent 2'(3')-O-(N-methylanthraniloyl)-tagged nucleotides alter hydrolysis and exchange kinetic, J. Biol. Chem. 285 (2010) 5132–5136.
[42] F. Shima, Y. Ijiri, S. Muraoka, J. Liao, M. Ye, M. Araki, K. Matsumoto, N. Yamamoto, T. Sugimoto, Y. Yoshikawa, T. Kumasaka, M. Yamamoto, A. Tamura, T. Kataoka, Structural basis for conformational dynamics of GTP-bound Ras protein, J. Biol. Chem. 285 (2010) 22696–22705.
[43] M. Araki, F. Shima, Y. Yoshikawa, S. Muraoka, Y. Ijiri, Y. Nagahara, T. Shirono, T. Kataoka, A. Tamura, Solution structure of the state 1 conformer of GTP-bound H-Ras protein and distinct dynamic properties between the state 1 and state 2 conformers, J. Biol. Chem. 286 (2011) 39644–39653.
[44] A. Iuga, M. Spoerner, H.R. Kalbitzer, E. Brunner, Solid-state ^{31}P NMR spectroscopy of microcrystals of the Ras protein and its effector loop mutants: comparison of solution and crystal structures, J. Mol. Biol. 342 (2004) 1033–1040.

[45] M. Spoerner, A. Wittinghofer, H.R. Kalbitzer, Perturbation of the conformational equilibria in Ras by selective mutations as studied by ^{31}P NMR spectroscopy, FEBS Lett. 578 (2004) 305–310.
[46] B. Ford, K. Skowronek, S. Boykevisch, D. Bar-Sagi, N. Nassar, Structure of the G60A mutant of Ras: implications for the dominant negative effect, J. Biol. Chem. 280 (2005) 25697–25705.
[47] M. Geyer, C. Herrmann, S. Wohlgemuth, A. Wittinghofer, H.R. Kalbitzer, Structure of the Ras-binding domain of RalGEF and implications for Ras binding and signaling, Nat. Struct. Biol. 4 (1997) 694–699.
[48] T. Linnemann, M. Geyer, B.K. Jaitner, C. Block, H.R. Kalbitzer, A. Wittinghofer, C. Herrmann, Thermodynamic and kinetic characterization of the interaction between the Ras binding domain of AF6 and members of the Ras subfamily, J. Biol. Chem. 274 (1999) 13556–13562.
[49] W. Gronwald, F. Huber, P. Grünewald, M. Spörner, S. Wohlgemuth, C. Herrmann, A. Wittinghofer, H.R. Kalbitzer, Solution structure of the Ras binding domain of the protein kinase Byr2 from Schizosaccharomyces pombe, Structure 9 (2001) 1029–1041.
[50] B. Ford, S. Boykevisch, C. Zhao, S. Kunzelmann, D. Bar-Sagi, C. Herrmann, N. Nassar, Characterization of a Ras mutant with identical GDP- and GTP-bound structures, Biochemistry 48 (2009) 11449–11457.
[51] T.I. Oprea, Property distribution of drug-related chemical databases, J. Comput. Aided Mol. Des. 14 (2000) 251–264.
[52] M. Mayer, B. Meyer, Group epitope mapping by saturation transfer difference NMR to identify segments of a ligand in direct contact with a protein receptor, J. Am. Chem. Soc. 123 (2001) 6108–6117.
[53] C. Dalvit, P. Pevarello, M. Tatò, M. Veronesi, A. Vulpetti, M. Sundström, Identification of compounds with binding affinity to proteins via magnetization transfer from bulk water, J. Biomol. NMR 18 (2000) 65–68.
[54] P. Schanda, E. Kupce, B. Brutscher, SOFAST-HMQC experiments for recording two-dimensional heteronuclear correlation spectra of proteins within a few seconds, J. Biomol. NMR 33 (2005) 199–211.
[55] A.G. Palmer, J. Cavanagh, P.E. Wright, M. Rance, Sensitivity improvement in proton-detected two-dimensional heteronuclear correlation NMR spectroscopy, J. Magn. Reson. 93 (1991) 151–170.
[56] F.H. Schumann, H. Riepl, T. Maurer, W. Gronwald, K.-P. Neidig, H.R. Kalbitzer, Combined chemical shift changes and amino acid specific chemical shift mapping of protein-protein interactions, J. Biomol. NMR 39 (2007) 275–289.
[57] C. Dominguez, R. Boelens, A.M.J.J. Bonvin, HADDOCK: a protein-protein docking approach based on biochemical or biophysical information, J. Am. Chem. Soc. 125 (2003) 1731–1737.
[58] P.A. Boriack-Sjodin, S.M. Margarit, D. Bar-Sagi, J. Kuriyan, The structural basis of the activation of Ras by Sos, Nature 394 (1998) 337–343.
[59] T. Sakamoto, A. Ojida, I. Hamachi, Molecular recognition, fluorescence sensing, and biological assay of phosphate anion derivatives using artificial Zn(II)-Dpa complexes, Chem. Commun. 2 (2009) 141–152.
[60] M. Geyer, R. Assheuer, C. Klebe, J. Kuhlmann, J. Becker, A. Wittinghofer, H.R. Kalbitzer, Conformational states of the nuclear GTP-binding protein Ran and its complexes with the exchange factor RCC1 and the effector protein RanBP1, Biochemistry 38 (1999) 11250–11260.
[61] S.M.G. Dias, R.A. Cerione, X-ray crystal structures reveal two activated states for RhoC, Biochemistry 46 (2007) 6547–6558.

[62] M.J. Phillips, G. Calero, B. Chan, S. Ramachandran, R.A. Cerione, Effector proteins exert an important influence on the signaling-active state of the small GTPase Cdc42, J. Biol. Chem. 283 (2008) 14153–14164.
[63] J. Liao, F. Shima, M. Araki, M. Ye, S. Muraoka, T. Sugimoto, M. Kawamura, N. Yamamoto, A. Tamura, T. Kataoka, Two conformational states of Ras GTPase exhibit differential GTP-binding kinetics, Biochem. Biophys. Res. Commun. 369 (2008) 327–332.
[64] R.B. Fenwick, S. Prasannan, L.J. Campbell, D. Nietlispach, K.A. Evetts, J. Camonis, H.R. Mott, D. Owen, Solution structure and dynamics of the small GTPase RalB in its active conformation: significance for effector protein binding, Biochemistry 48 (2009) 2192–2206.
[65] R.B. Fenwick, L.J. Campbell, K. Rajasekar, S. Prasannan, D. Nietlispach, J. Camonis, D. Owen, H.R. Mott, The RalB-RLIP76 complex reveals a novel mode of Ral-effector interaction, Structure 18 (2010) 985–995.
[66] T. Meierhofer, M. Eberhardt, M. Spoerner, Conformational states of the ADP ribosylation factor 1 complexed with different guanosine-triphosphates as studied by 31P NMR spectroscopy, Biochemistry 50 (2011) 6316–6327.
[67] T. Meierhofer, I.C. Rosnizeck, T. Graf, K. Reiss, B. König, H.R. Kalbitzer, M. Spoerner, Cu^{2+}-cyclen as probe to identify conformational states in guanine nucleotide binding proteins, J. Am. Chem. Soc. 133 (2011) 2048–2051.

Sugar-Based Inhibitors of Ras Activation: Biological Activity and Identification of Ras–Inhibitor Binding Interface

Alessandro Di Domizio, Francesco Peri[1]
Department of Biotechnology and Biosciences, University of Milano Bicocca, Milano, Italy
[1]Corresponding author: e-mail address: francesco.peri@unimib.it

Contents

Abstract

Inhibition of oncogenic Ras activation through small molecules is a promising approach to the pharmacologic treatment of human tumors. A common strategy to block Ras activation and signal transduction is based on molecules that interfere with the guanine exchange factors (GEF)-promoted nucleotide exchange. We developed several generations of small molecules active in inhibiting Ras activation at low micromolar concentrations. Some of these compounds are more active on cell lines expressing oncogenic Ras than on normal cells and are therefore good hit compounds for anticancer drug development. The molecules belonging to the last generation are soluble in water and allowed the identification of binding site on Ras by means of NMR experiments in deuterated water. The experimentally-determined Ras-binding site comprises residues belonging to the α-2 helix and the β-3 strand of the central β-sheet in the Switch 2 region. Synthetic molecules bind Ras in a region belonging to the more extended Ras/GEF-binding site, and a possible mechanism of Ras inhibition by these compounds can be the blockade of GEF-mediated nucleotide exchange.

The Enzymes, Volume 33
ISSN 1874-6047
http://dx.doi.org/10.1016/B978-0-12-416749-0.00005-1

1. INTRODUCTION

Small GTPases H-Ras, K-Ras, and N-Ras, the products of the ras proto-oncogene, function as molecular switches cycling from GDP-bound inactive to GTP-bound active forms in intracellular signaling pathways controlling cell growth, differentiation, and apoptosis [1]. The intracellular level of active Ras-GTP derives from the balance between GTP hydrolysis (mediated by GTPase activating proteins, GAPs) and GDP to GTP exchange (catalyzed by guanine nucleotide exchange factors, GEF, such as Son of sevenless, Sos) [2]. The oncogenic potential of Ras is activated by point mutations mainly at amino acids 12, 13, and 61 [3]. Most of these mutations impair the intrinsic GTPase activity and moreover render Ras insensitive to the GAP action, leading to constitutive activation of downstream effectors, to uncontrolled cellular growth, and finally to oncogenesis [4]. Oncogenic Ras mutants are found in 30% of human tumors, with great incidence in colon (50%) and pancreatic (90%) adenocarcinomas [5]. Furthermore, aberrant Ras signaling contributes to pathogenesis of some human developmental disorders, including neuro-cardio-facio-cutaneous syndromes [6].

A variety of approaches aimed at chemically silencing oncogenic Ras activity through intervention with drug-like small molecules (molecular weight below 500 Da) have been explored in preclinical and clinical trials. Different strategies have been adopted to target Ras activation through small molecules and are comprehensively described in this and the next volumes: (i) inhibition of K-Ras membrane localization [7], (ii) combination of farnesyl, geranylgeranyl, and palmitoyltransferases inhibitors [8,9], (iii) stabilization of the inactive Ras conformation that weakly binds effectors [10], (iv) inhibition of Ras–effector (Ras–Raf) interaction [11–13], and (v) inhibition of Ras–Sos (GEF) interaction [14–16].

In this chapter, we describe the rational design, synthesis, and biological characterization of small molecules derived from natural sugars developed by our group as inhibitors of nucleotide exchange on Ras. These molecules belong to the last class of compounds that inhibit Ras activation by interfering with the Ras–Sos interaction.

2. INITIAL COMPOUNDS AND THE CHARACTERIZATION OF Ras–GDP COMPLEX

Our project started in 2005, when we were inspired by the mechanism of action of a panel of small molecules developed by Schering–Plough that are able to block Ras–Sos interaction by binding to the Ras–GDP complex [17,18].

Some of these compounds, for example, SCH-54341, SCH-56407, SCH-53870, and SCH-54292 (Fig. 5.1), were found to be active in inhibiting NGF-induced neurite outgrowth on PC-12 cells with IC_{50} ranging from 10 to 20 μM. The biological readout indirectly indicated inhibition of Ras activation and signaling. By electronspray ionization mass spectrometry, it has been possible to detect for all molecules, after incubating with Ras–GDP, the formation of a ternary complex inhibitor–Ras–GDP [17].

A preliminary structure–activity study showed that in these compounds, the putative pharmacophore is formed by two aromatic moieties connected through a linear linker (Fig. 5.1). One of these moieties is an aryl(phenyl)-hydroxylamine group, the other group can vary and in general is a mono- or polycyclic aromatic group. Compounds SCH-56407, -53870, -54292 contain both the phenyl hydroxylamine and aromatic groups and show *in vitro*

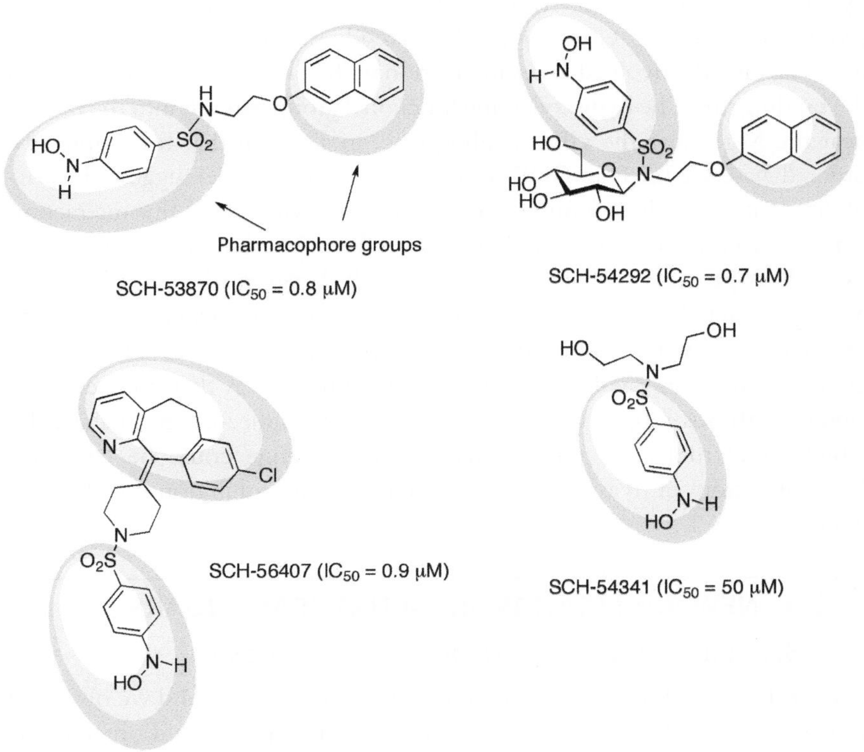

Pharmacophore groups

SCH-53870 (IC_{50} = 0.8 μM)

SCH-54292 (IC_{50} = 0.7 μM)

SCH-56407 (IC_{50} = 0.9 μM)

SCH-54341 (IC_{50} = 50 μM)

Figure 5.1 Chemical structures of Schering–Plough Ras inhibitors: the putative pharmacophore groups are outlined by a gray circle.

activities (IC_{50} less than 1 μM) higher than compound SCH-54341 (IC_{50} = 50 μM), lacking the aromatic group. The binding of these compounds to Ras was studied for SCH-54292 in which the complete pharmacophore (a β-naphthol linked to a phenyl hydroxylamine sulfonamide moiety) is covalently linked to the monosaccharide glucose [19]. SCH-54292 was incubated with Ras in the presence of EDTA that partially denatured the protein by chelating magnesium, then Mg^{2+} was added, and a stable complex in which the inhibitor is "locked" into the binding site of Ras:GDP:Mg^{2+} was formed. Even though this procedure does not reproduce the "real" inhibitor–Ras-binding process *in vitro* and *in vivo*, and somehow forces the inhibitor into Ras-binding site, NMR studies on the SCH-54292:Ras:GDP complex were performed using ^{15}N-enriched Ras protein. ^{1}H–^{15}N HSQC spectra were acquired, and a binding site on Ras was identified, based on the observed chemical shift changes in ^{1}H–^{15}N bidimensional peaks of Ras backbone induced by ligand binding [19]. In this NMR-derived binding model, the naphthyl group of SCH-54292 is positioned in a hydrophobic pocket formed by M72, I100, and V103, while the phenyl hydroxylamine group binds in proximity of the Q61, G60, Y96, and K16 residues. The glucose moiety seems not to be directly involved in the interaction with Ras, and points outside from the binding site. The hydroxylamine functional group extends into the pocket near the Mg^{2+} and the GDP β-phosphate, in a region that is occupied by ordered waters in the Ras–GDP crystal structure. The model suggests the possibility that the hydroxylamine could chelate the Mg^{2+}, and this interaction very likely contributes to ligand affinity for Ras. The binding site proposed by authors partially overlaps with Ras Switch 2 region (defined as residues 57–75). Switch 2 region changes conformation as a consequence of nucleotide exchange and, with the phosphate-binding P-loop (residues 10 – 17) and the Switch 1 (residues 25–40), is one of the regions of Ras that interact more closely with Sos [20]. Authors speculate that the molecular mechanism by which SCH molecules inhibit GDP–GTP exchange could be based on (i) stabilization of Ras GDP-bound state or (ii) inhibition of Ras–Sos interaction.

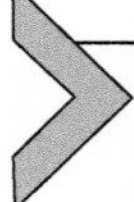

3. NEW Ras LIGANDS FROM NATURAL SUGARS

3.1. Ras Inhibitors derived from arabinose

We synthesized SCH-53870 and SCH-54292 compounds and we observed a very low solubility of both these compounds in aqueous and organic solvents. This negatively influences bioavailability and probably hampered the preclinical and clinical development of SCH molecules. We first investigated the possibility to modify the chemical structure of SCH-53870 to improve the solubility in water and the *in vitro* Ras-binding affinity. From a first screening of 20 compounds [21], we observed that the naphthyl group of SCH-53870 can be replaced by a phenyl, the sulfonamide with an amide, and the hydroxylamine with an hydrogen-bond acceptor/donor group such as the 3,4-dihydroxyphenyl (catechol) group, preserving the biological activity on Ras. Some of these replacements, in particular the introduction of the catechol moiety instead of phenylhydroxylamine, improved water solubility of synthetic molecules [21]. We then aimed at optimizing at the same time water solubility and Ras-binding properties by transferring the two aromatic moieties of the pharmacophore (Fig. 5.1) on polyhydroxylated scaffolds derived from natural monosaccharides. We rationally designed a variety of putative Ras ligands (Fig. 5.2) by replacing the linear (SCH-53870) or cyclic (SCH-56407) 5-atom linkers with conformationally constrained cyclic structures derived from natural monosaccharides. Type 1 compounds were obtained by functionalization of a bicyclic core derived from D-arabinofuranose with a phenylhydroxylamino and one or two benzyl groups. The D-arabinofuranose bicyclic scaffold was rationally designed to accommodate the two aromatic moieties composing the pharmacophore at relative distances similar to that found in SCH molecules. Type 2 compounds are formed by thioglucopyranose core linking the two aromatic groups. Type 3 compounds are formed by two aromatic moieties connected through a hydroxylamine linker that in turn is linked to a natural sugar (glucose, *N*-acetyl glucosamine, lactose). Worthy of note, in type 3 compounds, the phenyl hydroxylamine moiety has been replaced by a dihydroxyphenyl group that increases water solubility, allowing us to maintain biological activity, according to our observations on a panel of aromatic compounds [21].

The rational design of type 1 compounds was done in collaboration with Stephanie van Neuren and Matthias Stein (Anterio Consult & Research GmbH, Mannheim, Germany [22]). The chemical structure of the first panel of type 1 compounds that was rationally designed and synthesized by our group is composed of the arabinose-derived bicyclic scaffold bearing two benzyl and a phenylhydroxylamine group (Fig. 5.3A).

The absolute stereochemistry at C-2 (see Fig. 5.3A for bicycle numbering) and the type of linkage between the phenyl hydroxylamine group and the arabinose core (X = amide or sulfonamide, Fig. 5.3A) varied in the panel, but these differences in the chemical structures did not significantly affect the affinity for Ras and the biological properties in this group of compounds.

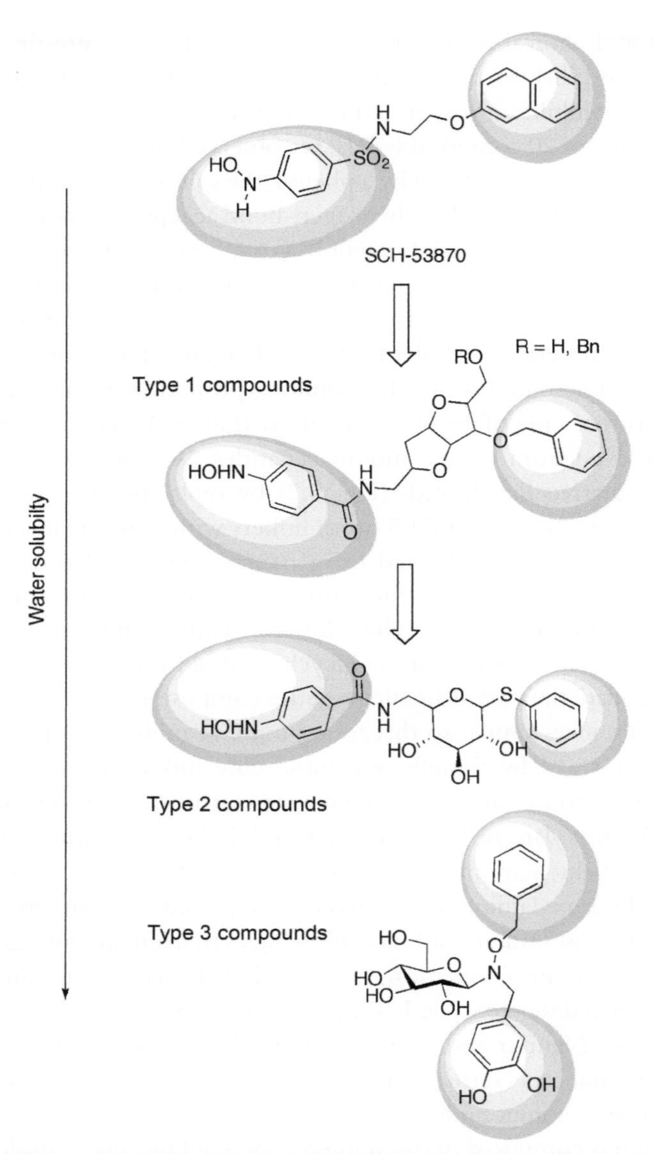

Figure 5.2 Rationally designed sugar-derived Ras inhibitors. The water solubility increases from the top to the bottom. Type 3 compounds are water soluble, while in our hands compound SCH-53870 was sparingly soluble in water.

A

Type 1 inhibitors

$X = SO_2$ or C=O
Absolute stereochemistry in C-2 (R) or (S)

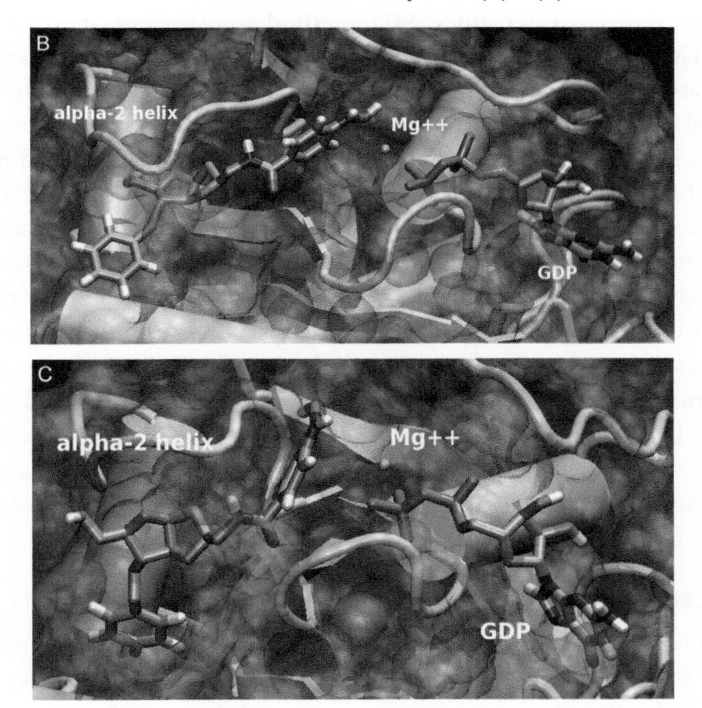

Figure 5.3 (A) Structure of the common bicyclic core of inhibitors type 1, first generation; (B) Ras–inhibitor type 1 binding pose as obtained using the Glide program for docking; (C) binding pose of the same inhibitor with Ras obtained by molecular dynamics. (See color plate.)

Before applying the virtual ligand docking, the X-ray crystal structure of the Ras–GDP complex (PDB ID: 4q21) was optimized and relaxed in three-step procedure, and the structure of small molecules was also submitted to conformational scan to find the most representative conformations for subsequent docking studies [22].

Using the Glide docking program (Glide 2.7, Schrodinger LLC, 2003), the orientation of the putative inhibitors within the binding site was found to be similar to that described by Schering–Plough [19], with the hydroxylamine group chelating the bivalent Mg^{2+} ion (Fig. 5.3B). Glide requires the generation of an initial grid as a first step; the boundaries of the box enclosing this grid were defined by the Mg^{2+} ion, the β-phosphate group of the bound GDP, and the two residues R68 and I100, previously identified by Schering–Plough researchers [19] as ligand contact points. This procedure is somehow biased toward the identification of a binding site localized in proximity of the GDP-binding region. We then adopted the steered/interactive molecular dynamics (MD) methodology to generate the binding conformation of type 1 compound. The molecule was positioned so that the oxygen atom of hydroxylamine is at a distance of 2.0 Å from the Mg^{2+} ion. An initial equilibration of the whole system was then performed by means of a 20.0 ns MD simulation in explicit water (NPT ensemble, 293.15 K), the distance between the hydroxylamine oxygen and Mg^{2+} being constrained to the value of 2.0 Å during all the dynamic simulation. Three different MD simulations of 10.0 ns in explicit water were then carried out with the system completely free, starting from the same conformational structure of the system, but using different initial velocities Maxwell distributions. The frame corresponding to the most representative conformation of the type 1 compound bound to Ras was then selected among all the generated conformations. In this binding pose (Fig. 5.3C), the small-molecule type 1 is situated in a more external binding site if compared to the one obtained by Glide (Fig. 5.3B). As a consequence, the hydroxylamine group of ligand does not chelate the magnesium ion.

3.2. Activity on purified Ras and k-Ras transformed cells

The activity of this first set of compounds in interfering with Sos-promoted GDP to GTP exchange on purified human Ras protein (p21h-Ras) and in inhibiting the Ras-dependent yeast growth and the growth of normal and k-Ras transformed NIH3T3 mouse fibroblast was investigated [22].

This work was carried out in collaboration with Prof. Enzo Martegani and Dr Sonia Colombo of our Department at the University of Milano Bicocca. The Sos-stimulated guanine nucleotide exchange was monitored by using the catalytically active form of Sos (the Cdc25Mm domain) in the presence of an excess of fluorescently labeled *N*-methylanthraniloyl GTP (MANT-GTP). The fluorescence of MANT-GTP proportionally increases upon binding to Ras and GDP exchange. Human p21h-Ras was incubated with MANT-GTP in the absence and the presence of increasing concentrations of the putative inhibitors and compound SCH-53870 was used as a positive control in the same experimental conditions.

IC_{50} values for four type 1 compounds and for SCH-53870 compound were found to be of the same order of magnitude, ranging from 35 to 76 μM [22]. Comparison of activities in inhibiting Sos-mediated GDP–GTP exchange and docking scores suggested that all compounds accommodate similarly into the Ras cavity and thus have very similar binding orientation and affinity. Finally, this first generation of type 1 compounds inhibited in a dose-dependent way the Ras-dependent yeast growth and k-Ras transformed NIH3T3 cell growth and proliferation [22].

We then developed a second generation of type 1 compounds in which the benzyl group on C-2′ has been removed so that these compounds present the bicyclic core with only one benzyl ether group (in C-4) and the phenylhydroxylamine group linked to C-1′ with amide or sulfonamide bonds (Fig. 5.4) [23].

These compounds were active in inhibiting *in vitro* Sos-promoted GDP–GTP exchange with an IC_{50} similar to that found for the first generation, and one of these compounds with the amide bond at C-1′ was found to be particularly active in inhibiting fluorescent GDP dissociation from Ras.

Interestingly, only amide-containing molecules were active in inhibiting cellular growth in both normal and k-Ras-transformed mammalian cells, while sulfonamides were inactive [23]. These data suggest that, while interacting strongly with purified Ras thus affording good *in vitro* activity, the more polar sulfonamide moiety can reduce uptake through cell membrane, thus decreasing inhibitor potency in cells [23].

Interestingly, type 1 compounds of third generation (Fig. 5.4), lacking also the benzyl ether group on C4, were more water soluble than other compounds because of the presence of two polar hydroxyl groups, but totally lack biological activity *in vitro* and on cells [23].

3.3. NMR binding experiments

We then aimed at characterizing the pharmacophore structure on inhibitors by means of NMR binding experiments with purified human Ras (p21h-Ras) [23]. This work has been done by Cristina Airoldi in collaboration with Prof. Jesús Jiménez-Barbero, of the CSIC, Madrid, Spain.

For ligands that exchange between the free and bound states at a moderately fast rate, trNOESY experiments provide a helpful means to study their conformations at the receptor-binding sites. trNOESY experiments

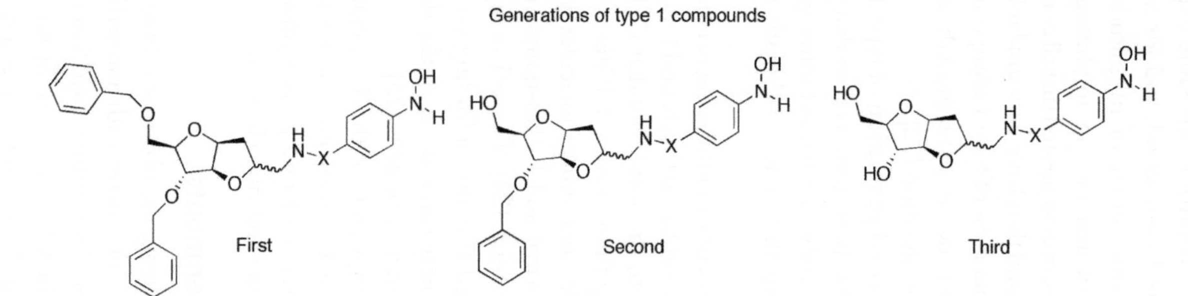

Figure 5.4 Structure of type 1 compounds: first, second, and third generations with, respectively, three, two, and one aromatic groups.

were performed to investigate the interactions in solution between second/third generation of type 1 compounds and the Ras–GDP complex [23]. We also studied ligand–Ras interaction by saturation-transfer difference (STD) experiments that are a very helpful means to detect ligand binding to receptors. In the STD experiment, after irradiating the protein in a region that does not contain signals of the ligand, magnetization is transferred from the protein to the ligand, and the part of the ligand having the strongest contact to the protein shows the most intense NMR signals, enabling the mapping of the ligand binding epitope [24].

STD experiments were performed on ligand–protein mixtures with the aim of confirming the interaction, and of studying which region of the ligand directly interacts with Ras. Data from both trNOESY and STD experiments converged to point out that the major interactions of inhibitors with Ras involved both the benzyl and the phenylhydroxylamine moieties of the ligands (Fig. 5.5) [23].

The backbone (core) of all inhibitors did not interact or interacted very weakly (compounds **2** and **3**) with Ras. Other generations of type 1 and 2 compounds were synthesized with a 3,4 dihydroxyphenyl (3,4 catechol) group replacing phenylhydroxylamine and STD experiments consistently indicated the involvement of benzyl ether and catechol groups in the interaction with Ras (Fig. 5.5) [25]. Interestingly, the interaction with Ras determined by STD/trNOESY for the compounds lacking the benzyl ether group (second generation, type 1, see Fig. 5.4) was totally absent, thus mirroring the lack of biological activity *in vitro* and in cells.

Taken together, these observations suggest that the pharmacophore of our sugar-derived Ras inhibitors is composed of two aromatic rings opportunely spaced, one bearing two hydroxyl or a hydroxylamine group, the other being a benzyl ether group.

3.4. Docking analysis

AutoDock docking analysis between Ras and compounds **1–5** (Fig. 5.5) showed that in high-scoring poses, all compounds bind Ras in a cleft near the Switch 2 region [25]. Compounds **1**, **2**, and **4** share the same pharmacophores, but have different scaffolds. Computed-binding energies for the best poses obtained by docking calculations (Fig. 5.5) indicate that compound **2** (arabinose scaffold) has a more favorable binding energy than compounds **1** (linear scaffold) and **4** (glucose scaffold). The (S)—but not the (R)—stereoisomer of compound **3** (arabinose scaffold) has a similar binding energy as compound **2**.

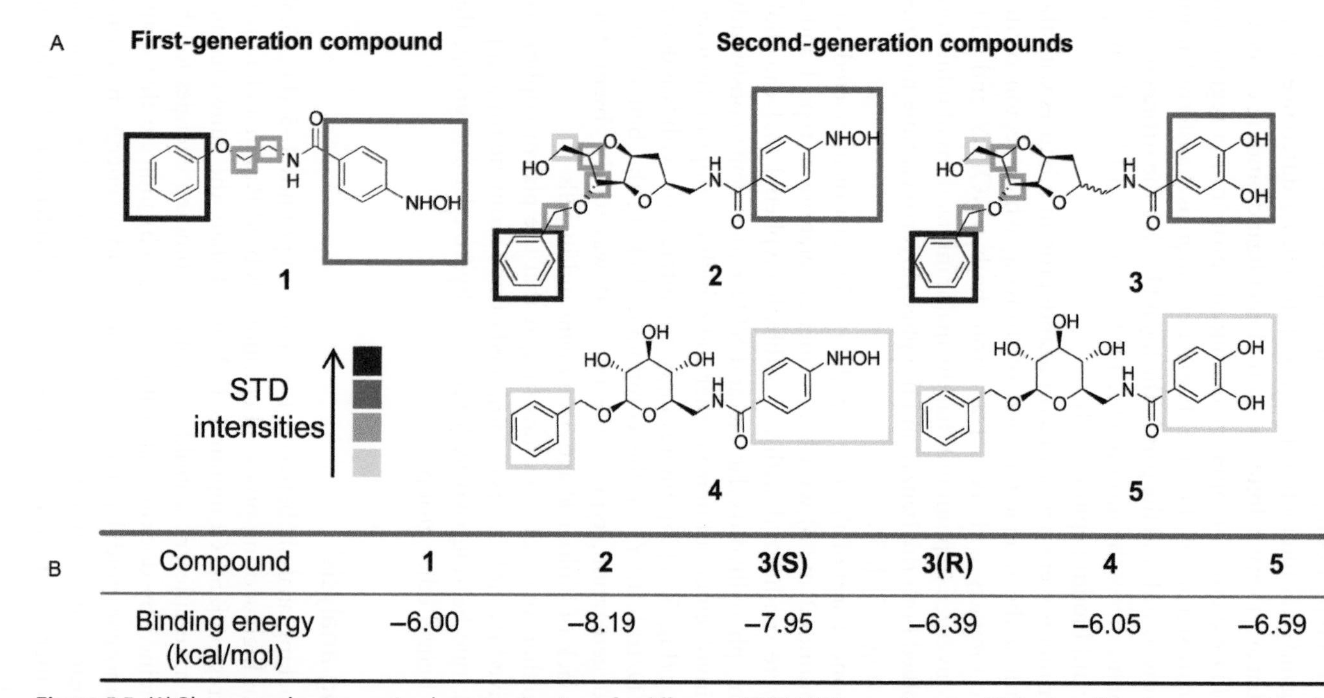

B

Compound	1	2	3(S)	3(R)	4	5
Binding energy (kcal/mol)	−6.00	−8.19	−7.95	−6.39	−6.05	−6.59

Figure 5.5 (A) Pharmacophore mapping by saturation transfer difference (STD) NMR experiments of Schering–Plough compound **1**, of compounds of type 1 (molecules **2**, **3**) and type 2 (molecules **4**, **5**). The intensity of STD signal (vertical arrow, gray scale) is proportional to the force of interaction between inhibitor groups and residues on Ras-binding site. (B) The binding energies calculated by Autodock-4 are reported in the table.

Docking analysis shows that the two aromatic pharmacophores are involved in aromatic stacking interactions with the aromatic rings of Y96 and Y64, while a further stabilizing contribution is given by the formation of several hydrogen bonds, where a key role is played by Q99 and some residues of the Switch 2 region, in particular R68 and Q62 [25]. NMR (STD) and trNOESY experiments in solution also point out for all compounds that aromatic groups are the stronger binding determinants, the phenyloxy or benzyloxy groups of compounds **1–3** interacting stronger with Ras than the phenylhydroxylamine or 3,4-dihydroxyphenyl groups (Fig. 5.5) [26]. Consistently with docking analysis, compound **2** was characterized by the strongest STD signals, indicating higher affinity for Ras (Fig. 5.5). In Fig. 5.5, the intensities of STD signals are reported for the different moieties of compounds **1–5** ranging from low (weak interaction with Ras, light gray square) to high (strong interaction with Ras, black square).

The binding region of molecules of type 1 and 2 is localized in the Ras Switch 2 region as described above and overlaps with Ras–Sos binding interface [20]. The interface between Ras and Sos is primarily hydrophilic and very extensive, with 3600 A^2 of surface area buried in the complex. At the heart of the interface between Ras and Sos is a cluster of three hydrophobic side chains from the Switch 2 region of Ras (Y64, M67, and Y71) which are buried into the hydrophobic core of Sos at the base of the binding site. Surrounding this hydrophobic anchor is an array of polar and charged interactions between Sos and Ras that results in almost every external side chain of Switch 2 being coordinated by Sos [20]. The capacity to interfere with Ras–Sos interaction was investigated for type 1 and 2 compounds by surface plasmon resonance (SPR) experiments in collaboration with prof. Marco Vanoni of our Department [27]. SPR experiments were performed by using N-terminal His-tagged p21hRas protein immobilized to an NTA sensor chip preactivated with $NiCl_2$. Sos binding to Ras was detected as an increase in resonance units. Pretreatment of the NTA-Ras chip with type 1 and 2 compounds (in the 0.5–10 μM range) decreased subsequent binding of Sos in a dose-dependent manner. Type 1 compounds were slightly more potent than type 2 in inhibiting Ras–Sos interaction.

The Ras-binding property of compounds type 2, with a glucose scaffold, was investigated in nucleotide exchange experiments on purified Ras [27]. These compounds showed higher water solubility than type 1 compound but weaker activity in inhibiting Sos-promoted GDP–GTP exchange *in vitro*, thus reflecting the less potent capacity to displace Sos from Ras in SPR experiments [27].

The NMR experiment (STD, trNOESY) for type 2 compounds confirmed the principal role of the aromatic moieties in the interaction with Ras [27].

3.5. Activity on cancer cells

In collaboration with Prof. Alberto Bardelli and Dr Federica di Nicolantonio (The Oncogenomics Center, IRCC University of Torino, Italy), we investigated the capacity of synthetic molecules of type 1 and 2 to induce selective cytotoxicity in human colorectal cancer cells expressing K-RasG13D [25]. The HCT-116 colorectal cancer cell line contains a heterozygous G38A mutation in the K-RAS gene that originates a glycine to aspartate (G13D) change. Derivative clones carrying either the wild-type (HKe-3 cells) or the K-RASG38A allele (HK2-6 cells) were generated by plasmid-mediated homologous recombination [28]. We tested the activity of compounds **2**, **3**, and **5** (Fig. 5.5) belonging to type 1 and type 2 inhibitor families on parental HCT-116 and on both HKe-3 and HK2-6 clones. Compound **5** with a glucose scaffold was inactive (nontoxic) on all cell types. Only compound **2** showed a selective toxicity effect at 50–100 μM concentration on cells expressing oncogenic RasG13D (both HCT-116 parental and HK2-6 clone). At the same concentration, the toxicity effect on HK-2 clone expressing only wild-type Ras is much lower. These data point to a selectivity of action of compound **2** on the oncogenic RasG13D protein versus normal Ras. To our knowledge, no other small molecules are known with selective action on Ras mutants. Moreover, the very low mortality of cells with wild-type Ras up to a 50 μM concentration indicates low toxicity of these compounds for normal cells [25].

In order to determine whether the selective cytotoxic effect of compound **2** on human colorectal cancer-derived cell lines was due to stronger inhibition of RasG13D, the ability of compounds **1–5** to interfere *in vitro* with Sos-catalyzed nucleotide exchange on wild-type and RasG13D proteins (used at 0.25 μM) was tested using the GEF catalytic domain of RasGRF1. The inhibitory effect of compounds **1–5** on the Sos-catalyzed nucleotide exchange on wild-type and RasG13D proteins was tested. Compounds **2** and **3** (arabinose scaffold) showed highest activity on GEF-mediated exchange of Ras ($IC_{50} < 100$ μM). Surprisingly, none of the compounds, including **2** that showed some selective effect on cells expressing the mutant RAS allele, showed any selective inhibitory effect on GEF-mediated nucleotide exchange of RasG13D [25].

The extensive biological characterization of type 1 and 2 compounds presented here pointed out a promising therapeutic potential of some molecules;

in particular, type 1 compounds with an arabinose-derived scaffold were selected as lead compounds for drug development. However, the low aqueous solubility of these synthetic compounds hampered meaningful NMR and crystallography studies and therefore most insights on their mechanism of action and Ras-binding mode were derived only from computational analysis. Indeed, numerous efforts by our group to cocrystallize the hydroxylamine-containing inhibitors with Ras were unsuccessful. H-Ras (1–166) bound to our compounds was purified and subjected to crystallization trials. Many commercially available sparse matrix conditions were tried to identify crystallizing conditions, but currently have not resulted in hit. Preformed H-Ras crystals were soaked in crystallizing mother liquor containing excess dissolved synthetic compounds of type 1 and type 2 and were found to dissolve and not reappear even after prolonged storage. The extremely low water solubility of all synthetic compounds necessitated the addition of at least 10% methanol that resulted in aggregation and precipitation of H-Ras, leading to a generalized shift of cross-peaks in 2D NMR spectra.

4. Ras INHIBITORS WITH IMPROVED WATER SOLUBILITY

The design of type 3 compounds differing for the sugar moiety [26,29] (Fig. 5.6) was mainly inspired by the necessity to increase water solubility while maintaining inhibitor activity. Indeed, the structure of *N*-glucosylated sulphonamide SCH-54292 previously synthesized by Schering–Plough laboratories and used in NMR experiments to evaluate binding to H-Ras is very similar to that of type 3 compounds [19]. Interestingly, we

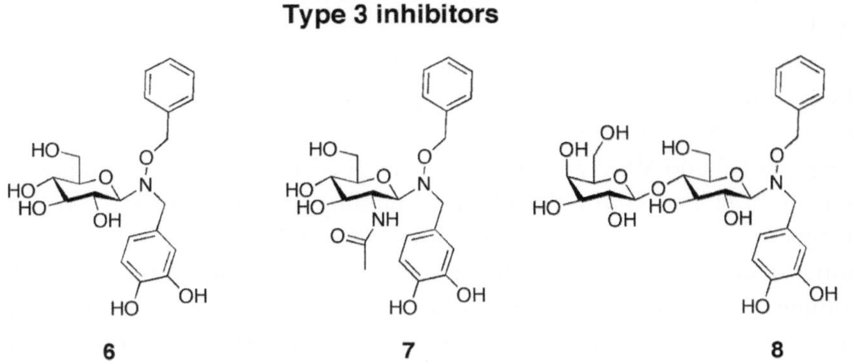

Type 3 inhibitors

6 7 8

Figure 5.6 Type 3 compounds differing in the sugar moiety: molecules **6** (glucose), **7** (*N*-acetylglucosamine), and **8** (lactose).

resynthesized compound SCH- 54292 and found it almost completely insoluble at the concentrations required for NMR-based binding experiments. To ensure water solubility, the hydrophilic sugar and dihydroxybenzyl moieties were included in the design of **6**, and the hydrophobic naphthyl of SCH-54292 was replaced by a benzyl group. The key step in the synthesis of **6** is the chemoselective glycosylation of the *O*-benzyl-*N*-(3,4-dihydroxybenzyl) hydroxylamine.

In contrast with compound SCH-54292 and with type 1 and 2 compounds, type 3 compound **6** turned out to be readily soluble in water and aqueous buffers used for biological and NMR characterization. This improved water solubility is due to the presence of the hydrophilic sugar moiety. Isothermal titration calorimetric (ITC) studies were performed by Dr Celestine Thomas (University of Montana, Missoula, USA) to quantify the affinity of compound **6** toward H-Ras (1–166). This molecule was able to bind to H-Ras with an affinity of 37 μM and with an enthalpy of binding of about 1964 ± 40 cal/mol (error was calculated on the basis of three independent titrations). Compound **6** exhibited exothermic heats in ITC profile. Interestingly, the stoichiometric value of the compound 1: H-Ras interaction was 1:1 (equimolar).

In collaboration with Dr Vadim Gaponenko and Dr Sherwin Abraham (University of Illinois at Chicago, USA), compound **6** was then tested for its ability to bind H-Ras–GDP by NMR experiments in solution. The titration of compound **6** into a solution of ^{15}N H-Ras (1–166) was followed by ^{15}N-edited HSQC experiments that allowed delineation of the Ras–ligand binding interface. A series of spectra were collected at different Ras:ligand molar ratios. The results of these experiments are depicted in Fig. 5.7. The residues exhibiting statistically significant chemical shift perturbations are marked on the graph.

The statistically significant chemical shift perturbations were observed in the β-3 strand (residues I55, T58, A59, and Y64) of the central β-sheet and α-2 helix of Switch 2 (residues M67, R68, Y71, M72, R73, T74, and G75). Residues K5 and V103 are adjacent to the β-3/α-2 region and complete a domain of H-Ras-GDP. The observed binding site covers Switch 2 region that is essential for Ras interactions with the Ras–GEF domain of RasGRF1, homologous to SOS, as described by the crystal structure of the Ras–SOS complex [20]. Therefore, it is possible that compound **6** interferes with H-Ras–GEF interaction, as previously outlined using SPR experiments [27]. The affinity of binding to H-Ras as calculated by ITC experiments is in the same order of magnitude of the affinity between Ras–GDP and GEF.

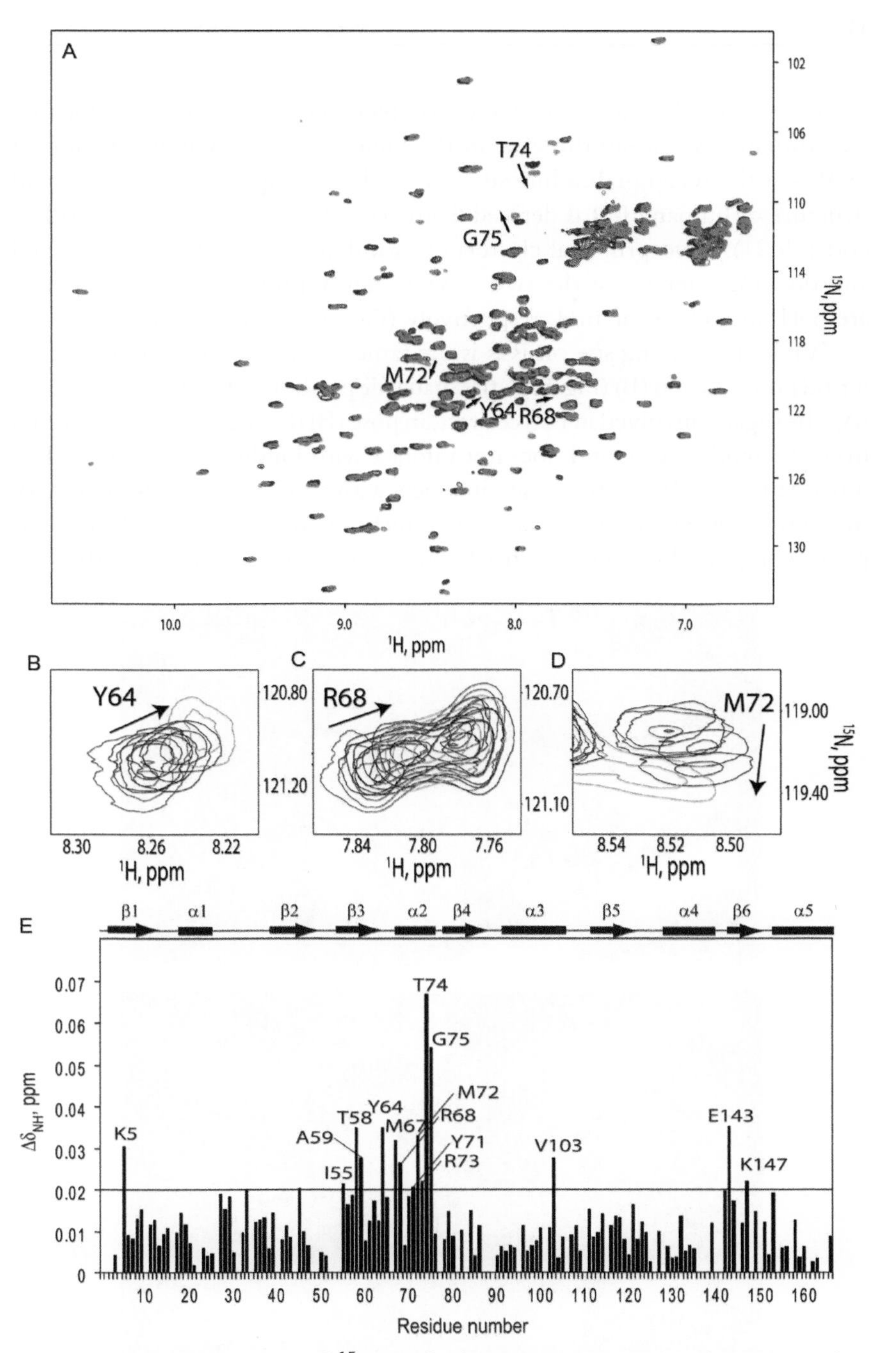

Figure 5.7 (A) NMR titration of ^{15}N-enriched H-Ras-GDP (1–166) with compound **6**. Enlarged view of crosspeaks of Ras residues Y6 (B), R68 (C), and M72 (D). E: Normalized HN chemical shift perturbations. The boxes and arrows above the graph represent helices and sheets in H-Ras-GDP (1–166). The horizontal line marks the average value of chemical shift perturbations plus one standard deviation. (See color plate.)

The NMR binding data were complemented by MD and docking experiments. Assuming the cleft in the vicinity of the Switch 2 region of H-Ras as the receptor-binding site, we docked compound **6** into the rigid structure of human H-Ras derived from crystallographic coordinates (PDB code: 4q21). Two principal clusters of ligand arrangements were obtained and docking energies of the two lowest-energy protein-ligand complexes are 6.41 and 5.91 kcal/mol, respectively (Fig. 5.8, poses A and B).

While the binding site on Ras is the same for the two minimum energy ligand poses (A) and (B) (Fig. 5.8), the spatial disposition of **6** is different: in pose (A), the sugar is involved in Ras contact, in pose (B) the sugar moiety protrudes from the binding cleft and does not interact with binding site residues. The inner part of the Ras-binding site interacts with a polar moiety in the ligand; the sugar hydroxyls in pose A and the catechol hydroxyls in pose B. In both poses, compound **6** does not chelate Mg^{2+}. Worthy of note, in both poses

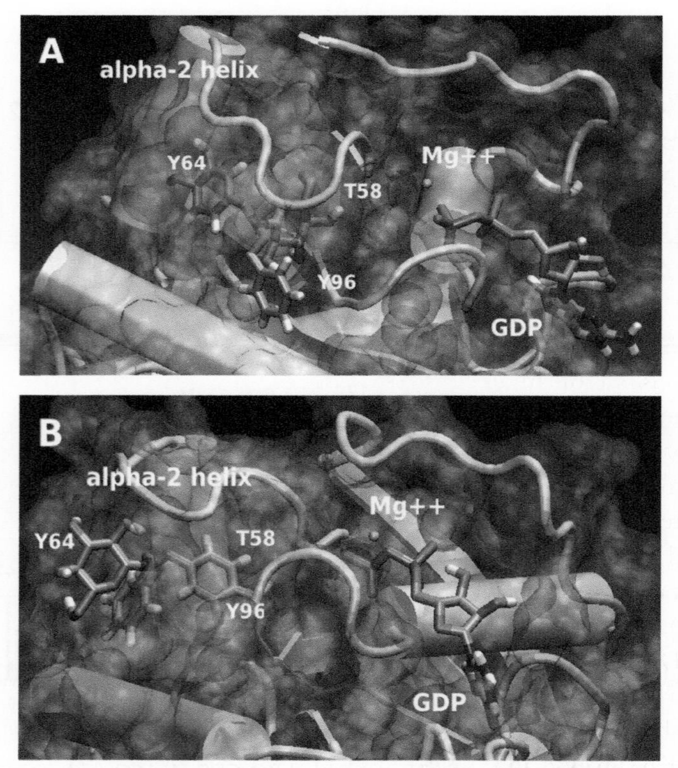

Figure 5.8 The two most representative lowest-energy ligand binding poses (A) and (B) as obtained from MD calculations on compound **6**. In pose (A), the sugar moiety is inserted into the Ras-binding pocket, while in (B) protrudes from the binding site. (See color plate.)

(A) and (B), the ligand occupies a binding region very similar to that identified by MD calculations on type 1 compounds and depicted in Fig. 5.3C.

Then, we aimed to verify if the Ras–ligand interaction model observed for **6** can be extended to water-soluble molecules **7** and **8** (Fig. 5.6) that share the aromatic moiety with **6** but with different sugars (respectively, GlcNAc and Lac). We investigated the interaction of **7** and **8** with Ras by means of STD experiments [29]. Ras residues perturbed by addition of the ligands include K5, R68, R73, T74, and G75 (perturbed by all compounds **6**, **7**, and **8**), Y71 (compounds **6** and **8**), and E76 and V112 (compounds **7** and **8**). Residues R68, Y71, R73, T74, G75, and E76 map to the β-3/α-2 region of Ras (Switch 2 region, residues 57–75). Residues K5 and V112 belonging to the β-1 and β-5 strand, respectively, are involved in Ras interaction for the three compounds. These results clearly point out that molecules **7** and **8**, similarly to **6**, bind to Switch 2 region of Ras, also involved in binding with effectors, such as GEFs.

5. CONCLUSIONS

It is interesting to compare the binding interface identified in our studies on glycosylated compounds **6–8** to the binding site recently detected by a combination of NMR and cocrystallization data for small molecules composed of aromatic moieties such as benzamidines, benzimidazoles, and indole derivatives (Fig. 5.9) [15,16].

Fesik and coworkers cocrystallized a panel of small-molecule ligands (all containing aromatic and indole rings) with both wild-type Ras and oncogenic G12V K-Ras [16]. In all cocrystals, the small aromatic molecules occupy a hydrophobic pocket located between the α-2 helix of Switch 2 and the central β sheet of the protein. Analysis of the ligand–protein cocrystal structures reveals that all the compounds bind to a pocket that is not readily observed in the ligand-free form but in an "open" form of the protein (Fig. 5.9). The pocket is created by a conformational change in which the α-2 helix moves away from the central β sheet [16]. Fang and coworkers soaked K-Ras with benzamidine (BZDN), benzimidazole (BZIM), and 4,6-dichloro-2-methyl-3-aminoethyl-indole (DCAI), thus obtaining cocrystals [15]. In all three cases, compounds bind to a similar site on K-Ras, that is, the region between helix α-2 and the core β-sheet, β-1–β-3. Residues surrounding the binding pocket in Ras include K5, L6, V7, I55, L56, and T74. This site overlaps with the binding interface determined by Fesik [16].

The aromatic small molecules (ASMs) studied by Fang [15] and Fesik [16] induce a conformational rearrangement on Ras generating an open

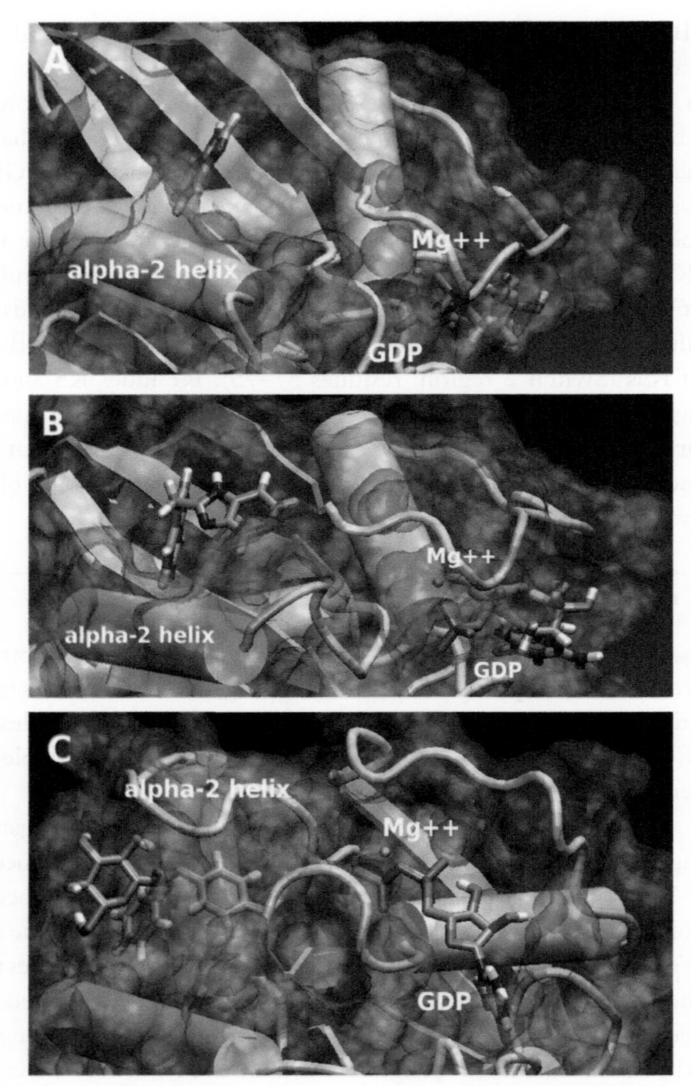

Figure 5.9 Ras-small molecules complexes: (A) Ras–BZI complex (PDB ID: 4dsu) as determined by NMR and X-ray analysis [15]; (B) Ras–0QX complex (PDB ID: 4epv), X-ray structure [16]; (C) Ras-molecule **6** (pose B, Fig. 5.8) as determined by NMR and molecular dynamics optimization [26]. (See color plate.)

conformation by displacing the α-2 helix. A binding site is then generated between helix α-2, the β-1, and the β-3 strands.

In conclusion, the sugar-derived synthetic molecules developed by our group bind Ras on part of the interface commonly involved in Ras–GAP

(Sos) interactions. The Ras residues that interact with small molecules belong to the α-2 helix and β-3 strand of the Switch 2 region and are close to Switch 1 [26]. This binding site does not match to that recently identified by NMR and cocrystallization studies for other ASMs [15,16]. However, both binding sites of our sugar-containing compounds and ASMs overlap with the large Ras–Sos-binding interface and therefore the mechanism of action of our compounds and ASM is assumed to be the same and relays on inhibition of Ras–Sos interaction.

The small molecules of type 1, with an arabinose-derived bicyclic core, are the most active in inhibiting GDP–GTP exchange *in vitro*, in inhibiting Ras signaling in cells, and present higher toxicity on the HCT-116 colorectal cancer cell lines expressing k-Ras than on HCT-116 with normal Ras. These compounds are therefore the most promising candidates for drug development. On the other hand, the less biologically active type 3 compounds have improved water solubility, allowing the experimental determination of Ras-binding site by NMR.

All compounds presented in this chapter are useful prototypes to develop efficient anticancer drugs as well as high affinity chemical probes to investigate the molecular aspect of Ras activation and signaling.

ACKNOWLEDGMENTS

F. Peri thanks all colleagues and coworkers that enthusiastically participated to the exciting adventure of the "Ras project;" the COST Action CM1102 MultiGlycoNano; Finlombarda, Regione Lombardia (Italy), project: "Network Enabled Drug Design" (NEDD 14546); the Italian Ministry of University and Research (MIUR), PRIN 2010–11, project: "Italian network for the development of multivalent nanosystems."

REFERENCES

[1] A.E. Karnoub, R.A. Weinberg, Ras oncogenes: split personalities, Nat. Rev. Mol. Cell Biol. 9 (7) (2008) 517–531.
[2] J. Bos, H. Rehmann, A. Wittinghofer, GEFs and GAPs: critical elements in the control of small G proteins, Cell 129 (5) (2007) 865–877.
[3] E. Diaz-Flores, K. Shannon, Targeting oncogenic Ras, Genes Dev. 21 (16) (2007) 1989–1992.
[4] H. Sun, et al., Expression of CD14 protein in U937 cells induced by vitamin D3, Xi Bao Yu Fen Zi Mian Yi Xue Za Zhi 21 (2) (2005) 155–158.
[5] L. Chin, et al., Essential role for oncogenic Ras in tumour maintenance, Nature 400 (6743) (1999) 468–472.
[6] S. Schubbert, K. Shannon, G. Bollag, Hyperactive Ras in developmental disorders and cancer, Nat. Rev. Cancer 7 (4) (2007) 295–308.
[7] D. van der Hoeven, et al., Fendiline inhibits K-Ras plasma membrane localization and blocks K-Ras signal transmission, Mol. Cell Biol. 33 (2) (2013) 237–251.

[8] R.B. Lobell, et al., Evaluation of farnesyl: protein transferase and geranylgeranyl:protein transferase inhibitor combinations in preclinical models, Cancer Res. 61 (24) (2001) 8758–8768.
[9] E.C. Lerner, et al., Inhibition of the prenylation of K-Ras, but not H- or N-Ras, is highly resistant to CAAX peptidomimetics and requires both a farnesyltransferase and a geranylgeranyltransferase I inhibitor in human tumor cell lines, Oncogene 15 (11) (1997) 1283–1288.
[10] I.C. Rosnizeck, et al., Stabilizing a weak binding state for effectors in the human ras protein by cyclen complexes, Angew. Chem. Int. Ed Engl. 49 (22) (2010) 3830–3833.
[11] O. Müller, et al., Identification of potent Ras signaling inhibitors by pathway-selective phenotype-based screening, Angew. Chem. Int. Ed Engl. 43 (4) (2004) 450–454.
[12] H. Waldmann, et al., Sulindac-derived Ras pathway inhibitors target the Ras-Raf interaction and downstream effectors in the Ras pathway, Angew. Chem. Int. Ed Engl. 43 (4) (2004) 454–458.
[13] J. Kato-Stankiewicz, et al., Inhibitors of Ras/Raf-1 interaction identified by two-hybrid screening revert Ras-dependent transformation phenotypes in human cancer cells, Proc. Natl. Acad. Sci. U.S.A. 99 (22) (2002) 14398–14403.
[14] A. Patgiri, et al., An orthosteric inhibitor of the Ras-Sos interaction, Nat. Chem. Biol. 7 (9) (2011) 585–587.
[15] T. Maurer, et al., Small-molecule ligands bind to a distinct pocket in Ras and inhibit SOS-mediated nucleotide exchange activity, Proc. Natl. Acad. Sci. U.S.A. 109 (14) (2012) 5299–5304.
[16] Q. Sun, et al., Discovery of small molecules that bind to K-Ras and inhibit Sos-mediated activation, Angew. Chem. Int. Ed Engl. 51 (25) (2012) 6140–6143.
[17] A. Ganguly, et al., Detection and structural characterization of ras oncoprotein-inhibitors complexes by electrospray mass spectrometry, Bioorg. Med. Chem. 5 (5) (1997) 817–820.
[18] A. Taveras, et al., Ras oncoprotein inhibitors: the discovery of potent, ras nucleotide exchange inhibitors and the structural determination of a drug-protein complex, Bioorg. Med. Chem. 5 (1) (1997) 125–133.
[19] A. Ganguly, et al., Interaction of a novel GDP exchange inhibitor with the Ras protein, Biochemistry 37 (45) (1998) 15631–15637.
[20] P. Boriack-Sjodin, et al., The structural basis of the activation of Ras by Sos, Nature 394 (6691) (1998) 337–343.
[21] S. Colombo, et al., Structure-activity studies on arylamides and arysulfonamides Ras inhibitors, Curr. Cancer Drug Targets 10 (2) (2010) 192–199.
[22] F. Peri, et al., Design, synthesis and biological evaluation of sugar-derived Ras inhibitors, Chembiochem 6 (10) (2005) 1839–1848.
[23] F. Peri, et al., Sugar-derived Ras inhibitors: group epitope mapping by NMR spectroscopy and biological evaluation, Eur. J. Org. Chem. 16 (2006) 3707–3720.
[24] M. Mayer, B. Meyer, Group epitope mapping by saturation transfer difference NMR to identify segments of a ligand in direct contact with a protein receptor, J. Am. Chem. Soc. 123 (25) (2001) 6108–6117.
[25] A. Palmioli, et al., Selective cytotoxicity of a bicyclic Ras inhibitor in cancer cells expressing K-Ras(G13D), Biochem. Biophys. Res. Commun. 386 (4) (2009) 593–597.
[26] A. Palmioli, et al., First experimental identification of Ras-inhibitor binding interface using a water-soluble Ras ligand, Bioorg. Med. Chem. Lett. 19 (15) (2009) 4217–4222.
[27] C. Airoldi, et al., Glucose-derived Ras pathway inhibitors: evidence of Ras-ligand binding and Ras-GEF (Cdc25) 14 interaction inhibition, Chembiochem 8 (12) (2007) 1376–1379.
[28] S. Shirasawa, et al., Altered growth of human colon cancer cell lines disrupted at activated Ki-ras, Science 260 (5104) (1993) 85–88.
[29] E. Sacco, et al., Binding properties and biological characterization of new sugar-derived Ras ligands, Medchemcomm 2 (5) (2011) 396–401.

CHAPTER SIX

Development of EHop-016: A Small Molecule Inhibitor of Rac

Suranganie Dharmawardhane[1], Eliud Hernandez, Cornelis Vlaar
Department of Biochemistry, University of Puerto Rico Medical Sciences Campus, San Juan, Puerto Rico, USA
[1]Corresponding author: e-mail address: su.d@upr.edu

Contents

Abstract

The Rac inhibitor EHop-016 was developed as a compound with the potential to inhibit cancer metastasis. Inhibition of the first step of metastasis, migration, is an important strategy for metastasis prevention. The small GTPase Rac acts as a pivotal binary switch that is turned "on" by guanine nucleotide exchange factors (GEFs) via a myriad of cell surface receptors, to regulate cancer cell migration, survival, and proliferation. Unlike the related GTPase Ras, Racs are not usually mutated, but overexpressed or overactivated in cancer. Therefore, a rational Rac inhibitor should block the activation of Rac by its upstream effectors, GEFs, and the Rac inhibitor NSC23766 was developed using this rationale. However, this compound is ineffective at inhibiting the elevated Rac activity of metastatic breast cancer cells. Therefore, a panel of small molecule compounds were derived from NSC23766 and screened for Rac activity inhibition in metastatic cancer cells. EHop-016 was identified as a compound that blocks the interaction of Rac with the GEF Vav in metastatic human breast cancer cells with an IC_{50} of ~1 μM. At higher concentrations (10 μM), EHop-016 inhibits the related Rho GTPase Cdc42, but not Rho,

The Enzymes, Volume 33
ISSN 1874-6047
http://dx.doi.org/10.1016/B978-0-12-416749-0.00006-3

and also reduces cell viability. Moreover, EHop-016 inhibits the activation of the Rac downstream effector p21-activated kinase, extension of motile actin-based structures, and cell migration. Future goals are to develop EHop-016 as a therapeutic to inhibit cancer metastasis, either individually or in combination with current anticancer compounds. The next generation of EHop-016-based Rac inhibitors is also being developed.

1. INTRODUCTION

EHop-016 (*N*4-(9-ethyl-9*H*-carbazol-3-yl)-*N*2-(3-morpholin-4-yl-propyl)-pyrimidine-2,4-diamine) is a small molecule compound that we recently characterized as an inhibitor of the small GTPase, Rac. The relevance of developing EHop-016, and related Rac inhibitors, stems from their potential as antimetastatic cancer therapeutics.

1.1. Relevance of developing targeted therapy for cancer metastasis

Despite recent progress in early detection and improved adjuvant therapy, the prognosis for cancer patients is still limited by distant metastases [1]. During metastatic progression, malignant cancer cells migrate from the primary tumor, invade the tumor microenvironment, and enter the circulatory system to establish secondary tumors at distant sites, thus complicating accurate diagnosis and treatment [2]. In addition to the dysregulated signaling in metastatic cancer cells, their interactions with the stromal cells in the tumor microenvironment that infiltrate the tumor tissue, such as macrophages, myeloid-derived suppressor cells, T lymphocytes, and adipocytes, increase tumor-associated inflammation and invasion. Cancer and stromal cell crosstalk signaling also promotes invasion and metastasis via cytokines, growth factors, and proteases that remodel the tumor microenvironment [3]. In addition, stromal cells in specific organs such as bone and lung (e.g., neutrophils and bone marrow-derived hematopoietic immune progenitor cells) signal to the homing cancer cells to promote establishment of the premetastatic niche [4]. Therefore, a viable antimetastasis therapeutic should inhibit invasive/migratory signaling in cancer cells, as well as those of the stromal cells that recruit them to invade the vasculature and establish secondary tumors at vital organ sites.

For breast cancer, preventing metastasis is critical, because breast cancer can be cured if detected at the early stages of the disease. Nevertheless, 30% of breast cancer patients can develop stage IV metastatic disease in bone, liver, and lung with a 5-year survival rate of 20% [5–7]. Although current systemic metastatic cancer therapies are generally active at the beginning of therapy, most patients develop resistance with time [5,8]. Therefore, improved targeted and combinatorial therapeutic strategies are required to effectively combat metastatic disease.

1.1.1 Regulating the actin cytoskeleton in cancer metastasis

Intravasation, the first step of metastasis, involves migration away from the primary tumor employing mechanisms similar to those used during normal cell migration [9]. Therefore, molecules that regulate cell migration may become dysregulated during metastasis and act as metastasis promoters [10]. Cell migration and invasion are guided by actin polymerization, rearrangement of the actin cytoskeleton to extend motile structures, and the modulation of cell–cell and cell–extracellular matrix adhesions. These activities require sophisticated molecular coordination and signaling orchestrated by both cancer cells and the stromal cells in the tumor microenvironment. The Rho family small GTPases have been implicated as key regulators of the spatial and temporal activities of metastatic cancer cells, as well as stromal cells, during invasion and directed migration [11–13].

1.2. The small GTPase Rac as a central regulator of cell signaling to the actin cytoskeleton

The Rho family, of which the most-studied members are Rho, Rac, and Cdc42, is a ubiquitiously expressed and evolutionarily conserved family of Ras-related small GTP-binding proteins that regulate actin dynamics and intracellular signaling. Rho GTPases control diverse cellular functions related to cancer development, including actin cytoskeleton organization, invasion and metastasis, transcriptional regulation, cell cycle progression, apoptosis, vesicle trafficking, and cell-to-cell and cell-to-extracellular matrix adhesions. Rho GTPases are activated during signaling from cell surface receptors that regulate GTP/GDP cycling via a number of accessory proteins: Rho guanine nucleotide dissociation inhibitors, Rho guanine nucleotide exchange factors (GEFs), and Rho guanine nucleotide activating proteins (GAPs) [11,14,15].

Of the Rho family of GTPases, Rac proteins (Racs 1 and 3 expressed in nonhematopoietic cells and Rac2 in hematopoietic cells) have been specifically implicated in organization of the actin cytoskeleton into cell surface protrusions called lamellipodia that control directed cell migration during leukocyte chemotaxis [16], as well as cancer cell invasion, and thus metastasis

[17–29]. The Rac homolog Cdc42 also modulates the actin cytoskeleton during migration/invasion via *de novo* actin polymerization and extension of motile actin structures called filopodia, and has been implicated in breast cancer malignancy [30,31].

Racs are also essential for Ras and other oncogene-mediated transformation via regulation of Ras/mitogen activated protein kinase (MAPK) signaling [32–34]. Hyperactive Rac1 and Rac3 have been implicated with increased survival, proliferation, and invasion of breast cancer, gliomas, melanomas, and leukemia [20,35–41]. Wild-type Rac1 overexpression has been associated with a range of human cancers: breast, brain, gastric, and pancreatic cancers, as well as ulcerative colitis [36,37,42–46]. Studies have also demonstrated a cancer-promoting role for the constitutively active Rac1b splice variant that is overexpressed in breast and colorectal cancers [39,47–50].

Although functionally relevant Rac1 mutations are rare, activating Rac1 mutations have been found in melanoma [51]. A "Rac1 risk allele" has also been reported from patients at risk for developing colon cancer [52]. More recently, fast cycler mutations of Rac, with transformative ability, were reported from a range of human cancer cell lines [53]. However, given their low frequency, and because these mutations were identified from cancer cell lines that have been in culture for a long period, the importance of Rac mutants in human carcinogenesis remains to be validated.

Rac and the close homolog Cdc42 are ideal therapeutic targets for metastatic cancer prevention, especially in breast cancer for a number of reasons. Rac is a key downstream effector of ErbB/epidermal growth factor receptors (EGFRs) that are often overexpressed in metastatic breast cancer [9,26]. Overexpression of human epidermal growth factor receptor 2 (HER2) in mammary epithelial cells increased Rac1 activity, implicating Rac signaling in the malignant phenotype of HER2-type breast cancer [54]. Moreover, Rac1 was recently shown to regulate breast cancer cell proliferation and estrogen receptor (ER)α levels, thus implicating Rac in modulating ER function in breast cancer [55]. Racs have also been implicated with reversal of growth factor receptor (GFR) targeted therapy resistance signaling pathways [56].

The malignant phenotype of Rac overexpression has been associated with the activity of the Rac downstream effector p21-activated kinase (PAK) [57,58]. Moreover, elevated HER2 expression in human breast cancer specimens, an indicator of poor prognosis, has been associated with PAK levels [59]. PAKs are central activators of several cancer pathways that include not only actin cytoskeletal changes during migration, but also cell adhesion, survival, and proliferation [60–62]. PAK and other downstream effectors of Racs regulate cell proliferation, survival, angiogenesis, cell polarity, epithelial to mesenchymal transition, cell–extracellular matrix adhesion, as well as migration/invasion via a number of signaling sequelae [22,26,59,62–68]. Although Rac-mediated production of reactive oxygen species (ROS) via NADPH oxidase activity is part of the innate immune response, Rac-mediated ROS production has also been shown to regulate the invasive potential of cancer cells [69–72].

Rac action is implicated with cancer progression and acquisition of therapy resistance via multiple pathways that include signal transduction to Rac. GEFs from integrin, G protein-coupled receptors (GPCR), GFR/receptor tyrosine kinases (RTK), and cytokine/janus kinase/signal transducer and activator of transcription (STAT) receptors. These cell surface receptors regulate a myriad of cancer-promoting signal cascades that have also been implicated with Rac/PAK activity. These pathways include: phosphoinositide 3-kinase (PI3-K)/Akt/mammalian target of rapamycin (mTOR); MAPKs: extracellular regulated kinase, jun kinase (JNK), and p38 MAPK; protein kinas C ε, and STATs (Fig. 6.1) [22,26,34,61–67,73–80]. Recent studies have implicated Rac activity in mTOR signaling, where Rac regulates both mTOR complex 1 (mTORC1) and mTORC2 activation during cancer cell migration, cell size control, epithelial to mesenchymal transition, and metastasis [66–68]. Since mTOR signaling has emerged as a central regulator of cancer progression and acquisition of therapy resistance [81], these studies suggest a key role for Rac in regulation of cancer malignancy.

Because the malignant phenotype of Rac is closely associated with activation of its direct downstream effectors PAKs [57,58], much effort has been focused on the development of PAK inhibitors as anticancer therapeutics [82–84]. However, in addition to PAK, Racs have multiple downstream effectors, such as WASP family verprolin-homologous protein (WAVE) and mammalian-enabled (Mena)/vasodilator-stimulated phosphoprotein, that contribute to cancer [85,86]. Therefore, targeting Rac activation is a more viable approach for the development of anticancer drugs [87].

1.3. Rac.GEFs in cancer metastasis

So far, over 60 potential Rac.GEFs have been identified [88–90]. A large subset of the Rho.GEFs is characterized by a Dbl homology (DH) domain, followed by a pleckstrin homology domain that forms the structural basis for the guananine nucleotide exchange activity [91]. DH domains are present in

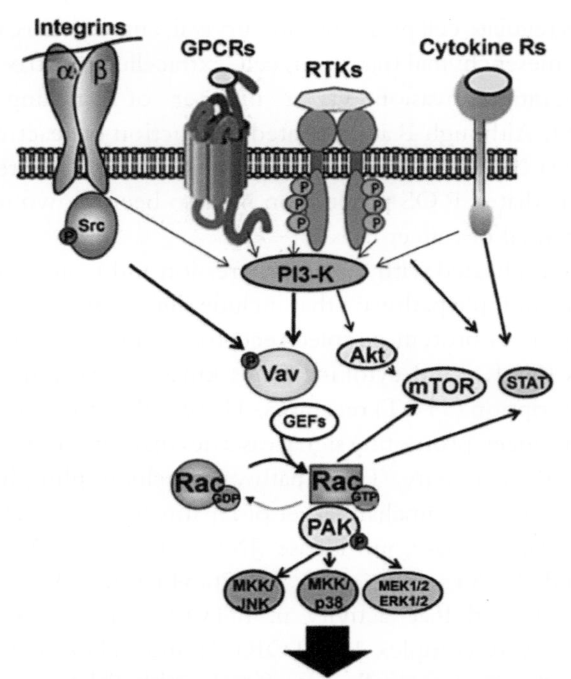

Figure 6.1 Simplified scheme of the major signaling pathways of Rac activation in human cancer. Cell surface integrin receptors, G protein coupled receptors (GPCR), receptor tyrosine kinases (RTK), cytokine receptors, and nonreceptor tyrosine kinases activate Ras/mitogen activated protein kinase (MAPK), phosphoinositide 3-kinases (PI3-K), and signal transducers and activator of transcription (STAT) pathways. PI3-K phosphorylates phosphatidyl inositol biphosphate (PIP2) to form the signaling intermediate PIP3 that recruits phospholipid-dependent kinase and Akt/protein kinase B. Akt activates the mammalian target of rapamycin (mTOR), a central regulator of protein synthesis and metabolism. PI3-K signaling also modulates Vav and other Rac.GEFs such as p-REX that exchange the GDP on Rac for GTP, thus activating Rac. Rac action is implicated with signaling to PAK, mTOR complex1 (mTORC1), mTORC2, MAPKs: extracellular regulated kinases (ERK1/2), p38 MAPK, Jun kinase (JNK), and STATs to control cancer progression to metastasis. (See color plate.)

a number of proto-oncogenes such as T-cell invasion and metastasis gene product (Tiam-1), Trio, Vav, and PIP3-dependent Rac exchanger 1 (p-Rex1) that have been implicated in cancer progression [24,26,92–97].

1.3.1 Vav/Rac signaling in cancer metastasis

Of the Rho GEFs, Vav is of note due to its importance in both hematopoietic (Vav1) and nonhematopoietic (Vav2, 3) signaling to activate Rac2 (in immune cells) and Rac1 and Rac3 in cancer cells. Vav isoforms have also been implicated in tumor growth, angiogenesis, and metastasis in a number of cancers including breast cancer [24,93–95,98–105], as well as immune and stromal cell signaling relevant for recruitment of cancer cells in the tumor microenvironment [106–108]. Although Vav2 can act as a GEF for RhoA, Rac1, and Cdc42 *in vitro*, its transforming activity has been ascribed primarily to its ability to activate Rac1 [93]. Therefore, cell surface receptor-activated or oncogenic Vav has been shown to regulate tumor progression, invasion, and angiogenesis via Rac-regulated activation of downstream cancer-promoting molecules such as PI3K, PAK, p38 MAPK, and JNK [93–95,105,109–113] (Fig. 6.1). Moreover, Rac and Cdc42 have been shown to be necessary for Vav-induced cell transformation, migration, and metastasis [93,112,114–120]. Vav/Rac signaling is also significant during the development of breast cancer therapy resistance [54,121–123]. Recent studies have shown that Vav2 and Vav3 regulate a lung metastasis-specific gene signature in breast cancer cells, thus implicating Vav/Rac signaling in the control of specific steps of metastasis to the lung at the transcription level [24,98]. Significantly, the current literature on cell signaling during cancer metastasis strongly implicates Vav/Rac signaling in promoting cancer malignancy both via signaling in cancer cells as well as in stromal and immune cells in the tumor microenvironment. Thus, our recent characterization of EHop-016 as a specific inhibitor of the Vav/Rac interaction has direct implications for its further development as an anticancer metastasis therapeutic [124].

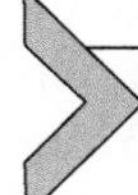

2. DEVELOPMENT OF Rac INHIBITORS TO IMPEDE METASTATIC CANCER PROGRESSION

Using the rationale that a structure–function-based design to block the signaling step of Rac activation by GEFs is a viable strategy for inhibiting Rac functions, Gao et al. [125] identified a small chemical compound from the National Cancer Institute chemical database, NSC23766, as a Rac-specific inhibitor. This compound fits into a surface groove of Rac1 known to be critical for GEF specification, and was shown to inhibit Rac1 binding and activation by a subset of Rac-specific GEFs, that is, Trio and T-cell invasion and metastasis gene product (Tiam1), in a dose-dependent manner. The interaction between NSC23766 and Rac is specific and does not affect Cdc42 or RhoA binding or activation by their respective GEFs, interaction of Rac1 and GAPs, or other downstream effectors [126,127]. NSC23766 has been used to demonstrate the significance of Rac activity in cancer cell proliferation, migration, invasion, metastasis, and therapy resistance, as well as platelet aggregation and hematopoietic cell migration [122,126,128–134]. However, the high effective concentrations ($IC_{50} > 75$ μM) of NSC23766 limit its use as a therapeutic agent [125].

Other known Rac inhibitors, including NSC23766 derivatives, have IC_{50}s of 10–50 μM [135,136]. Virtual screening of NSC23766 in a ZINC database identified improved structures with IC_{50}s from 12 to 57 μM [135]. EHT 1864, a recently described inhibitor that selectively blocks the interaction of Rac with its downstream effectors, is also effective at 10–50 μM and is, therefore, more useful than NSC23766 at inhibiting a number of Rac functions [136,137]. Accordingly, EHT 1864 has been successfully used to demonstrate the role of Rac1 in growth and inflammatory responses in human endothelial cells, medulloblastoma migration, ER expression and spread of breast cancer, platelet activation, and lymphoma development [22,55,138–141].

Selective inhibitors for the Rac1B GTPase, which is an alternatively spliced constitutively active form Rac1, have also been designed with specificity for Rac1B inhibition over Rac1 or Cdc42 [142]. However, the Rac1B-specific inhibitors are not universally applicable to inhibit invasive cancers with overactive Rac or Cdc42.

2.1. Design and synthesis of NSC23766 derivatives

A rational Rac inhibitor should be effective at physiologically relevant concentrations (in the nM range), be specific to Rac activation by a range of Rac.GEFs, have high solubility and bioavailability, and be nontoxic in cell and animal models. To determine the effective concentration of NSC23766 in invasive cancer cells, we first tested the effect of this compound on the Rac activity of a highly metastatic human cancer cell line, MDA-MB-435. Using metastatic variants of this MDA-MB-435 cell line, we previously reported that hyperactive Rac is associated with high invasive and metastatic efficiency [36]. In MDA-MB-435 cells, the IC_{50} for Rac activity inhibition by NSC23766 is as high as ~100 μM [124,143]. Therefore, we used NSC23766 as a lead structure for the synthesis of new derivatives with the potential for improved Rac1 inhibitory activity.

2.1.1 Synthesis of NSC23766 derivatives

As described in our strategy for the development of Rac1 inhibitors with increased potency, we utilized NSC23766 as a template in which the three major chemical building blocks were substituted with commercially available heteroarylamines, dichloropyrimidines, and primary or secondary aliphatic amines with tail-end amino-substituents [143]. The central building block of NSC23766, the pyrimidine core (B1–B2) that binds a critical tryptophan residue (Trp56) of Rac1 [125], was maintained in our design (Fig. 6.2). The second group of building blocks (C1–C8) consisted of heterobicyclic arylamino groups that were connected to the 4-position of the pyrimidine ring, thereby mimicking the substituted aminoquinoline group of NSC23766. The third group of building blocks contained primary or secondary aliphatic amine with a tail-end amino group (A1–A8), mimicking the substituent on the 2-position of the pyrimidine ring of NSC23766. The new NSC23766 derivatives were prepared via a facile two-step synthesis by reacting dichloropyrimidines (B1–B2) with (hetero)-arylamines (C1–C8) to provide a 3:1 to 5:1 mixture of the 4-substituted and 2-substituted products. After silica gel chromatography, the pure 4-substituted product was reacted with amines (A2–A8) to produce the desired novel derivatives (Fig. 6.2). Detailed synthetic methods are described in Ref. [143]. Out of the 128 possible combinations, 32 compounds were randomly selected for synthesis and subsequently screened for activity.

2.1.2 Initial screening of novel Rac inhibitors

To screen for the relative efficiency of the new NSC23766 derivatives, we used The G-LISA Rac Activation Assay (Cytoskeleton, Inc., Denver, CO), as described in Ref. [124,143]. This assay detects the activated (GTP bound) Rac1, 2, and 3 isoforms from cell lysates by their specific interaction with the Cdc42 and Rac interactive binding domain (CRIB) from PAK [144].

In the highly metastatic MDA-MB-435 cells, NSC23766 inhibited Rac with an IC_{50} of 95 μM. Therefore, we compared the effect of the 32 new compounds at 50 μM, following a 24-h incubation in MDA-MB-435 cells, in the presence of serum. At this concentration, NSC23766 inhibits Rac by 20% compared to vehicle controls. In Fig. 6.2, the range of Rac inhibitory activities for compounds with the C1–C8 building blocks is provided. Each range represents at least two compounds with these fragments that in addition have different building blocks selected from B1 to B2 and A1 to A8. In the series of compounds with C2 and C4 blocks, the most active compounds are approximately 50% more active than NSC23766 (30–31% versus 20% Rac inhibition, respectively). In addition, the most active compound in the series with C3 is two times

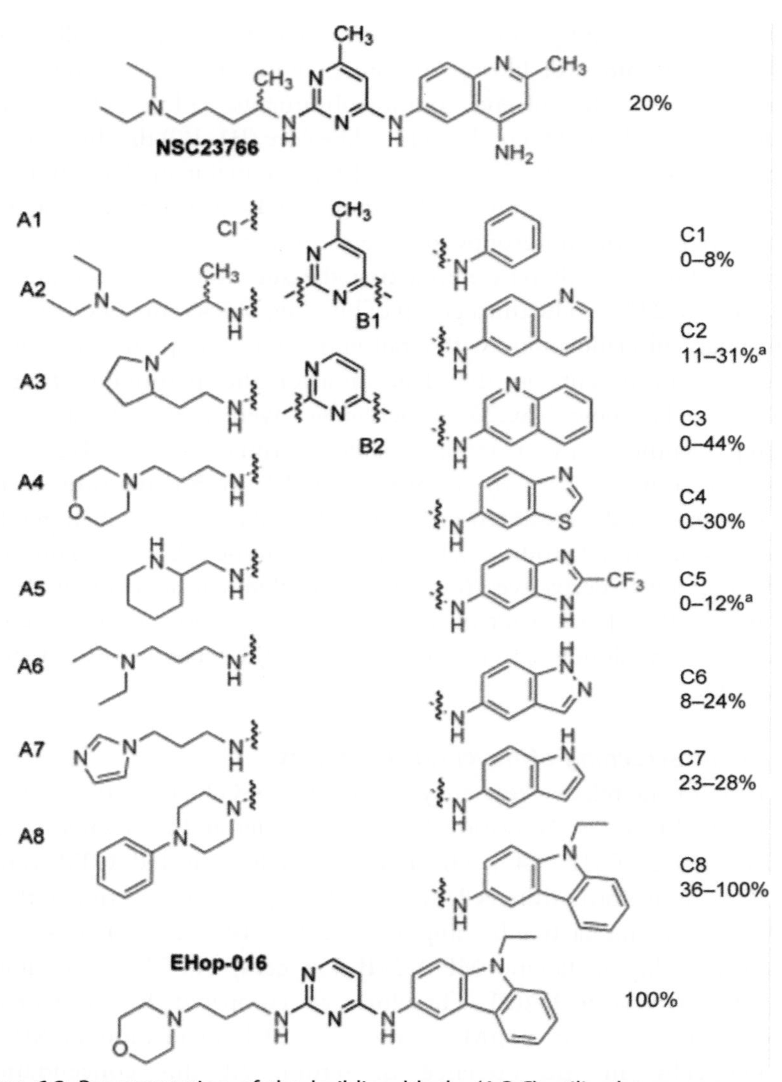

Figure 6.2 Representation of the building blocks (A,B,C) utilized to prepare a small library of NSC23766 derivatives. Indicated in italics are the percentages of Rac inhibition of a group of compounds containing units C1–C8. (For color version of this figure, the reader is referred to the online version of this chapter.)

more active than NSC23766 (44% versus 20% Rac inhibition). Significantly, all of the four compounds in the series with the carbazole fragment C8 were found to inhibit Rac at a higher efficiency than NSC23766. Initially, the Rac inhibitory potential of these compounds was thought to result partially from interference of cytotoxicity and (or) inhibition of Rac-induced effects on cell viability. Nevertheless, upon lowering the assay concentration of the most active compound (EHop-016) to noncytotoxic levels, this compound was indeed found to inhibit Rac with an $IC_{50} \sim 1$ μM [124].

Since the preliminary screening experiments indicated that derivatives with fragment C8 were cytotoxic, we initially focused our attention on the compounds with C2, C3, or C4 building blocks. The most active compounds of these series inhibited Rac approximately 1.5–2 times more potently than NSC23766, and demonstrated higher efficiency at reducing cell spreading, extension of lamellopodia, and directed migration toward serum at 50 μM concentrations [143]. In addition, although Rac is only inhibited by 30–44% at this concentration, cell migration was reduced by 80–90%. Therefore, it may be hypothesized that partial inhibition of Rac, or localization at different compartments, is sufficient to produce a dramatic effect on Rac-regulated cell functions. Nevertheless, preliminary data also demonstrated that at least one of the compounds inhibits Cdc42 to a similar extent as Rac. Thus, Cdc42 inhibition may also contribute to the effects of NSC23766 derivatives on cell functions relevant for metastasis. Although selectivity for specific GTPases is important for the development of biochemical probes, for therapeutic purposes, it will be more advantageous to develop dual inhibitors, where dual Rac–Cdc42 inhibitors are expected to be synergistic and interfere with multiple steps in the metastatic process. Consequently, among the future goals in our laboratories is the development of prospective dual Rac–Cdc42 inhibitors.

2.2. Identification of EHop-016 as a Rac inhibitor

From the screening results described above, EHop-016 appeared to demonstrate complete inhibition of Rac1 at a concentration of 50 μM. As shown in Fig. 6.3, the decreased Rac activity in response to 50 μM EHop-016 may at least be partially due to the reduced MDA-MB-435 cell viability (~40%) at this concentration. Therefore, we subsequently determined the Rac inhibitory activity of EHop-016 at concentrations at which EHop-016 was not cytotoxic (<5 μM). Serendipitously, EHop-016 specifically inhibited Rac activity at these lower concentrations, and the IC_{50} for Rac inhibition in this cell line was subsequently determined to be ~1 μM [124]. In conclusion, we identified EHop-016 as a novel inhibitor of Rac, at 100-fold more potency than the parent compound NSC23766 and 10–50-fold more effective than other available Rac inhibitors [135,137,145].

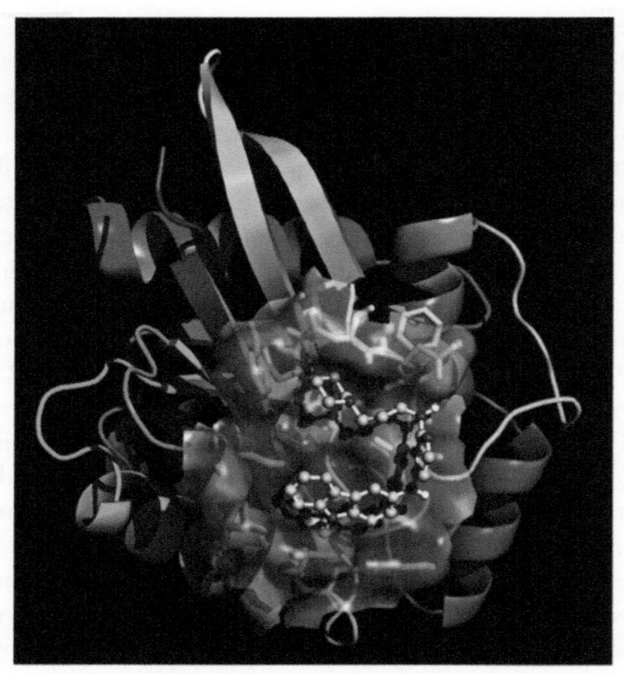

Figure 6.3 EHop-016 docked into the GEF-binding pocket of Rac1. AutodockTools were used to prepare the GEF-interacting region of Rac and EHop-016 for docking, centered on the original position of NSC23766 on the crystal structure of the GEF-interacting surface of Rac. In this configuration, Ehop-016 interacts with residues that have been shown to interact with the Rac guanine nucleotide exchange factor Vav. (See color plate.)

Figure 6.2 demonstrates that all compounds containing C8 (four derivatives were tested) were more potent Rac inhibitors than the parent compound, indicating that this carbazole group is key to increased Rac inhibitory activity. To explain the marked difference in inhibition potency between the carbazole unit and the other arylamine fragments C1–C7, molecular docking studies were conducted utilizing the crystal structure of the Rac–NSC23766 complex [126,146] as a template. Figure 6.3 demonstrates that the energetically most favorable position of EHop-016 in Rac1 is in the area around Trp56, which has been shown to be critical for binding of Rac to its GEFs, similar to the region of NSC23766 interaction. However, the crystal structure of the Rac effector region and NSC23766 demonstrates that NSC23766 is stretched linearly on the surface of Rac [125,132]. Our calculations with EHop-016 in the same Rac effector region indicate a bent U-shape that places the carbazole group (C8) of EHop-016 (or its analogs) with relatively lipophilic residues in a more efficient configuration. In addition, unit A4 of EHop-016 provides additional binding energy via hydrogen bonding with Asn39 [124].

In light of our finding that EHop-016 inhibits the Vav2–Rac interaction [124], it is notable that Ehop-016 interacts with residues Thr35, Val36, Asn39, Ala59, and Tyr64, which have been demonstrated to be relevant for binding of the closely related GEF Vav1 to Rac1. Attempts to further elucidate binding interactions via cocrystallization of EHop-016 with Rac are ongoing.

2.3. Functional characterization of EHop-016

2.3.1 EHop-016 inhibits Rac activity at 0.5–5 μM and Cdc42 at concentrations >5 μM

As shown in Fig. 6.4, the 100% inhibition of Rac activity in response to 50 μM EHop-016 may partially be due to the reduced MDA-MB-435 cell viability (~40%) at this concentration. We subsequently determined that in this highly metastatic HER2 overexpressing cancer cell line, EHop-016 inhibited Rac activity at low physiologically relevant concentrations (<5 μM) with an IC50 of ~1 μM [124]. Therefore, EHop-016 is 100-fold more potent than the parent compound NSC23766 and 10–50-fold more effective than other available Rac inhibitors [135,137,145]. EHop-016 also inhibited the Rac activity of the HER2 (−) (triple negative) metastatic breast cancer cell line MDA-MB-231; however, the IC_{50} for Rac inhibition by EHop-016 in this cell line was higher (~3 μM) [124]. Additionally, we have determined that the MDA-MB-435 cell line expresses three to four times more Vav2 than MDA-MB-231 cells (data not shown). Therefore, the increased EHop-016 potency in the MDA-MB-435 cells may indicate that EHop-016 action depends on inhibition of HER2 and (or) Vav signaling.

2.3.2 EHop-016 reduces cell viability at concentrations >5 μM

At concentrations <5 μM, EHop-016 is specific for Rac, and has no effect on Rho or the close homolog Cdc42. However, at concentrations >5 μM, EHop-016 inhibits Cdc42 activity with a 75% inhibition at 10 μM [124]. Interestingly, MDA-MB-435 cell viability was not significantly affected at the concentrations that inhibited Rac activity (<5 μM), while significant effects on cell viability were observed at EHop-016 concentrations that also

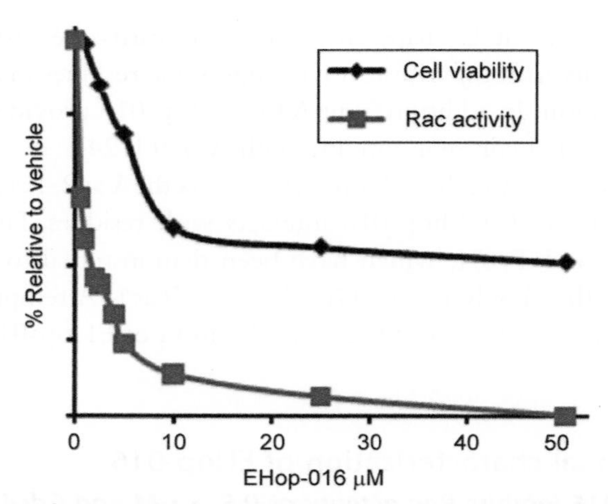

Figure 6.4 Effect of EHop-016 on cell viability and Rac activity. MDA-MB-435 cells were treated with vehicle (0.1% DMSO), or varying concentrations of EHop-016 (0–10 μM) for 24 h. Cell viability was measured using the MTT cell survival and proliferation kit (Millipore, Inc). Rac activity was measured from cell lysates by a G-LISA Rac1 Activation Assay (Cytoskeleton, Inc., Denver, CO). The mean values ± SEM ($N = 3$) are presented relative to vehicle (100%). All data points for Rac activity >0.5 μM EHop-016, and for viability at >5.0 μM EHop-016, were statistically significant compared to vehicle controls ($p \leq 0.05$). (For color version of this figure, the reader is referred to the online version of this chapter.)

inhibit Cdc42 activity (Fig. 6.4), suggesting that the reduced cell viability in response to EHop-016 may be due to inhibition of Cdc42 activity or a combination of reduced Rac and Cdc42 activities.

2.3.3 EHop-016 inhibits the interaction of Vav2 and Rac

To identify upstream effectors of Rac that are inhibited by EHop-016, MDA-MB-435 cell lysates were incubated with glutathione beads coupled to a GST-tagged nucleotide-free form of Rac1, Rac1(G15A), with a high affinity for Rac GEFs. Vav2 specifically associated with Rac1(G15A), while neither Tiam-1 or Trio-bound Rac1(G15A) from MDA-MB-435 cells in serum [124]. EHop-016 (4 μM) inhibited the association of Vav2 with Rac1 (G15A) by 50% (Fig. 6.5). Since 4 μM EHop-016 results in ~100% inhibition of Rac activity [124], EHop-016 may also inhibit the interaction of GEFs that have yet to be identified. Moreover, EHop-016 at >40 μM inhibited the interaction of a purified active domain of Tiam-1 with Rac1(G15A) ([124], demonstrating that at high concentrations, EHop-016 may inhibit other GEFs.

Vav is a GEF for RhoA, Rac1, and Cdc42; however, Vav-mediated cell transformation has been ascribed mainly to its ability to activate Rac1 [93].

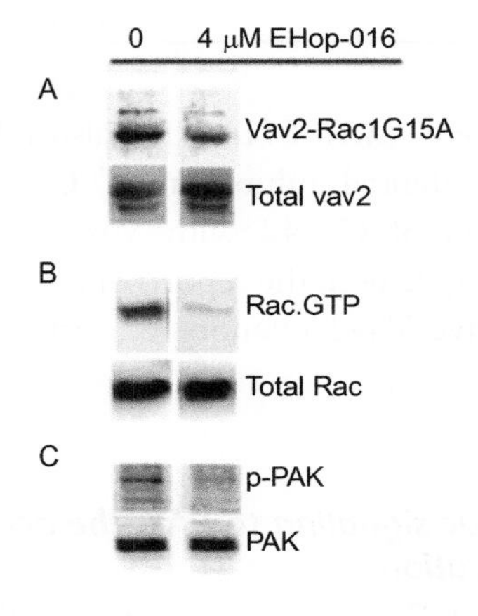

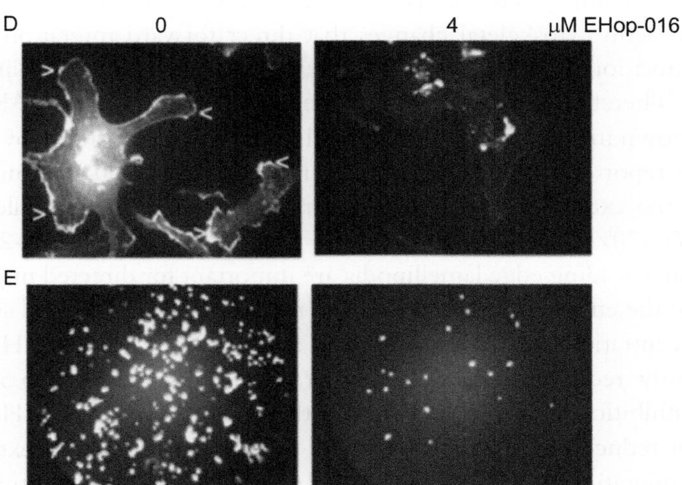

Figure 6.5 Effect of EHop-016 on Rac activation and downstream activities in highly metastatic MDA-MB-435 cells. (A) GST-Rac1(G15A) (mutant form of Rac that selectively binds active GEFs) beads were preincubated with vehicle (0), or 4 μM EHop-016 prior to incubation with MDA-MB-435 cell lysates. A representative western blot ($N=3$) immunostained for Vav2 is shown. Top row, pulldown; bottom row, total cell lysate. (B) MDA-MB-435 cells treated with vehicle (0) or EHop-016 (4 μM) for 24 h were lysed and subjected to a pulldown assay for Rac.GTP using a GST-CRIB domain of PAK and western blotted with a pan Rac antibody (Rac1,2,3). Representative western blot ($N=3$) is shown. Top row, pulldown; bottom row, total cell lysate. (C) MDA-MB-435 cells were treated with vehicle (0) or 4 μM EHop-016 for 24 h and the cells lysed and western blotted for active P-$^{thr\ 423}$ PAK (upper band) or total PAK (lower band). Representative western blot ($N=3$) is shown. (D) MDA-MB-435 metastatic breast cancer cells were treated with vehicle or EHop-016 at 4 μM for 24 h. Cells were fixed, permeabilized, and stained with Rhodamine phalloidin to visualize F-actin. Representative micrographs are shown. Arrows indicate lamellipodia. (E) MDA-MB-435 cells treated with vehicle (0) or EHop-016 (4 μM) for 24 h were subjected to a Transwell migration assay. The cells that migrated to the underside of the membrane of the top well (through 8 μm diameter pores) were stained with propidium iodide and quantified. Representative micrographs of cells that migrated for each treatment are shown.

At higher concentrations (≥ 10 μM), EHop-016 also inhibits Cdc42, but not RhoA, indicating a potential inhibition of Cdc42 activity either via blocking the interaction of Cdc42 with Vav or alternative GEFs. Taken together, our results agree with the reports of Rac1 and Cdc42 mediating the constitutively active Vav2 phenotype, with Rho A playing an antagonistic role [147].

2.3.4 EHop-016 inhibits Rac signaling to PAK, the actin cytoskeleton, and cancer cell migration

The downstream effector of Rac, PAK, is a key regulator of the Rac-mediated actin cytoskeletal changes that direct forward migration during cancer invasion, as well as cell adhesion, survival, and proliferation [60–62]. Therefore, we determined the effect of EHop-016 on PAK activity. As shown in Fig. 6.5, 4 μM EHop-016 reduced PAK activity by ~80%. We also reported that at similar concentrations, EHop-016 significantly reduced the extension of Rac-directed motile actin structures called lamellipodia (~70%) in both MDA-MB-435 (Fig. 6.5) and MDA-MB-231 cells [124]. Since leading edge lamellipodia are important for directed migration, we tested the effect of EHop-016 on cell migration by a Transwell assay. At the concentrations of EHop-016 that inhibit Rac activity, EHop-016 significantly reduced directed cell migration by ~60% (Fig. 6.5) Therefore, inhibition of Vav interaction with Rac and Cdc42 by EHop-016 results in reduced metatstaic cancer cell viability, lamellipodia extension, and cell migration, implicating EHop-016 as a viable antimetastatic cancer therapeutic.

2.3.5 Summary

The working model that the Rac inhibitor EHop-016 can impede breast cancer metastasis is based on our data that EHop-016 inhibits the interaction of Vav with Rac, PAK activity, and decreases invasive actin structures and migration [124] (Fig. 6.6). Inhibition of the Vav/Rac interaction may not only impede PAK signaling to the actin cytoskeleton, it also has the potential to block effects of PAK on cell survival and proliferation [60–62]. Moreover, our data suggest that EHop-016 may block GEFs other than Vav. These may be oncogenic GEFs, such as p-REX that is regulated via GPCR-mediated activation of PI3-K that has been implicated in Rac-mediated cell migration/invasion, tumorigenesis, and metastasis [23,25].

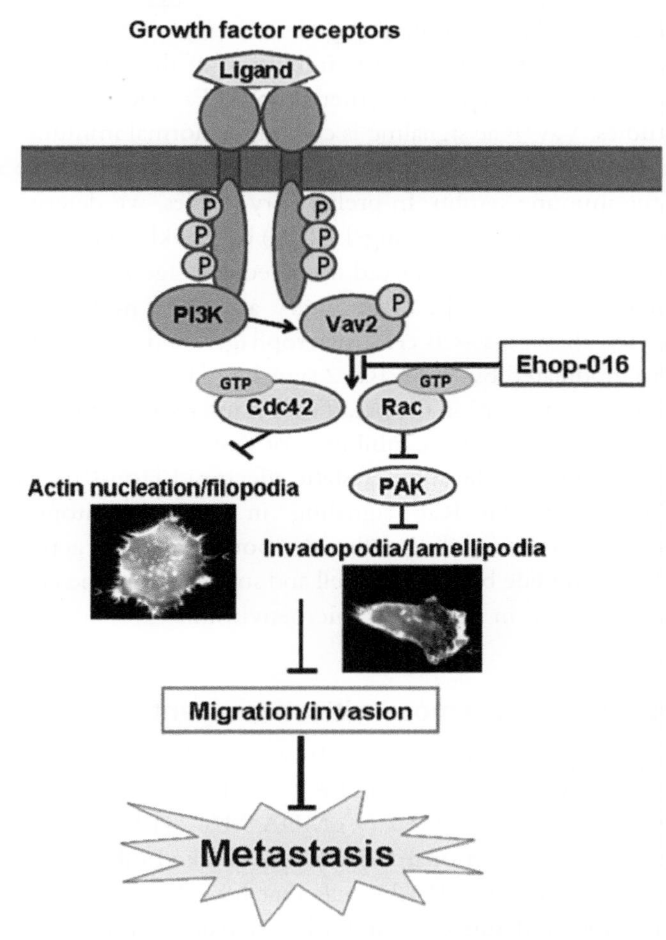

Figure 6.6 Effects of EHop-016 on Rac signaling. EHop-016 inhibits the Vav2/Rac interaction, Rac and PAK activities, lamellipodia formation, and metastatic cancer cell migration. Thus, EHop-016 holds promise as an anticancer metastatic agent. (For color version of this figure, the reader is referred to the online version of this chapter.)

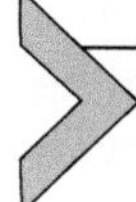

3. FUTURE DIRECTIONS

3.1. EHop-016 in metastatic cancer therapy

Future directions include the identification of the role of Vav and other GEFs that may be inhibited by EHop-016, and a comprehensive analysis of all of the molecular mechanisms that are inhibited by EHop-016. As shown in Fig. 6.1, EHop-016 may impede cancer progression via multiple mechanisms as a result of the specific inhibition of the interaction of Vav with Racs and Cdc42, as well as other Rho.GEFs that will be identified in future studies. Vav/Rac signaling is critical for normal immune function [148], and EHop-016 administration for long periods may have adverse effects on the immune system. In preliminary studies, we determined that administration of EHop-016 (10 mg/kg BW) by weekly oral gavage to athymice nude mice for ~2 months did not affect average mouse weight (see supplementary data [124]). Although athymic nude mice lack T lymphocytes, they possess B cells, macrophages, and neutrophils, whose activities are modulated by Vav1/Rac2 signaling [149–152]. Since macrophages and neutrophils within the breast tumor microenvironment promote cancer cell invasion [3,153,154], inhibition of Vav/Rac signaling in immune cells may in fact enhance the antimetastatic effects of EHop-016. Therefore, the importance of Vav/Rac signaling in both hematopoietic and nonhematopoietic cell signaling offers the potential for targeted delivery of EHop-016 to impede both cancer cell and stromal/immune cell crosstalk signaling and invasion in the tumor microenvironment.

3.2. EHop-016 as a chemosensitization agent

Our initial characterization of EHop-016 was conducted with two human metastatic breast cancer cell lines: the ER (−) HER2 (++) MDA-MB-435 high metastatic cells and the ERα (−) MDA-MB-231 low metastatic cells that do not express a functional HER2. The marked inhibitory effects of EHop-016 on the MDA-MB-435 cells that have high intrinsic Rac and Vav activities [36,124] suggest that HER2 signaling to Vav and Rac, or oncogenic Vav activity of the MDA-MB-435 cell line is inhibited by EHop-016.

EGFR1 and HER2 are often upregulated in metastatic breast cancers and HER2 amplification is a prognostic factor for metastatic breast cancer [155,156]. Such overexpression of HER2 occurs in 30% of breast cancer patients and leads to increased metastasis and reduced disease-free survival [157–159]. Intriguingly, a number of studies have linked HER2 signaling with Rac activity in metastatic cancer. Rac activity is associated with trastuzumab resistance and inhibition of Rac has been shown to reverse trastuzumab resistance [122,123]. Moreover, overexpression of HER2 in mammary epithelial cells increased Rac1 activity [54]. HER2 expression has also been correlated with the activity of the Rac downstream effector PAK in human breast cancer specimens [59].

The current therapies for HER2 type breast cancer are the monoclonal antibody trastuzumab (Herceptin) and the tyrosine kinase inhibitor lapatinib [160]. Unfortunately, patients often present with intrinsic or acquired resistance to these anti-EGFR therapeutics [161,162]. Trastuzumab resistance has been attributed to many factors, including bypass signaling through other receptors and activation of downstream signaling pathways independent of HER2 [157,163]. Therefore, a viable strategy to overcome resistance to HER2-directed therapy is combinatorial therapies targeting HER2, as well as downstream resistance pathways [164].

Elevated downstream signaling in trastuzumab-resistant breast cancers has been attributed to MAPK and PI3-K/Akt/mTOR signaling pathways [121,163,165–168]. As discussed in Section 1, PI3-K signaling activates Rac GEFs, such as Vav, p-REX-1, and Tiam-1, that have been implicated in increased metastatic properties and survival of HER-2 overexpressing cells [114,115,117] (Fig. 6.1). Moreover, receptor-activated or oncogenic Vavs regulate tumor progression, invasion, and angiogenesis via Rac-mediated activation of PAK, p38 MAPK, and JNK [109–113].

Our initial studies to test the potential of EHop-016 to sensitize the trastuzumab-resistant MDA-MB-435 cell line to trastuzumab therapy are promising. As shown in Fig. 6.2A, following a 24-h treatment, trastuzumab did not affect MDA-MB-435 cell viability at concentrations as high as 20 μg/ml. As previously reported by us [124] and shown in Fig. 6.3, MDA-MB-435 cell numbers decreased ~20% and 50%, respectively, in response to 5 and 10 μM EHop-016. Interestingly, MDA-MB-435 cell viability decreased to a higher extent, that is, ~40% and 80%, respectively, when 5 or 10 μM of EHop-016 was combined with 5 or 10 μg/ml trastuzumab (Fig. 6.7). These data indicate a synergistic effect of EHop-016 and trastuzumab on inhibition of MDA-MB-435 cell viability.

A recent paradigm shift in effective aggressive breast cancer therapy is to mitigate the high prevalence of intrinsic and acquired resistance to single-agent regimens by dual anti-HER2 therapy of trastuzumab and other inhibitors [169,170]. Therefore, our characterization of EHop-016 offers a timely alternative to effective combinatorial therapy. We are currently testing the pharmacodynamics, pharmacokinetics, toxicity, and efficiency of EHop-016 in mouse models.

3.3. Development of next-generation inhibitors

EHop-016 fits some of the criteria for a specific inhibitor of metastasis by demonstrating an IC_{50} of 1 μM and being specific for the activation of Rac by Vav. Moreover, EHop-016 does not affect cell viability at

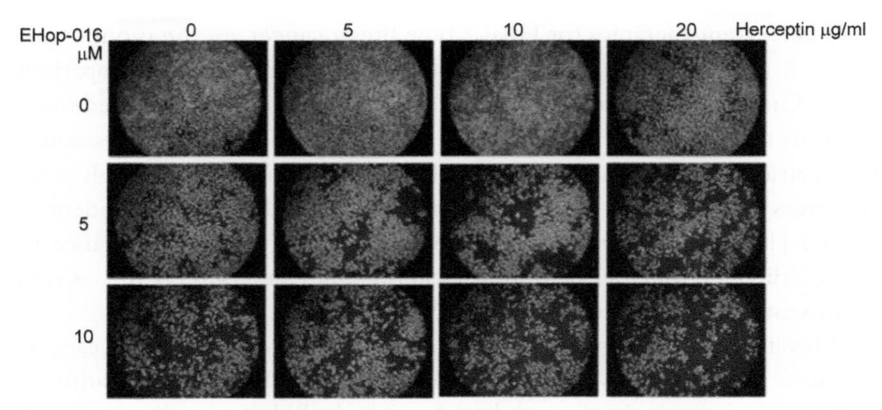

Figure 6.7 Effect of EHop-016 and trastuzumab on MDA-MB-435 cell viability. Trastuzumab-resistant MDA-MB-435 cells were treated with 0, 5, 10 μM EHop-016 (rows) or 0, 5, 10 μg/ml trastuzumab (columns) for 24 h, fixed and stained with propidium iodide. Representative micrographs from duplicates are shown. (See color plate.)

<10 μM, and is relatively soluble in aqueous solutions. However, the micromolar effective concentrations of EHop-016 may prove to be physiologically insufficient for development as a viable cancer therapeutic. Therefore, plans are underway to develop the next generation of EHop-016 compounds that are expected to be effective at lower (nanomloar) concentrations with a tighter and more energetically favorable binding at the GEF interaction site of Rac.

ACKNOWLEDGMENTS

We thank the following members of the Dharmawardhane laboratory who characterized the efficacy of EHop-016 in metastatic cancer cells: Linette Castillo-Pichardo, Ph.D.; Alina De La Mota-Peynado; Tessa Humphries-Bickley; and Brendaliz Montalvo-Ortiz, Ph.D. This work was supported by DoD/US Army BCRPW81XWH-0701-0330 and NIH/NIGMS SC3GM084824-02S1 (to S. Dharmawardhane), RCMI G12RR03051 to UPR MSC, and RCMI G12RR03035 to Universidad Central del Caribe, PR.

REFERENCES

[1] S. Riethdorf, K. Pantel, Disseminated tumor cells in bone marrow and circulating tumor cells in blood of breast cancer patients: current state of detection and characterization, Pathobiology 75 (2008) 140–148.

[2] E.C. Woodhouse, R.F. Chuaqui, L.A. Liotta, General mechanisms of metastasis, Cancer 80 (1997) 1529–1537.

[3] J. Condeelis, J.W. Pollard, Macrophages: obligate partners for tumor cell migration, invasion, and metastasis, Cell 124 (2006) 263–266.

[4] H.A. Smith, Y. Kang, The metastasis-promoting roles of tumor-associated immune cells, J. Mol. Med. 91 (4) (2013) 411–429.

[5] A.M. Gonzalez-Angulo, F. Morales-Vasquez, G.N. Hortobagyi, Overview of resistance to systemic therapy in patients with breast cancer, Adv. Exp. Med. Biol. 608 (2007) 1–22.
[6] A.J. Minn, G.P. Gupta, P.M. Siegel, P.D. Bos, W. Shu, D.D. Giri, A. Viale, A.B. Olshen, W.L. Gerald, J. Massague, Genes that mediate breast cancer metastasis to lung, Nature 436 (2005) 518–524.
[7] G.R. Mundy, Metastasis to bone: causes, consequences and therapeutic opportunities, Nat. Rev. Cancer 2 (2002) 584–593.
[8] J. Gligorov, J.P. Lotz, Optimal treatment strategies in postmenopausal women with hormone-receptor-positive and HER2-negative metastatic breast cancer, Breast Cancer Res. Treat. 112 (2008) 53–66.
[9] J. Condeelis, R.H. Singer, J.E. Segall, The great escape: when cancer cells hijack the genes for chemotaxis and motility, Annu. Rev. Cell Dev. Biol. 21 (2005) 695–718.
[10] P.S. Steeg, Metastasis suppressors alter the signal transduction of cancer cells, Nat. Rev. Cancer 3 (2003) 55–63.
[11] F.M. Vega, A.J. Ridley, Rho GTPases in cancer cell biology, FEBS Lett. 582 (2008) 2093–2101.
[12] A.J. Ridley, Regulation of macrophage adhesion and migration by Rho GTP-binding proteins, J. Microsc. 231 (2008) 518–523.
[13] S.J. Heasman, A.J. Ridley, Mammalian Rho GTPases: new insights into their functions from in vivo studies, Nat. Rev. Mol. Cell Biol. 9 (2008) 690–701.
[14] A.J. Ridley, Rho GTPases and actin dynamics in membrane protrusions and vesicle trafficking, Trends Cell Biol. 16 (2006) 522–529.
[15] R. Rathinam, A. Berrier, S.K. Alahari, Role of Rho GTPases and their regulators in cancer progression, Front. Biosci. 17 (2012) 2561–2571.
[16] S. Dharmawardhane, G.M. Bokoch, Rho GTPases and leukocyte cytoskeletal regulation, Curr. Opin. Hematol. 4 (1997) 12–18.
[17] A. Hall, Rho GTPases and the control of cell behaviour, Biochem. Soc. Trans. 33 (2005) 891–895.
[18] M. Lin, K.L. van Golen, Rho-regulatory proteins in breast cancer cell motility and invasion, Breast Cancer Res. Treat. 84 (2004) 49–60.
[19] C.G. Kleer, K.A. Griffith, M.S. Sabel, G. Gallagher, K.L. van Golen, Z.F. Wu, S.D. Merajver, RhoC-GTPase is a novel tissue biomarker associated with biologically aggressive carcinomas of the breast, Breast Cancer Res. Treat. 93 (2005) 101–110.
[20] A.Y. Chan, S.J. Coniglio, Y.Y. Chuang, D. Michaelson, U.G. Knaus, M.R. Philips, M. Symons, Roles of the Rac1 and Rac3 GTPases in human tumor cell invasion, Oncogene 24 (2005) 7821–7829.
[21] P. Burbelo, A. Wellstein, R.G. Pestell, Altered Rho GTPase signaling pathways in breast cancer cells, Breast Cancer Res. Treat. 84 (2004) 43–48.
[22] E. Katz, A.H. Sims, D. Sproul, H. Caldwell, M.J. Dixon, R.R. Meehan, D.J. Harrison, Targeting of Rac GTPases blocks the spread of intact human breast cancer, Oncotarget 3 (2012) 608–619.
[23] X.R. Bustelo, Intratumoral stages of metastatic cells: a synthesis of ontogeny, Rho/Rac GTPases, epithelial-mesenchymal transitions, and more, Bioessays 34 (2012) 748–759.
[24] L. Barrio-Real, M.G. Kazanietz, Rho GEFs and cancer: linking gene expression and metastatic dissemination, Sci. Signal. 5 (2012) e43.
[25] M. Parri, P. Chiarugi, Rac and Rho GTPases in cancer cell motility control, Cell Commun. Signaling 8 (2010) 23.
[26] E. Wertheimer, A. Gutierrez-Uzquiza, C. Rosemblit, C. Lopez-Haber, S.M. Soledad, M.G. Kazanietz, Rac signaling in breast cancer: a tale of GEFs and GAPs, Cell. Signal. 24 (2011) 353–362.

[27] N.A. Mack, H.J. Whalley, S. Castillo-Lluva, A. Malliri, The diverse roles of Rac signaling in tumorigenesis, Cell Cycle 10 (2011) 1571–1581.
[28] X.R. Bustelo, V. Sauzeau, I.M. Berenjeno, GTP-binding proteins of the Rho/Rac family: regulation, effectors and functions in vivo, Bioessays 29 (2007) 356–370.
[29] S.Y. Pai, C. Kim, D.A. Williams, Rac GTPases in human diseases, Dis. Markers 29 (2010) 177–187.
[30] H.Y. Shi, L.J. Stafford, Z. Liu, M. Liu, M. Zhang, Maspin controls mammary tumor cell migration through inhibiting Rac1 and Cdc42, but not the RhoA GTPase, Cell Motil. Cytoskeleton 64 (May 2007) 338–346.
[31] D.S. Hirsch, W.J. Wu, Cdc42: an effector and regulator of ErbB1 as a strategic target in breast cancer therapy, Expert Rev. Anticancer Ther. 7 (2007) 147–157.
[32] R.G. Qiu, J. Chen, D. Kirn, F. McCormick, M. Symons, An essential role for Rac in Ras transformation, Nature 374 (1995) 457–459.
[33] M.W. Renshaw, E. Lea-Chou, J.Y. Wang, Rac is required for v-Abl tyrosine kinase to activate mitogenesis, Curr. Biol. 6 (1996) 76–83.
[34] Z. Wang, E. Pedersen, A. Basse, T. Lefever, K. Peyrollier, S. Kapoor, Q. Mei, R. Karlsson, A. Chrostek-Grashoff, C. Brakebusch, Rac1 is crucial for Ras-dependent skin tumor formation by controlling Pak1-Mek-Erk hyperactivation and hyperproliferation in vivo, Oncogene 29 (2010) 3362–3373.
[35] N.G. Azios, L. Krishnamoorthy, M. Harris, L.A. Cubano, M. Cammer, S.F. Dharmawardhane, Estrogen and resveratrol regulate Rac and Cdc42 signaling to the actin cytoskeleton of metastatic breast cancer cells, Neoplasia 9 (2007) 147–158.
[36] P.J. Baugher, L. Krishnamoorthy, J.E. Price, S.F. Dharmawardhane, Rac1 and Rac3 isoform activation is involved in the invasive and metastatic phenotype of human breast cancer cells, Breast Cancer Res. 7 (2005) R965–R974.
[37] J.P. Mira, V. Benard, J. Groffen, L.C. Sanders, U.G. Knaus, Endogenous, hyperactive Rac3 controls proliferation of breast cancer cells by a p21-activated kinase-dependent pathway, Proc. Natl. Acad. Sci. U.S.A. 97 (2000) 185–189.
[38] T. Yoshida, Y. Zhang, L.A. Rivera Rosado, J. Chen, T. Khan, S.Y. Moon, B. Zhang, Blockade of Rac1 activity induces G1 cell cycle arrest or apoptosis in breast cancer cells through downregulation of cyclin D1, survivin, and X-linked inhibitor of apoptosis protein, Mol. Cancer Ther. 9 (2010) 1657–1668.
[39] A. Schnelzer, D. Prechtel, U. Knaus, K. Dehne, M. Gerhard, H. Graeff, N. Harbeck, M. Schmitt, E. Lengyel, Rac1 in human breast cancer: overexpression, mutation analysis, and characterization of a new isoform, Rac1b, Oncogene 19 (2000) 3013–3020.
[40] V. Walf-Vorderwulbecke, J. de Boer, S.J. Horton, R. van Amerongen, N. Proost, A. Berns, O. Williams, Frat2 mediates the oncogenic activation of Rac by MLL fusions, Blood 120 (2012) 4819–4828.
[41] L.E. Dalton, J. Kamarashev, I. Barinaga-Rementeria Ramirez, G. White, A. Malliri, A. Hurlstone, Constitutive Rac activation is not sufficient to initiate melanocyte neoplasia but accelerates malignant progression, J. Invest. Dermatol. 133 (2013) 1572–1581.
[42] E. Wertheimer, M.G. Kazanietz, Rac1 takes center stage in pancreatic cancer and ulcerative colitis: quantity matters, Gastroenterology 141 (2011) 427–430.
[43] I. Heid, C. Lubeseder-Martellato, B. Sipos, P.K. Mazur, M. Lesina, R.M. Schmid, J.T. Siveke, Early requirement of Rac1 in a mouse model of pancreatic cancer, Gastroenterology 141 (2011) 719–730.
[44] C.M. Morris, L. Haataja, M. McDonald, S. Gough, D. Markie, J. Groffen, N. Heisterkamp, The small GTPase RAC3 gene is located within chromosome band 17q25.3 outside and telomeric of a region commonly deleted in breast and ovarian tumours, Cytogenet. Cell Genet. 89 (2000) 18–23.

[45] Y. Pan, F. Bi, N. Liu, Y. Xue, X. Yao, Y. Zheng, D. Fan, Expression of seven main Rho family members in gastric carcinoma, Biochem. Biophys. Res. Commun. 315 (2004) 686–691.
[46] S.L. Hwang, J.H. Chang, T.S. Cheng, W.D. Sy, A.S. Lieu, C.L. Lin, K.S. Lee, S.L. Howng, Y.R. Hong, Expression of Rac3 in human brain tumors, J. Clin. Neurosci. 12 (2005) 571–574.
[47] A. Singh, A.E. Karnoub, T.R. Palmby, E. Lengyel, J. Sondek, C.J. Der, Rac1b, a tumor associated, constitutively active Rac1 splice variant, promotes cellular transformation, Oncogene 23 (2004) 9369–9380.
[48] M. Stallings-Mann, D. Radisky, Matrix metalloproteinase-induced malignancy in mammary epithelial cells, Cells Tissues Organs 185 (2007) 104–110.
[49] P. Jordan, R. Brazao, M.G. Boavida, C. Gespach, E. Chastre, Cloning of a novel human Rac1b splice variant with increased expression in colorectal tumors, Oncogene 18 (1999) 6835–6839.
[50] P. Matos, P. Jordan, Increased Rac1b expression sustains colorectal tumor cell survival, Mol. Cancer Res. 6 (2008) 1178–1184.
[51] E. Hodis, I.R. Watson, G.V. Kryukov, S.T. Arold, M. Imielinski, J.P. Theurillat, E. Nickerson, D. Auclair, L. Li, C. Place, D. Dicara, A.H. Ramos, M.S. Lawrence, K. Cibulskis, A. Sivachenko, D. Voet, G. Saksena, N. Stransky, R.C. Onofrio, W. Winckler, K. Ardlie, N. Wagle, J. Wargo, K. Chong, D.L. Morton, K. Stemke-Hale, G. Chen, M. Noble, M. Meyerson, J.E. Ladbury, M.A. Davies, J.E. Gershenwald, S.N. Wagner, D.S. Hoon, D. Schadendorf, E.S. Lander, S.B. Gabriel, G. Getz, L.A. Garraway, L. Chin, A landscape of driver mutations in melanoma, Cell 150 (2012) 251–263.
[52] A.M. Muise, T. Walters, W. Xu, G. Shen-Tu, C.H. Guo, R. Fattouh, G.Y. Lam, V.M. Wolters, J. Bennitz, J. van Limbergen, P. Renbaum, Y. Kasirer, B.Y. Ngan, D. Turner, L.A. Denson, P.M. Sherman, R.H. Duerr, J. Cho, C.W. Lees, J. Satsangi, D.C. Wilson, A.D. Paterson, A.M. Griffiths, M. Glogauer, M.S. Silverberg, J.H. Brumell, Single nucleotide polymorphisms that increase expression of the guanosine triphosphatase RAC1 are associated with ulcerative colitis, Gastroenterology 14 (2011) 633–641.
[53] M. Kawazu, T. Ueno, K. Kontani, Y. Ogita, M. Ando, K. Fukumura, A. Yamato, M. Soda, K. Takeuchi, Y. Miki, H. Yamaguchi, T. Yasuda, T. Naoe, Y. Yamashita, T. Katada, Y.L. Choi, H. Mano, Transforming mutations of RAC guanosine triphosphatases in human cancers, Proc. Natl. Acad. Sci. U.S.A. 110 (2013) 3029–3034.
[54] Y. Ueda, S. Wang, N. Dumont, J.Y. Yi, Y. Koh, C.L. Arteaga, Overexpression of HER2 (erbB2) in human breast epithelial cells unmasks transforming growth factor beta-induced cell motility, J. Biol. Chem. 279 (2004) 24505–24513.
[55] A.E. Rosenblatt, M.I. Garcia, L. Lyons, Y. Xie, C. Maiorino, L. Desire, J. Slingerland, K.L. Burnstein, Inhibition of the Rho GTPase, Rac1, decreases estrogen receptor levels and is a novel therapeutic strategy in breast cancer, Endocr. Relat. Cancer 18 (2011) 207–219.
[56] T. Vu, F.X. Claret, Trastuzumab: updated mechanisms of action and resistance in breast cancer, Front Oncol. 2 (2012) 62.
[57] A. Whale, F.N. Hashim, S. Fram, G.E. Jones, C.M. Wells, Signalling to cancer cell invasion through PAK family kinases, Front. Biosci. 16 (2011) 849–864.
[58] B. Dummler, K. Ohshiro, R. Kumar, J. Field, Pak protein kinases and their role in cancer, Cancer Metastasis Rev. 28 (2009) 51–63.
[59] L.E. Arias-Romero, O. Villamar-Cruz, A. Pacheco, R. Kosoff, M. Huang, S.K. Muthuswamy, J. Chernoff, A Rac-Pak signaling pathway is essential for ErbB2-mediated transformation of human breast epithelial cancer cells, Oncogene 29 (2010) 5839–5849.

[60] D.Z. Ye, J. Field, PAK signaling in cancer, Cell Logist. 2 (2012) 105–116.
[61] C.W. Menges, E. Sementino, J. Talarchek, J. Xu, J. Chernoff, J.R. Peterson, J.R. Testa, Group I p21-activated kinases (PAKs) promote tumor cell proliferation and survival through the AKT1 and Raf-MAPK pathways, Mol. Cancer Res. 10 (2012) 1178–1188.
[62] L.E. Rias-Romero, J. Chernoff, p21-Activated kinases in Erbb2-positive breast cancer: a new therapeutic target? Small Gtpases. 1 (2010) 124–128.
[63] L. Adam, R. Vadlamudi, S.B. Kondapaka, J. Chernoff, J. Mendelsohn, R. Kumar, Heregulin regulates cytoskeletal reorganization and cell migration through the p21-activated kinase-1 via phosphatidylinositol-3 kinase, J. Biol. Chem. 273 (1998) 28238–28246.
[64] A.M. Brumby, K.R. Goulding, T. Schlosser, S. Loi, R. Galea, P. Khoo, J.E. Bolden, T. Aigaki, P.O. Humbert, H.E. Richardson, Identification of novel Ras-cooperating oncogenes in Drosophila melanogaster: a RhoGEF/Rho-family/JNK pathway is a central driver of tumorigenesis, Genetics 188 (2011) 105–125.
[65] R. Samaga, J. Saez-Rodriguez, L.G. Alexopoulos, P.K. Sorger, S. Klamt, The logic of EGFR/ErbB signaling: theoretical properties and analysis of high-throughput data, PLoS Comput. Biol. 5 (2009) e1000438.
[66] A. Saci, L.C. Cantley, C.L. Carpenter, Rac1 regulates the activity of mTORC1 and mTORC2 and controls cellular size, Mol. Cell 42 (2011) 50–61.
[67] P. Gulhati, K.A. Bowen, J. Liu, P.D. Stevens, P.G. Rychahou, M. Chen, E.Y. Lee, H.L. Weiss, K.L. O'Connor, T. Gao, B.M. Evers, mTORC1 and mTORC2 regulate EMT, motility, and metastasis of colorectal cancer via RhoA and Rac1 signaling pathways, Cancer Res. 71 (2011) 3246–3256.
[68] I. Bracho-Valdes, P. Moreno-Alvarez, I. Valencia-Martinez, E. Robles-Molina, L. Chavez-Vargas, J. Vazquez-Prado, mTORC1- and mTORC2-interacting proteins keep their multifunctional partners focused, IUBMB Life 63 (2011) 880–898.
[69] G.M. Bokoch, B. Diebold, J.S. Kim, D. Gianni, Emerging evidence for the importance of phosphorylation in the regulation of NADPH oxidases, Antioxid. Redox Signal. 11 (2009) 2429–2441.
[70] S.J. Park, Y.T. Kim, Y.J. Jeon, Antioxidant dieckol downregulates the Rac1/ROS signaling pathway and inhibits Wiskott-Aldrich syndrome protein (WASP)-family verprolin-homologous protein 2 (WAVE2)-mediated invasive migration of B16 mouse melanoma cells, Mol. Cells 33 (2012) 363–369.
[71] M. Nieborowska-Skorska, P.K. Kopinski, R. Ray, G. Hoser, D. Ngaba, S. Flis, K. Cramer, M.M. Reddy, M. Koptyra, T. Penserga, E. Glodkowska-Mrowka, E. Bolton, T.L. Holyoake, C.J. Eaves, S. Cerny-Reiterer, P. Valent, A. Hochhaus, T.P. Hughes, H. van der Kuip, M. Sattler, W. Wiktor-Jedrzejczak, C. Richardson, A. Dorrance, T. Stoklosa, D.A. Williams, T. Skorski, Rac2-MRC-cIII-generated ROS cause genomic instability in chronic myeloid leukemia stem cells and primitive progenitors, Blood 119 (2012) 4253–4263.
[72] W.S. Wu, The signaling mechanism of ROS in tumor progression, Cancer Metastasis Rev. 25 (2006) 695–705.
[73] C. Costa, G. Germena, E. Hirsch, Dissection of the interplay between class I PI3Ks and Rac signaling in phagocytic functions, Sci. World J. 10 (2010) 1826–1839.
[74] H.W. Yang, M.G. Shin, S. Lee, J.R. Kim, W.S. Park, K.H. Cho, T. Meyer, H.W. Do, Cooperative activation of PI3K by Ras and Rho family small GTPases, Mol. Cell 47 (2012) 281–290.
[75] E.K. Kim, S.J. Yun, J.M. Ha, Y.W. Kim, I.H. Jin, J. Yun, H.K. Shin, S.H. Song, J.H. Kim, J.S. Lee, C.D. Kim, S.S. Bae, Selective activation of Akt1 by mammalian target of rapamycin complex 2 regulates cancer cell migration, invasion, and metastasis, Oncogene 30 (2011) 2954–2963.

[76] M.C. Caino, C. Lopez-Haber, J.L. Kissil, M.G. Kazanietz, Non-small cell lung carcinoma cell motility, rac activation and metastatic dissemination are mediated by protein kinase C epsilon, PLoS One 7 (2012) e31714.
[77] E. Aksamitiene, S. Achanta, W. Kolch, B.N. Kholodenko, J.B. Hoek, A. Kiyatkin, Prolactin-stimulated activation of ERK1/2 mitogen-activated protein kinases is controlled by PI3-kinase/Rac/PAK signaling pathway in breast cancer cells, Cell. Signal. 23 (2011) 1794–1805.
[78] V.S. Ramgolam, S.D. DeGregorio, G.K. Rao, M. Collinge, S.S. Subaran, S. Markovic-Plese, R. Pardi, J.R. Bender, T cell LFA-1 engagement induces HuR-dependent cytokine mRNA stabilization through a Vav-1, Rac1/2, p38MAPK and MKK3 signaling cascade, PLoS One 5 (2010) e14450.
[79] Y. Zhang, L.A. Rivera Rosado, S.Y. Moon, B. Zhang, Silencing of D4-GDI inhibits growth and invasive behavior in MDA-MB-231 cells by activation of Rac-dependent p38 and JNK signaling, J. Biol. Chem. 284 (2009) 12956–12965.
[80] R. Arulanandam, M. Geletu, H. Feracci, L. Raptis, Activated Rac1 requires gp130 for Stat3 activation, cell proliferation and migration, Exp. Cell Res. 316 (2010) 875–886.
[81] M. Laplante, D.M. Sabatini, mTOR signaling in growth control and disease, Cell 149 (2012) 274–293.
[82] J.V. Kichina, A. Goc, B. Al-Husein, P.R. Somanath, E.S. Kandel, PAK1 as a therapeutic target, Expert Opin. Ther. Targets 14 (2010) 703–725.
[83] C. Yi, J. Maksimoska, R. Marmorstein, J.L. Kissil, Development of small-molecule inhibitors of the group I p21-activated kinases, emerging therapeutic targets in cancer, Biochem. Pharmacol. 80 (2010) 683–689.
[84] C.C. Ong, A.M. Jubb, P.M. Haverty, W. Zhou, V. Tran, T. Truong, H. Turley, T. O'Brien, D. Vucic, A.L. Harris, M. Belvin, L.S. Friedman, E.M. Blackwood, H. Koeppen, K.P. Hoeflich, Targeting p21-activated kinase 1 (PAK1) to induce apoptosis of tumor cells, Proc. Natl. Acad. Sci. U.S.A. 108 (2011) 7177–7182.
[85] A.L. Bishop, A. Hall, Rho GTPases and their effector proteins, Biochem. J. 348 (2000) 241–255.
[86] S. Baranwal, S.K. Alahari, Rho GTPase effector functions in tumor cell invasion and metastasis, Curr. Drug Targets 12 (2011) 1194–1201.
[87] S.W. Deacon, J.R. Peterson, Chemical inhibition through conformational stabilization of Rho GTPase effectors, Handb. Exp. Pharmacol. 186 (2008) 431–460.
[88] K.L. Rossman, C.J. Der, J. Sondek, GEF means go: turning on RHO GTPases with guanine nucleotide-exchange factors, Nat. Rev. Mol. Cell Biol. 6 (2005) 167–180.
[89] A. Schmidt, A. Hall, Guanine nucleotide exchange factors for Rho GTPases: turning on the switch, Genes Dev. 16 (2002) 1587–1609.
[90] A. Bernards, GAPs galore! A survey of putative Ras superfamily GTPase activating proteins in man and Drosophila, Biochim. Biophys. Acta 1603 (2003) 47–82.
[91] G.R. Hoffman, R.A. Cerione, Signaling to the Rho GTPases: networking with the DH domain, FEBS Lett. 513 (2002) 85–91.
[92] H.C. Adams III., R. Chen, Z. Liu, I.P. Whitehead, Regulation of breast cancer cell motility by T-cell lymphoma invasion and metastasis-inducing protein, Breast Cancer Res. 12 (2010) R69.
[93] T.R. Palmby, K. Abe, A.E. Karnoub, C.J. Der, Vav transformation requires activation of multiple GTPases and regulation of gene expression, Mol. Cancer Res. 2 (2004) 702–711.
[94] S.L. Miller, J.E. DeMaria, D.O. Freier, A.M. Riegel, C.V. Clevenger, Novel association of Vav2 and Nek3 modulates signaling through the human prolactin receptor, Mol. Endocrinol. 19 (2005) 939–949.

[95] M.E. Minard, L.S. Kim, J.E. Price, G.E. Gallick, The role of the guanine nucleotide exchange factor Tiam1 in cellular migration, invasion, adhesion and tumor progression, Breast Cancer Res. Treat. 84 (2004) 21–32.
[96] M.S. Sosa, C. Lopez-Haber, C. Yang, H. Wang, M.A. Lemmon, J.M. Busillo, J. Luo, J.L. Benovic, A. Klein-Szanto, H. Yagi, J.S. Gutkind, R.E. Parsons, M.G. Kazanietz, Identification of the Rac-GEF P-Rex1 as an essential mediator of ErbB signaling in breast cancer, Mol. Cell 40 (2010) 877–892.
[97] J.C. Montero, S. Seoane, A. Ocana, A. Pandiella, P-Rex1 participates in Neuregulin-ErbB signal transduction and its expression correlates with patient outcome in breast cancer, Oncogene 30 (2011) 1059–1071.
[98] C. Citterio, M. Menacho-Marquez, R. Garcia-Escudero, R.M. Larive, O. Barreiro, F. Sanchez-Madrid, J.M. Paramio, X.R. Bustelo, The rho exchange factors vav2 and vav3 control a lung metastasis-specific transcriptional program in breast cancer cells, Sci. Signal. 5 (2012) ra71.
[99] I. Hornstein, E. Pikarsky, M. Groysman, G. Amir, N. Peylan-Ramu, S. Katzav, The haematopoietic specific signal transducer Vav1 is expressed in a subset of human neuroblastomas, J. Pathol. 199 (2003) 526–533.
[100] G. Denicola, D.A. Tuveson, VAV1: a new target in pancreatic cancer? Cancer Biol. Ther. 4 (2005) 509–511.
[101] G. Lazer, Y. Idelchuk, V. Schapira, E. Pikarsky, S. Katzav, The haematopoietic specific signal transducer Vav1 is aberrantly expressed in lung cancer and plays a role in tumourigenesis, J. Pathol. 219 (2009) 25–34.
[102] R.A. Bartolome, I. Molina-Ortiz, R. Samaniego, P. Sanchez-Mateos, X.R. Bustelo, J. Teixido, Activation of Vav/Rho GTPase signaling by CXCL12 controls membrane-type matrix metalloproteinase-dependent melanoma cell invasion, Cancer Res. 66 (2006) 248–258.
[103] K.T. Lin, J. Gong, C.F. Li, T.H. Jang, W.L. Chen, H.J. Chen, L.H. Wang, Vav3-Rac1 signaling regulates prostate cancer metastasis with elevated Vav3 expression correlating with prostate cancer progression and posttreatment recurrence, Cancer Res. 72 (2012) 3000–3009.
[104] F. Valderrama, S. Thevapala, A.J. Ridley, Radixin regulates cell migration and cell-cell adhesion through Rac1, J. Cell Sci. 125 (2012) 3310–3319.
[105] S.K. Sastry, Z. Rajfur, B.P. Liu, J.F. Cote, M.L. Tremblay, K. Burridge, PTP-PEST couples membrane protrusion and tail retraction via VAV2 and p190RhoGAP, J. Biol. Chem. 281 (2006) 11627–11636.
[106] L. Ilan, S. Katzav, Human Vav1 expression in hematopoietic and cancer cell lines is regulated by c-Myb and by CpG methylation, PLoS One 7 (2012) e29939.
[107] M. Groysman, I. Hornstein, A. Alcover, S. Katzav, Vav1 and Ly-GDI two regulators of Rho GTPases, function cooperatively as signal transducers in T cell antigen receptor-induced pathways, J. Biol. Chem. 277 (2002) 50121–50130.
[108] M.J. Oberley, D.S. Wang, D.T. Yang, Vav1 in hematologic neoplasms, a mini review, Am. J. Blood Res. 2 (2012) 1–8.
[109] D.M. Brantley-Sieders, G. Zhuang, D. Vaught, T. Freeman, Y. Hwang, D. Hicks, J. Chen, Host deficiency in Vav2/3 guanine nucleotide exchange factors impairs tumor growth, survival, and angiogenesis in vivo, Mol. Cancer Res. 7 (2009) 615–623.
[110] T.A. Garrett, J.D. Van Buul, K. Burridge, VEGF-induced Rac1 activation in endothelial cells is regulated by the guanine nucleotide exchange factor Vav2, Exp. Cell Res. 313 (2007) 3285–3297.
[111] B.P. Liu, K. Burridge, Vav2 activates Rac1, Cdc42, and RhoA downstream from growth factor receptors but not beta1 integrins, Mol. Cell Biol. 20 (2000) 7160–7169.
[112] V. Patel, H.M. Rosenfeldt, R. Lyons, J.M. Servitja, X.R. Bustelo, M. Siroff, J.S. Gutkind, Persistent activation of Rac1 in squamous carcinomas of the head and

neck: evidence for an EGFR/Vav2 signaling axis involved in cell invasion, Carcinogenesis 28 (2007) 1145–1152.

[113] J.M. Servitja, M.J. Marinissen, A. Sodhi, X.R. Bustelo, J.S. Gutkind, Rac1 function is required for Src-induced transformation. Evidence of a role for Tiam1 and Vav2 in Rac activation by Src, J. Biol. Chem. 278 (2003) 34339–34346.

[114] C. Yang, E.A. Klein, R.K. Assoian, M.G. Kazanietz, Heregulin beta1 promotes breast cancer cell proliferation through Rac/ERK-dependent induction of cyclin D1 and p21Cip1, Biochem. J. 410 (2008) 167–175.

[115] A. De Laurentiis, O.E. Pardo, A. Palamidessi, S.P. Jackson, S.M. Schoenwaelder, E. Reichmann, G. Scita, A. Arcaro, The catalytic class I(A) PI3K isoforms play divergent roles in breast cancer cell migration, Cell. Signal. 23 (2010) 529–541.

[116] P. Sachdev, L. Zeng, L.H. Wang, Distinct role of phosphatidylinositol 3-kinase and Rho family GTPases in Vav3-induced cell transformation, cell motility, and morphological changes, J. Biol. Chem. 277 (2002) 17638–17648.

[117] S.E. Wang, I. Shin, F.Y. Wu, D.B. Friedman, C.L. Arteaga, HER2/Neu (ErbB2) signaling to Rac1-Pak1 is temporally and spatially modulated by transforming growth factor beta, Cancer Res. 66 (2006) 9591–9600.

[118] L. Zeng, P. Sachdev, L. Yan, J.L. Chan, T. Trenkle, M. McClelland, J. Welsh, L.H. Wang, Vav3 mediates receptor protein tyrosine kinase signaling, regulates GTPase activity, modulates cell morphology, and induces cell transformation, Mol. Cell Biol. 20 (2000) 9212–9224.

[119] N. Marcoux, K. Vuori, EGF receptor mediates adhesion-dependent activation of the Rac GTPase: a role for phosphatidylinositol 3-kinase and Vav2, Oncogene 22 (2003) 6100–6106.

[120] J.L. Wilsbacher, S.L. Moores, J.S. Brugge, An active form of Vav1 induces migration of mammary epithelial cells by stimulating secretion of an epidermal growth factor receptor ligand, Cell Commun. Signal. 4 (2006) 5.

[121] B. Weigelt, A.T. Lo, C.C. Park, J.W. Gray, M.J. Bissell, HER2 signaling pathway activation and response of breast cancer cells to HER2-targeting agents is dependent strongly on the 3D microenvironment, Breast Cancer Res. Treat. 122 (2010) 35–43.

[122] Y. Zhao, Z. Wang, Y. Jiang, C. Yang, Inactivation of Rac1 reduces Trastuzumab resistance in PTEN deficient and insulin-like growth factor I receptor overexpressing human breast cancer SKBR3 cells, Cancer Lett. 313 (2011) 54–63.

[123] M. Dokmanovic, D.S. Hirsch, Y. Shen, W.J. Wu, Rac1 contributes to trastuzumab resistance of breast cancer cells: Rac1 as a potential therapeutic target for the treatment of trastuzumab-resistant breast cancer, Mol. Cancer Ther. 8 (2009) 1557–1569.

[124] B.L. Montalvo-Ortiz, L. Castillo-Pichardo, E. Hernandez, T. Humphries-Bickley, L.M.-P. De, L.A. Cubano, C.P. Vlaar, S. Dharmawardhane, Characterization of EHop-016, novel small molecule inhibitor of Rac GTPase, J. Biol. Chem. 287 (2012) 13228–13238.

[125] Y. Gao, J.B. Dickerson, F. Guo, J. Zheng, Y. Zheng, Rational design and characterization of a Rac GTPase-specific small molecule inhibitor, Proc. Natl. Acad. Sci. U.S.A. 101 (2004) 7618–7623.

[126] N. Nassar, J. Cancelas, J. Zheng, D.A. Williams, Y. Zheng, Structure-function based design of small molecule inhibitors targeting Rho family GTPases, Curr. Top. Med. Chem. 6 (2006) 1109–1116.

[127] Y. Gao, J. Xing, M. Streuli, T.L. Leto, Y. Zheng, Trp(56) of rac1 specifies interaction with a subset of guanine nucleotide exchange factors, J. Biol. Chem. 276 (2001) 47530–47541.

[128] A. Gastonguay, T. Berg, A.D. Hauser, N. Schuld, E. Lorimer, C.L. Williams, The role of Rac1 in the regulation of NF-kB activity, cell proliferation, and cell migration in non-small cell lung carcinoma, Cancer Biol. Ther. 13 (2012) 647–656.

[129] B. Mizukawa, J. Wei, M. Shrestha, M. Wunderlich, F.S. Chou, A. Griesinger, C.E. Harris, A.R. Kumar, Y. Zheng, D.A. Williams, J.C. Mulloy, Inhibition of Rac GTPase signaling and downstream prosurvival Bcl-2 proteins as combination targeted therapy in MLL-AF9 leukemia, Blood 118 (2011) 5235–5245.
[130] Q.Y. Chen, L.Q. Xu, D.M. Jiao, Q.H. Yao, Y.Y. Wang, H.Z. Hu, Y.Q. Wu, J. Song, J. Yan, L.J. Wu, Silencing of Rac1 modifies lung cancer cell migration, invasion and actin cytoskeleton rearrangements and enhances chemosensitivity to antitumor drugs, Int. J. Mol. Med. 28 (2011) 769–776.
[131] M. Hamalukic, J. Huelsenbeck, A. Schad, S. Wirtz, B. Kaina, G. Fritz, Rac1-regulated endothelial radiation response stimulates extravasation and metastasis that can be blocked by HMG-CoA reductase inhibitors, PLoS One 6 (2011) e26413.
[132] H. Akbar, J. Cancelas, D.A. Williams, J. Zheng, Y. Zheng, Rational design and applications of a Rac GTPase-specific small molecule inhibitor, Methods Enzymol. 406 (2006) 554–565.
[133] E.K. Thomas, J.A. Cancelas, H.D. Chae, A.D. Cox, P.J. Keller, D. Perrotti, P. Neviani, B.J. Druker, K.D. Setchell, Y. Zheng, C.E. Harris, D.A. Williams, Rac guanosine triphosphatases represent integrating molecular therapeutic targets for BCR-ABL-induced myeloproliferative disease, Cancer Cell 12 (2007) 467–478.
[134] M.G. Binker, A.A. Binker-Cosen, H.Y. Gaisano, L.I. Cosen-Binker, Inhibition of Rac1 decreases the severity of pancreatitis and pancreatitis-associated lung injury in mice, Exp. Physiol. 93 (2008) 1091–1103.
[135] N. Ferri, A. Corsini, P. Bottino, F. Clerici, A. Contini, Virtual screening approach for the identification of new Rac1 inhibitors, J. Med. Chem. 52 (2009) 4087–4090.
[136] A. Shutes, C. Onesto, V. Picard, B. Leblond, F. Schweighoffer, C.J. Der, Specificity and mechanism of action of EHT 1864, a novel small molecule inhibitor of Rac family small GTPases, J. Biol. Chem. 282 (2007) 35666–35678.
[137] C. Onesto, A. Shutes, V. Picard, F. Schweighoffer, C.J. Der, Characterization of EHT 1864, a novel small molecule inhibitor of Rac family small GTPases, Methods Enzymol. 439 (2008) 111–129.
[138] A.C. Montezano, D. Burger, T.M. Paravicini, A.Z. Chignalia, H. Yusuf, M. Almasri, Y. He, G.E. Callera, G. He, K.H. Krause, D. Lambeth, M.T. Quinn, R.M. Touyz, Nicotinamide adenine dinucleotide phosphate reduced oxidase 5 (Nox5) regulation by angiotensin II and endothelin-1 is mediated via calcium/calmodulin-dependent, rac-1-independent pathways in human endothelial cells, Circ. Res. 106 (2010) 1363–1373.
[139] M.A. Davare, T. Saneyoshi, T.R. Soderling, Calmodulin-kinases regulate basal and estrogen stimulated medulloblastoma migration via Rac1, J. Neurooncol. 104 (2011) 65–82.
[140] L. Stefanini, Y. Boulaftali, T.D. Ouellette, M. Holinstat, L. Desire, B. Leblond, P. Andre, P.B. Conley, W. Bergmeier, Rap1-Rac1 circuits potentiate platelet activation, Arterioscler. Thromb. Vasc. Biol. 32 (2012) 434–441.
[141] A. Colomba, S. Giuriato, E. Dejean, K. Thornber, G. Delsol, H. Tronchere, F. Meggetto, B. Payrastre, F. Gaits-Iacovoni, Inhibition of Rac controls NPM-ALK-dependent lymphoma development and dissemination, Blood Cancer J. 1 (2011) e21.
[142] E. Beausoleil, C. Chauvignac, T. Taverne, S. Lacombe, L. Pognante, B. Leblond, D. Pallares, C.D. Oliveira, F. Bachelot, R. Carton, H. Peillon, S. Coutadeur, V. Picard, N. Lambeng, L. Desire, F. Schweighoffer, Structure-activity relationship of isoform selective inhibitors of Rac1/1b GTPase nucleotide binding, Bioorg. Med. Chem. Lett. 19 (2009) 5594–5598.

[143] E. Hernandez, L.M.-P. De, S. Dharmawardhane, C.P. Vlaar, Novel inhibitors of Rac1 in metastatic breast cancer, P. R. Health Sci. J. 29 (2010) 348–356.
[144] V. Benard, G.M. Bokoch, Assay of Cdc42, Rac, and Rho GTPase activation by affinity methods, Methods Enzymol. 345 (2002) 349–359.
[145] Z. Surviladze, A. Waller, Y. Wu, E. Romero, B.S. Edwards, A. Wandinger-Ness, L.A. Sklar, Identification of a small GTPase inhibitor using a high-throughput flow cytometry bead-based multiplex assay, J. Biomol. Screen. 15 (2010) 10–20.
[146] Y. Zheng, N. Nassar, K.R. Skowronek. U.S.Patent No. 17,826,982, (2010).
[147] L. Duan, G. Chen, S. Virmani, G. Ying, S.M. Raja, B.M. Chung, M.A. Rainey, M. Dimri, C.F. Ortega-Cava, X. Zhao, R.J. Clubb, C. Tu, A.L. Reddi, M. Naramura, V. Band, H. Band, Distinct roles for Rho versus Rac/Cdc42 GTPases downstream of Vav2 in regulating mammary epithelial acinar architecture, J. Biol. Chem. 285 (2010) 1555–1568.
[148] M. Turner, D.D. Billadeau, VAV proteins as signal integrators for multi-subunit immune-recognition receptors, Nat. Rev. Immunol. 2 (2002) 476–486.
[149] S. Malhotra, S. Kovats, W. Zhang, K.M. Coggeshall, Vav and Rac activation in B cell antigen receptor endocytosis involves Vav recruitment to the adapter protein LAB, J. Biol. Chem. 284 (2009) 36202–36212.
[150] P.J. Bhavsar, E. Vigorito, M. Turner, A.J. Ridley, Vav GEFs regulate macrophage morphology and adhesion-induced Rac and Rho activation, Exp. Cell Res. 315 (2009) 3345–3358.
[151] W. Ming, S. Li, D.D. Billadeau, L.A. Quilliam, M.C. Dinauer, The Rac effector p67phox regulates phagocyte NADPH oxidase by stimulating Vav1 guanine nucleotide exchange activity, Mol. Cell Biol. 27 (2007) 312–323.
[152] C. Kim, C.C. Marchal, J. Penninger, M.C. Dinauer, The hemopoietic Rho/Rac guanine nucleotide exchange factor Vav1 regulates N-formyl-methionyl-leucyl-phenylalanine-activated neutrophil functions, J. Immunol. 171 (2003) 4425–4430.
[153] J.B. Wyckoff, Y. Wang, E.Y. Lin, J.F. Li, S. Goswami, E.R. Stanley, J.E. Segall, J.W. Pollard, J. Condeelis, Direct visualization of macrophage-assisted tumor cell intravasation in mammary tumors, Cancer Res. 67 (2007) 2649–2656.
[154] J.J. Zhao, K. Pan, W. Wang, J.G. Chen, Y.H. Wu, L. Lv, J.J. Li, Y.B. Chen, D.D. Wang, Q.Z. Pan, X.D. Li, J.C. Xia, The prognostic value of tumor-infiltrating neutrophils in gastric adenocarcinoma after resection, PLoS One 7 (2012) e33655.
[155] S.K. Chan, M.E. Hill, W.J. Gullick, The role of the epidermal growth factor receptor in breast cancer, J. Mammary Gland Biol. Neoplasia 11 (2006) 3–11.
[156] C.L. Arteaga, C.I. Truica, Challenges in the development of anti-epidermal growth factor receptor therapies in breast cancer, Semin. Oncol. 31 (2004) 3–8.
[157] L.M. Bender, R. Nahta, Her2 cross talk and therapeutic resistance in breast cancer, Front. Biosci. 13 (2008) 3906–3912.
[158] S. Menard, S. Fortis, F. Castiglioni, R. Agresti, A. Balsari, HER2 as a prognostic factor in breast cancer, Oncology 61 (2001) 67–72.
[159] R. Nahta, S. Shabaya, T. Ozbay, D.L. Rowe, Personalizing HER2-targeted therapy in metastatic breast cancer beyond HER2 status: what we have learned from clinical specimens, Curr. Pharmacogenomics Pers. Med. 7 (2009) 263–274.
[160] A. Jeyakumar, T. Younis, Trastuzumab for HER2-positive metastatic breast cancer: clinical and economic considerations, Clin. Med. Insights Oncol. 6 (2012) 179–187.
[161] E. Lantz, I. Cunningham, G.M. Higa, Targeting HER2 in breast cancer: overview of long-term experience, Int. J. Womens Health 1 (2010) 155–171.
[162] J.T. Garrett, C.L. Arteaga, Resistance to HER2-directed antibodies and tyrosine kinase inhibitors: mechanisms and clinical implications, Cancer Biol. Ther. 11 (2011) 793–800.

[163] J.A. Wilken, N.J. Maihle, Primary trastuzumab resistance: new tricks for an old drug, Ann. N. Y. Acad. Sci. 1210 (2010) 53–65.
[164] B.N. Rexer, C.L. Arteaga, Intrinsic and acquired resistance to HER2-targeted therapies in HER2 gene-amplified breast cancer: mechanisms and clinical implications, Crit. Rev. Oncog. 17 (2012) 1–16.
[165] L. Wang, Q. Zhang, J. Zhang, S. Sun, H. Guo, Z. Jia, B. Wang, Z. Shao, Z. Wang, X. Hu, PI3K pathway activation results in low efficacy of both trastuzumab and lapatinib, BMC Cancer 11 (2011) 248.
[166] N. Normanno, L.A. De, M.R. Maiello, M. Campiglio, M. Napolitano, M. Mancino, A. Carotenuto, G. Viglietto, S. Menard, The MEK/MAPK pathway is involved in the resistance of breast cancer cells to the EGFR tyrosine kinase inhibitor gefitinib, J. Cell. Physiol. 207 (2006) 420–427.
[167] C. Brunner-Kubath, W. Shabbir, V. Saferding, R. Wagner, C.F. Singer, P. Valent, W. Berger, B. Marian, C.C. Zielinski, M. Grusch, T.W. Grunt, The PI3 kinase/mTOR blocker NVP-BEZ235 overrides resistance against irreversible ErbB inhibitors in breast cancer cells, Breast Cancer Res. Treat. 29 (2010) 387–400.
[168] N.E. Hynes, G. MacDonald, ErbB receptors and signaling pathways in cancer, Curr. Opin. Cell Biol. 21 (2009) 177–184.
[169] G.E. Konecny, Emerging strategies for the dual inhibition of HER2-positive breast cancer, Curr. Opin. Obstet. Gynecol. 25 (2013) 55–65.
[170] S.A. Hurvitz, L. Dirix, J. Kocsis, G.V. Bianchi, J. Lu, J. Vinholes, E. Guardino, C. Song, B. Tong, V. Ng, Y.W. Chu, E.A. Perez, Phase II randomized study of trastuzumab emtansine versus trastuzumab plus docetaxel in patients with human epidermal growth factor receptor 2-positive metastatic breast cancer, J. Clin. Oncol. 31 (2013) 1157–1163.

CHAPTER SEVEN

Aptamer-Derived Peptide Inhibitors of Rho Guanine Nucleotide Exchange Factors

Susanne Schmidt[1], Anne Debant[1]

Centre de Recherche en Biochimie Macromoléculaire, CNRS-UMR 5237, Universités Montpellier I et II, 1919 Route de Mende, Montpellier, France

[1]Corresponding authors: e-mail address: susanne.schmidt@crbm.cnrs.fr; anne.debant@crbm.cnrs.fr

Contents

Abstract

Small G proteins of the Rho family and their activators the guanine nucleotide exchange factors (RhoGEFs) regulate essential cellular functions and their deregulation has been associated with an amazing variety of human disorders, including cancer, inflammation, vascular diseases, and mental retardation. Rho GTPases and RhoGEFs therefore represent important targets for inhibition, not only in basic research but also for therapeutic purposes, and strategies to inhibit their function are actively being sought. Our lab has been very active in this field and has used the peptide aptamer technology to develop the first RhoGEF inhibitor, using the RhoGEF Trio as a model. Trio function has been described mainly in cell motility and axon growth in the nervous system via Rac1 GTPase activation, but recent findings suggest it to play also a role in the aggressive phenotype of various cancers, making it an attractive target for drug discovery.

The object of this chapter is to demonstrate that targeting a RhoGEF using the peptide aptamer technology represents a valid and efficient approach to inhibit cellular processes in which Rho GTPase activity is upregulated. This is illustrated here by the first description of a peptide inhibitor of the oncogenic RhoGEF Tgat, TRIP^{E32G}, which is functional *in vivo*. On a long-term perspective, these peptide inhibitors can also serve as therapeutic tools or as guides for the discovery of small-molecule drugs, using an aptamer displacement screen.

The Enzymes, Volume 33
ISSN 1874-6047
http://dx.doi.org/10.1016/B978-0-12-416749-0.00007-5

1. INTRODUCTION

Rho GTPases constitute a subfamily of the Ras superfamily of small GTPases and comprise more than 20 members in mammals. The family of Rho GTPases regulates various cellular processes, including polarity, proliferation, differentiation, cell shape and movement in different systems, by controlling actin cytoskeleton remodeling, vesicular trafficking, gene regulation and cell cycle progression [1]. Monomeric GTPases function as molecular switches, oscillating between an inactive, GDP-bound state and an active, GTP-bound state. They are activated by guanine nucleotide exchange factors (GEFs) that stimulate the GDP-to-GTP exchange rate [2]. The RhoGEFs are proteins that belong to two families of proteins, which often show tissue- and/or developmental-specific distribution: the recently identified family of DOCK proteins [3] and the Dbl family (comprising more than 60 members in mammals). The Dbl proteins are complex proteins defined by an invariant module, which contains a catalytic DH domain (Dbl homology, referring to Dbl, the founding member of the family), followed by a PH domain (Pleckstrin homology) that targets the GEF to the plasma membrane and/or regulates guanine nucleotide exchange [4–8]. In addition to this module, they usually harbor various protein–protein interaction domains, suggesting that they may participate in several signaling networks [2]. Indeed, the GEF proteins are directly responsible for Rho GTPase activation following various extracellular stimuli, and in this respect they constitute complex signaling hubs that transduce signaling to different pathways.

Deregulation of Rho GTPase function has been associated with various human disorders, including cancer and metastasis, and also cardiovascular and hepatic diseases, bacterial and viral pathogenesis, and developmental disorders such as mental retardation and neurodegenerative diseases [9,10]. Consistently, many Dbl family RhoGEFs have been isolated based on their oncogenic potency, which often results from a truncation of the protein, leading to uncontrolled GEF activity and subsequent aberrant Rho GTPase activation [11–15]. In addition, mutations in members of the Rho GTPase signaling, including the RhoGEF βPIX, have been associated with mental retardation, probably due to an alteration of actin cytoskeleton remodeling leading to aberrant formation and function of neuronal dendrites and spines [16]. All together, these features reveal RhoGEFs as key components to temporally and spatially control Rho GTPase activities during several biological processes. RhoGEFs therefore represent promising targets for inhibition not only in terms of understanding their function but also in pathology.

In this context, the RhoGEF Trio is an excellent paradigm illustrating the complexity of the structural organization of RhoGEFs, the multiple signaling pathways in which RhoGEFs are involved, and the central role of these proteins in human disease. Trio is a complex, multidomain protein harboring two distinct GEF domains (GEF1 and GEF2), a kinase domain, two SH3 motifs, a CRAL–Trio/Sec14 motif, and several spectrin repeats [17]. The GEF1 domain activates the GTPases Rac1 and RhoG, and the GEF2 domain acts specifically on RhoA [18,19]. The RhoGEF Trio is a master gene that plays key roles in cell migration, axon outgrowth, and guidance in invertebrates and vertebrates, mainly through the activation of Rac1 by its first GEF domain [20–23]. Trio mediates the effect of different extracellular signals including NGF, netrin-1, and cadherin-11 on axon outgrowth and guidance [20,24,25]. Numerous lines of evidence suggest that Trio could also be involved in oncogenesis [26]: (i) the amplification or overexpression of the *trio* gene has been described in different types of tumors, such as breast cancer, glioblastoma, and soft tissue sarcoma [27–34]; (ii) the *trio* gene can produce many splice variants [35], including Tgat, an oncogenic variant of Trio that has been identified in ATL (adult T cell leukemia) patients and that harbors only the RhoA-activating GEF2 domain [36–38]. Of note, Tgat-induced RhoA activation is essential for the oncogenic activity of the protein, and Tgat expression induces the formation of tumors in nude mice; (iii) mutations in the *trio* gene have been identified by deep sequencing of tumors [39]; (iv) recently, a synthetic-biology approach coupled to a genome-wide RNAi screen revealed Trio as an important mediator for normal and aberrant cell growth elicited by the activation of G protein-coupled receptors [40].

Trio thus appears as an attractive target for inhibition, not only in human pathology, but also to better understand its physiological function. This chapter will present the strategy we have successfully undertaken to address the challenging issue of identifying RhoGEF inhibitors.

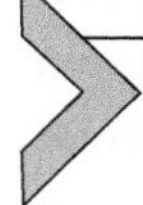

2. THE PEPTIDE APTAMER TECHNOLOGY: A BRIEF OVERVIEW

Strategies to inhibit the activation of Rho GTPases by their GEFs have been actively sought over the last decade, but so far only few successful examples have been described [41,42]. Indeed these proteins represent challenging targets for inhibition due to two main obstacles that have to be overcome: first, when trying to inhibit a specific RhoGEF, a high degree of specificity has to be achieved within such a complex family of more than 60 related proteins in mammals; and second, it is technically challenging to target efficiently complex GEF–GTPase interactions that are not yet well characterized.

We have successfully devised a peptide aptamer screening strategy to find peptide inhibitors of the RhoGEF Trio. Peptide aptamers are short peptides, generally composed of 20 random amino acids, which are conformationally constrained by a scaffold protein such as bacterial Thioredoxin A (TrxA), and which are commonly used as disrupters of protein–protein interactions [43]. The unique property of peptide aptamers is that they display a double constrained conformation in the N- and C-termini that has proven to generate binding to targets with high affinities [44]. Libraries of peptide aptamers are generally screened using the classical two-hybrid system in yeast, against the bait of interest. This technology has been widely used over the past decade to identify inhibitors of various intracellular targets, including a growing panel of oncotargets such as heat shock proteins, the human papillomavirus oncoprotein HPV16 E6 [45], cell cycle control proteins such as Cdk2 [43], and cell survival proteins, and also of antiviral and antibacterial agents [46,47]. Peptide aptamer inhibitors have also been applied to dissect and inhibit intracellular signaling pathways that regulate cellular proliferation, such as EGFR or Ras signaling [48].

The power of the strategy we used here relies on different features that are characteristic of peptide aptamers: (i) the screening of a highly combinatorial aptamer library, generating immense possibilities of random peptides; (ii) the unbiased approach in contrast to rational design of active sites, such as in the case of kinases where most of chemical inhibitors target the ATP-binding cavity; (iii) the high degree of binding specificity, which enables peptide aptamers to discriminate between closely related proteins within a functional family; (iv) the cell-based screening, which gives a direct read-out for toxicity and is more stringent; (v) the fact that peptide aptamers do not mimic cellular targets, which could have undesired effects in cells; (vi) finally, as compared to an approach by complete knockdown of the protein (such as siRNA), peptide aptamers represent an attractive alternative as they recognize selectively a single domain of a protein, thus interfering with one specific function, without affecting others. This is of great advantage when targeting complex proteins harboring numerous domains.

Using this unbiased strategy, we discovered the first RhoGEF inhibitor at the time, the peptide aptamer TRIPα, which blocks the activation of RhoA by the GEF2 domain of Trio *in vitro* and in cells [49]. More recently, we have devised an optimization screen based on the TRIPα peptide, which allowed us to discover a series of TRIP-like peptides with increased inhibitory properties toward Trio GEF2 activity, as compared to TRIPα. This is of particular importance for the use of the peptides in *in vivo* systems. Most importantly, the peptide $TRIP^{E32G}$ inhibits the GEF activity of Trio toward RhoA at a concentration lying in the low micromolar range and in a highly specific manner, and it significantly reduces the effect of Tgat, the oncogenic isoform of Trio, on tumor formation in nude mice [50].

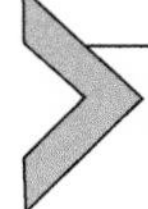

3. THE PEPTIDE APTAMER TECHNOLOGY APPLIED TO THE IDENTIFICATION OF RhoGEF INHIBITORS: EXAMPLE OF TRIPα APTAMER TARGETING Trio

In order to identify inhibitors targeting Trio, we adopted an innovative technique at the time, consisting in yeast two-hybrid screening of a peptide aptamer library to identify aptamers that bound to Trio GEF domains and were thus potential inhibitors. In this paragraph we will present the study we conducted that led to the identification of a specific peptide inhibitor of the second, RhoA-specific GEF domain of Trio [49].

The GEF2 domain of Trio, fused to the LexA DNA-binding domain, was expressed in yeast strain EGY048, which was subsequently transformed with the library of TrxA-constrained peptide aptamers [43], fused to the GAL4 transactivation domain. This library had a complexity of approximately 2×10^9, covering all possible combinations of random 20 amino acid peptides. Two million yeast transformants were screened for interaction with the bait, using appropriate selective medium, as illustrated in Fig. 7.1. Three aptamers binding to Trio GEF2, called TRIAPα, -β, and -γ (for *T*rio *i*nhibitory *ap*tamer), were selected for further study and tested for their binding specificity. Indeed, GEF domains of Dbl family GEFs are well conserved and the sequence of their catalytic DH domain is relatively similar. Binding specificity to Trio GEF2 was tested by yeast two-hybrid assay, using the DH–PH domains of the following RhoA-specific GEFs: Dbl, Vav1, Kalirin GEF2, p115 RhoGEF, and PDZ RhoGEF. This showed that the binding of the selected aptamers was indeed specific, since they bound mostly to Trio GEF2 (Fig. 7.2A). Only TRIAPα bound also, although to a lesser extent, to Kalirin GEF2, which is a paralog of Trio

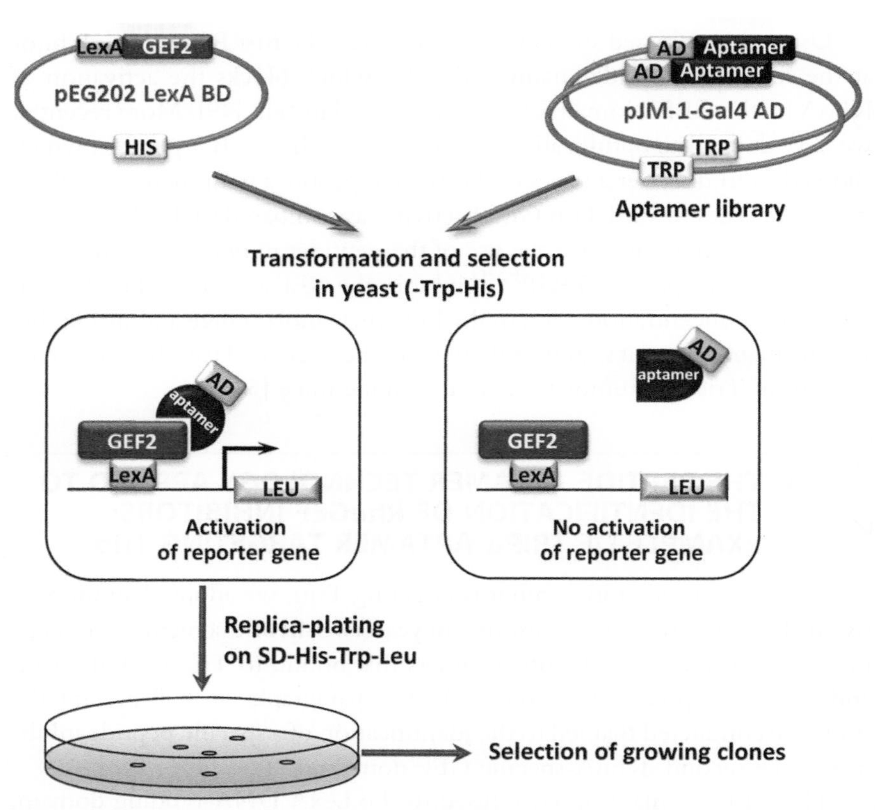

Figure 7.1 Principle of the yeast two-hybrid screening using an aptamer library as prey. The bait cDNA (Trio GEF2) fused to the LexA DNA-binding domain is inserted into the yeast pEG202 vector, bearing a histidine selection marker. The peptide aptamer library [43], fused to the GAL4 activation domain, is inserted into the pJM1 vector, carrying the tryptophan selection marker. Upon transformation into yeast strain EGY048 and plating on selective medium, only yeast clones containing both plasmids will grow. These clones will then be replica plated onto medium lacking leucine, to select for those colonies in which an aptamer/GEF2 interaction has occurred.

and shares 65% identity with Trio GEF2. TRIAPα and -γ aptamers also bound very weakly to Trio GEF1, which shares 39% identity to Trio GEF2. It should be emphasized that, at this stage, only binding of the aptamers to the GEF2 domain was assessed, which did not preclude for their inhibitory potential.

The next step consisted in testing whether binding of the aptamers to GEF2 could inhibit its catalytic activity toward RhoA. This was measured in an *in vitro* GEF assay, after production of recombinant RhoA, GEF2, and

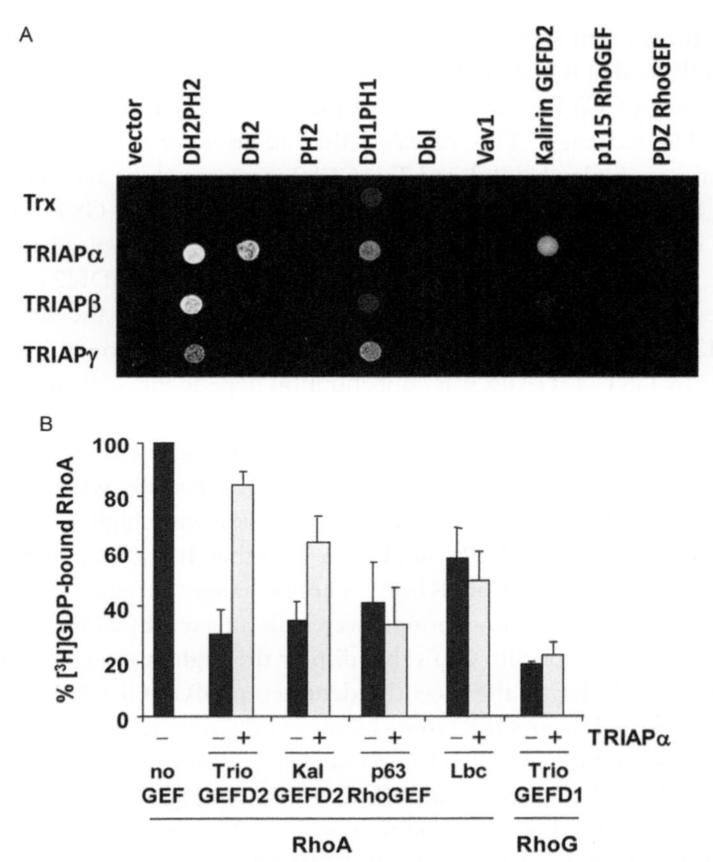

Figure 7.2 TRIAPα is a specific inhibitor of Trio GEF2. (A) Isolation of aptamers (TRIAPα, TRIAPβ, and TRIAPγ) binding to TrioGEF2. The selected aptamers binding to TrioGEF2 were tested for specific interaction against different baits. Aptamers were cotransformed in yeast strain EGY048 with either empty vector, Trio domain constructs (DH2PH2, DH2, PH2, or DH1PH1), various RhoA-specific GEFs (Dbl, Vav1, Kalirin GEF2, p115 RhoGEF, and PDZ RhoGEF), spotted on selective medium and monitored for growth. Trx represents the Thioredoxin protein with no inserted peptide. (B) To test the specificity of TRIAPα inhibition *in vitro*, exchange experiments were performed with different recombinant RhoA-specific GEFs: Trio GEF2 (0.4 μM), Kalirin GEF2 (KalGEFD2, 1.7 μM), p63RhoGEF (0.4 μM), Lbc (0.4 μM), and with the RhoG-specific GEF domain of Trio (TrioGEFD1, 0.1 μM). Different concentrations of RhoGEFs were used in order to yield similar nucleotide exchange efficiency. Recombinant GEFs were preincubated with or without a 20-fold molar excess of GST-aptamer TRIAPα, before adding 0.3 μM of [^{3}H]GDP-loaded GST-RhoA. The exchange activity was monitored by the decrease of [^{3}H]GDP-bound RhoA after 15 min. The amount of [^{3}H]GDP-bound RhoA incubated without GEF was defined as 100%. For each exchange factor, three independent experiments were performed in the absence (black bar) or the presence (gray bar) of TRIAPα.

peptide aptamers in bacteria. The GEF activity of TrioGEF2 was tested on [^{3}H]-GDP-loaded RhoA in the presence of nonlabeled GTP, by measuring the decrease of GTPase-associated radioactivity over time, due to [^{3}H]-GDP/GTP exchange. This revealed that addition of GST-TRIAPα to the reaction completely inhibited TrioGEF2 *in vitro* catalytic activity toward RhoA (Fig. 7.2B), whereas addition of GST-TRIAPβ or GST-TRIAPγ had no effect (see data in Ref. [49]), even though they bound to GEF2 in yeast. GST-TRIAPα inhibited both TrioGEF2 and TrioDH2 exchange activities [49], which is consistent with the fact that TRIAPα also recognized the DH2 catalytic domain alone in yeast (Fig. 7.2A). Moreover, inhibition of GEF2 by GST-TRIAPα was concentration dependent, with an apparent half-inhibitory concentration of around 4 μM [49].

When working with peptide inhibitors with the intent to use them *in vivo*, an essential issue to address is specificity, especially when targeting proteins that, like RhoGEFs, belong to families with high homologies. We therefore tested TRIAPα inhibition on other RhoGEFs that display *in vitro* exchange activity on RhoA. The exchange activities of Lbc and Dbl (Fig. 7.2B and data not shown) were not affected by GST-TRIAPα. Among the RhoA-specific GEFs that display the highest sequence identity with TrioGEF2, that is, the recently identified p63RhoGEF (71.5% identity) [51] and the Kalirin GEF2 (65% identity) [52], only Kalirin GEF2 activity was weakly inhibited by GST-TRIAPα. In addition, GST-TRIAPα did not block TrioGEF1 activity on RhoG or Rac1 (Fig. 7.2B and data not shown), although it slightly bound to it in yeast. Thus, the *in vitro* inhibitory effect of TRIAPα was selective for Trio GEF2.

As mentioned in Section 1, peptide aptamers are short peptides whose conformation is constrained by the presence of a scaffold protein, in our case bacterial TrxA. The variable moiety of TRIAPα consisted in 42 amino acids, which is a stretch long enough to adopt its own conformation, without the TrxA scaffold (Fig. 7.5A). To test whether such an unconstrained peptide retained the inhibitory properties of its constrained counterpart, a synthetic TRIPα (*Tr*io *i*nhibitory *p*eptide) peptide was added to the *in vitro* Trio GEF2 assay on RhoA. Fig. 7.3A shows that the TRIPα peptide had the same inhibitory properties as the TRIAPα aptamer and completely blocked the exchange activity of GEF2 toward RhoA.

The TRIPα peptide being specific and effective for Trio GEF2 inhibition *in vitro*, it was then tested for its effects in intact cells. First, binding of the peptide to GEF2 was confirmed by coimmunoprecipitation studies in COS-7 cells [49]. The degree of inhibition by TRIPα of GEF2-mediated activation of endogenous RhoA was then assessed by pull-down assay of

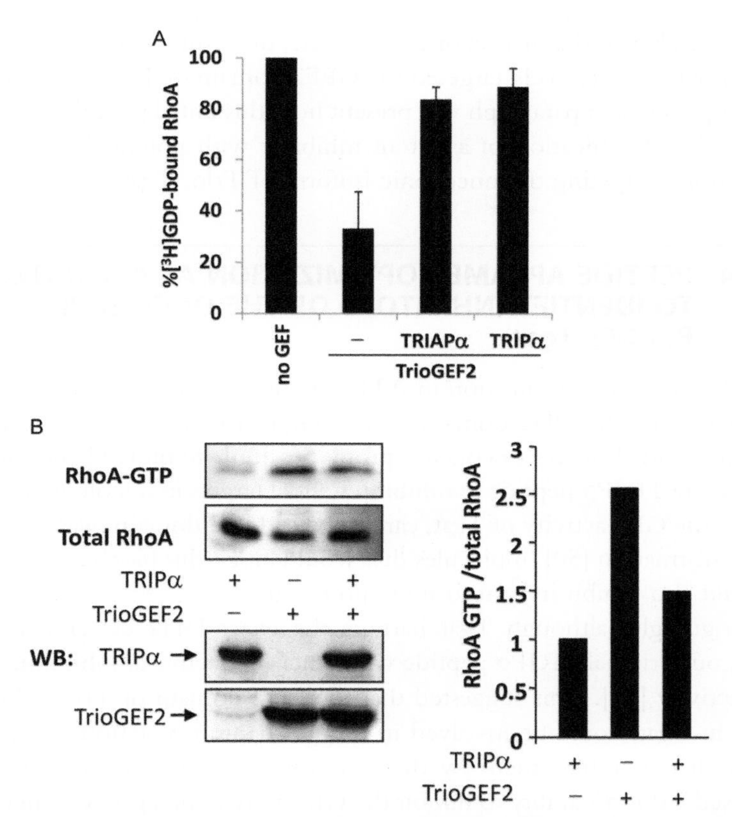

Figure 7.3 Characterization of the inhibitory properties of the TRIPα peptide *in vitro* and in cells. (A) TRIPα shows the same inhibitory properties as TRIAPα *in vitro*. A synthetic 42-amino acid peptide, TRIPα, corresponding to the variable moiety of TRIAPα, was tested in exchange assays for its ability to block Trio GEF2 activity on RhoA. GEF assays were performed as described in Fig. 7.2B. This showed that TRIPα had the same inhibitory potency as the TrxA-constrained TRIAPα. (B) Inhibitory effect of TRIPα on Trio GEF2-mediated RhoA activation in intact cells using the RhoA activity assay. COS-7 cells were transfected with TrioGEF2, GFP-TRIPα, or both, as indicated. Cell lysates were subjected to GST-pull-down using the recombinant RBD fragment of the RhoA-specific effector Rhotekin. The presence of the GTP-bound form of RhoA and of the total RhoA protein was detected using a monoclonal anti-RhoA antibody and is represented in the upper two panels. GFP-TRIPα and TrioGEF2 expression levels in the cell lysates are shown in the lower two panels. Quantification of the RhoA activity assay is shown in the histogram on the right part of the figure.

RhoA-GTP, using the RhoA-binding domain (RBD) of its effector Rhotekin fused to GST (Fig. 7.3B). This revealed that activation of endogenous RhoA by GEF2 was significantly impaired when coexpressing GFP-TRIPα, showing that this peptide was able to inhibit Trio GEF2 activity in intact cells.

This study was the first report, at the time, of an inhibitor of a RhoGEF protein, which effectively targeted the GEF2 domain of Trio *in vitro* and in cells [49]. The next paragraph will present how this initial peptide paved the way to the identification of a potent inhibitor with potential therapeutic applications, targeting the oncogenic isoform of Trio, Tgat.

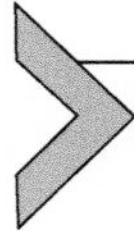

4. PEPTIDE APTAMER OPTIMIZATION AS A STRATEGY TO IDENTIFY INHIBITORS OF THE ONCOGENIC RhoGEF Tgat

The recent identification in ATL patient cells of Tgat, an isoform of Trio harboring the DH2 domain plus a unique 15 amino acid C-terminal extension and showing oncogenic properties [36], reinforced the interest in using the TRIPα peptide to inhibit potential targets in oncology. Indeed, because the GEF activity of Tgat, carried by its DH2 domain, was necessary for transformation [50], molecules that would block this biochemical activity could also inhibit its transforming potential.

Intriguingly, although Tgat harbors the exact DH2 domain of Trio GEF2, our original TRIPα peptide was rather ineffective at inhibiting Tgat GEF activity [50]. This suggested that the PH2 domain of Trio, which is absent in Tgat, could be involved in the mechanism of action of TRIPα, and/or that its replacement by the C-terminal 15 amino acid extension decreased TRIPα's ability to inhibit the GEF activity of Tgat. We therefore sought to optimize TRIPα inhibition efficiency. The rationale for this optimization strategy was that peptides that would bind stronger to Trio GEF2 domain might also be better at inhibiting its GEF activity and that of Tgat. We therefore decided to create a new library of peptide aptamers, based on the sequence of TRIPα (Fig. 7.5A), which was randomly mutated by PCR. The resulting peptides in the library harbored a mean of three point mutations per aptamer (between one and six mutations). Fig. 7.4 represents the flowchart of the optimization strategy we adopted to screen for novel peptide inhibitors targeting Tgat. Briefly, the new TRIPα-derived aptamer library we created was screened in yeast against Trio GEF2 as bait, using a two-hybrid system in which the interaction stringency could be modulated by increasing concentration of the 3-AT (3-aminotriazole) drug. We chose a concentration (80 and 120 mM) at which no interaction between GEF2 and the original TRIPα was detected anymore. Among the 35 clones selected for stronger binding to Trio GEF2, 11 were also stronger inhibitors of its catalytic activity, as assessed by the *in vitro* [^{3}H]-GDP dissociation assay

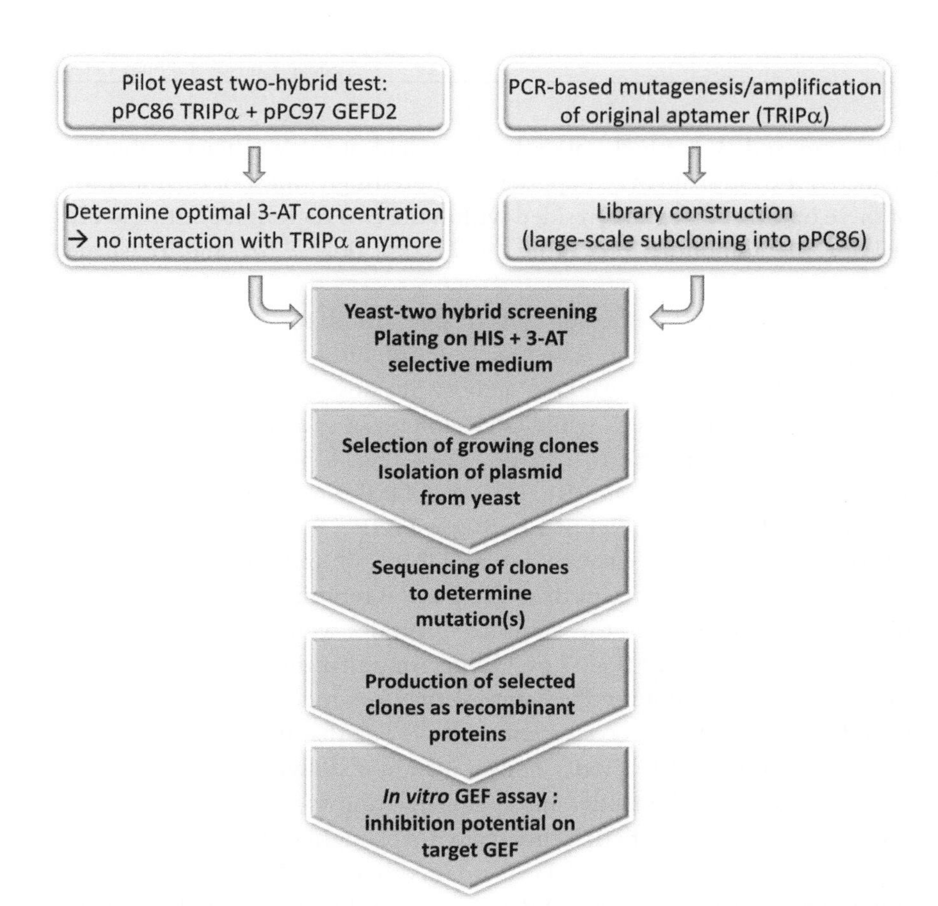

Figure 7.4 Flowchart of the peptide aptamer optimization protocol. Top right white-boxed text: Construction of a mutant TRIPα peptide aptamer library of preys (see Ref. [53] for further details). Random PCR-based mutagenesis was performed on the initial TRIPα aptamer, using PCR conditions that gave a mutation rate of about three mutations per clone. These mutated PCR fragments were then amplified and inserted into the appropriate yeast vector (pPC86), fused to the GAL4 transactivation domain and bearing a tryptophan selection marker, in order to constitute a library. The expected complexity of the library was about 5.8×10^6. Top left white-boxed text: Pilot yeast two-hybrid experiment to set up experimental conditions stringent enough to select for strong-binding aptamers. A two-hybrid system was chosen in which the threshold of interaction detection could be modulated by the concentration of the 3-aminotriazole (3-AT) drug [72]. This was done by subcloning the bait (GEF2) fused to the GAL4 DNA-binding domain (BD) into the pPC97 yeast vector, which harbors the leucine selection marker, and the original TRIPα aptamer fused to the GAL4 transactivation domain into yeast vector (pPC86), which bears the tryptophan selection marker. The histidine marker was chosen as a reporter gene, because its selection can be modulated by growing the yeast on 3-AT. This pilot yeast two-hybrid test served to determine the 3-AT concentration at which no binding to the original TRIPα was detected anymore, that is 80–120 mM 3-AT. Gray-boxed text: Sequence of experimental steps performed to obtain peptide aptamers with a good inhibition potential toward Tgat (see Ref. [53] for details).

(Fig. 7.5A and B; see Ref. [50,53]). This yield of more efficient inhibitors validated the rationale of our approach. Interestingly, most of the mutations of these optimized peptides resided within the two regions identified as crucial for the inhibitory properties of TRIPα (Fig. 7.5A and B, shaded residues).

The two most efficient TRIP-like inhibitors, $TRIP^{E32G}$ and $TRIP^{T16M/L17S}$, were then tested for their inhibitory activity toward Tgat. Both peptides inhibited Tgat GEF activity in a dose-dependent manner in a kinetics fluorescence assay, while GST alone (not shown) or GST-TRIPα, at the same concentrations, had no effect (Fig. 7.5C). Accordingly, the apparent inhibition constant (Ki_{app}) [50] of TRIPα toward Tgat was 89 μM (±33), and decreased to 7.4 μM (±5) for $TRIP^{E32G}$ and 5.1 μM (±4) for $TRIP^{T16M/L17S}$. These data show that $TRIP^{E32G}$ and $TRIP^{T16M/L17S}$ are about 15 times more efficient than TRIPα at inhibiting the exchange activity of Tgat. Interestingly, the optimized peptides were equally efficient on Tgat and on Trio DH2 (Fig. 7.5C), suggesting that the unique C-terminal extension of Tgat is not involved in the inhibitory mechanism of these optimized peptides. This was consistent with the fact that this sequence did not interfere with the GEF activity *in vitro* (data not shown).

At this stage of the study, and before moving to *in vivo* studies, it was essential to determine whether these two novel, highly efficient peptides still retained full specificity toward Tgat and no other RhoA-specific GEF. This was tested *in vitro* by mant-GTP fluorescence kinetics, and showed that $TRIP^{E32G}$ and $TRIP^{T16M/L17S}$ had no inhibitory effect on the exchange activities of the closely related RhoGEF/Rho GTPase tandems p115RhoGEF/RhoA, Lbc/RhoA, Dbl/RhoA, and not even on the very closely Trio-related p63RhoGEF/RhoA (70% identity with Tgat), or Trio GEF1/RhoG (40% identity with Tgat) (see data in Ref. [50]). Taken together, these data show that the optimized TRIP peptides were highly specific for Tgat and Trio DH2.

$TRIP^{E32G}$ and $TRIP^{T16M/L17S}$ were then tested for their inhibitory effects on Tgat in intact cells. The degree of inhibition of Tgat-mediated RhoA activation by both peptides was assessed by pull-down assays of RhoA-GTP, using the RBD of its effector Rhotekin fused to GST, as described in Fig. 7.3B. This revealed that activation of endogenous RhoA by Tgat was significantly impaired when coexpressing GFP-$TRIP^{E32G}$, showing that this peptide was able to inhibit Tgat activity in intact cells (Fig. 7.6A). Although $TRIP^{E32G}$ and $TRIP^{T16M/L17S}$ inhibited the *in vitro* GEF activity of Tgat to a similar extent, $TRIP^{T16M/L17S}$ was less efficient

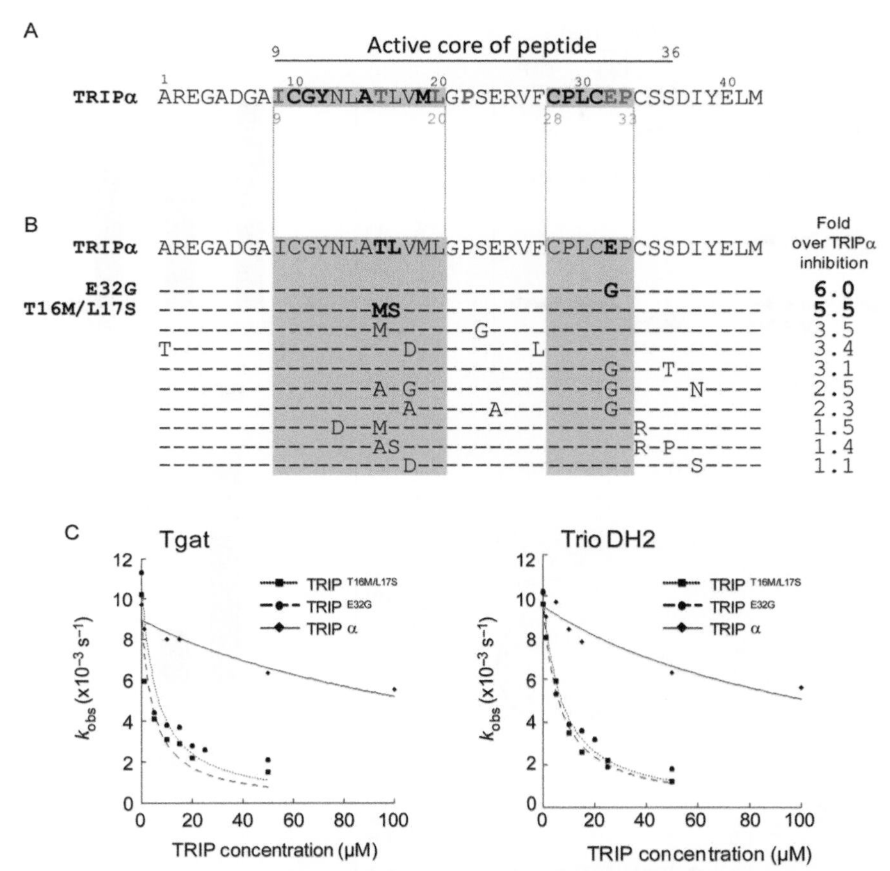

Figure 7.5 Identification and characterization of optimized TRIP-like inhibitory peptides. (A) The active core of the TRIPα peptide was determined by alanine-scanning, as described in Ref. [50]. Black bold letters indicate residues that are strictly required for inhibition, whereas gray bold letters represent residues that are needed for full inhibition. Shaded residues are regions (aa 9–20 and 28–33) that emerge as being essential for TRIPα activity. (B) Amino acid sequence of the optimized TRIP-like peptides. Inhibition efficiency of each new peptide was measured on Trio GEF2 and compared with the inhibition potential of the original TRIPα peptide. "Fold over TRIPα inhibition" means stronger inhibition at the same concentration of inhibitor (inhibition by TRIPα was set to 1). (C) Inhibition of Tgat GEF activity by $TRIP^{E32G}$ and $TRIP^{T16M/L17S}$ *in vitro*. FRET fluorescence exchange assays were performed using constant concentrations of RhoA (1 μM), equal amounts of Tgat (left panel) or Trio DH2 (right panel), and increasing concentrations of GST-TRIP peptides, up to 100 μM. Results are expressed as k_{obs} values plotted as a function of the indicated TRIP inhibitor concentration.

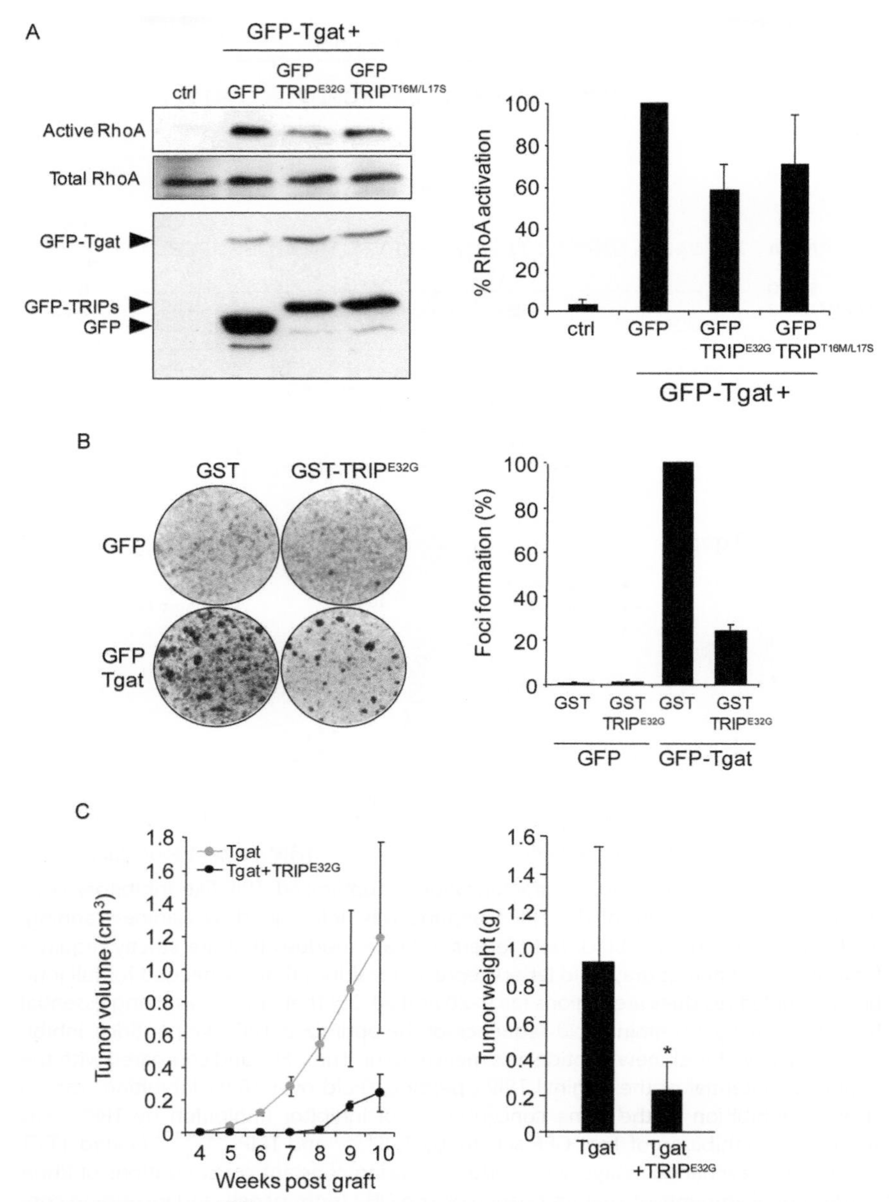

Figure 7.6 The TRIP^{E32G} peptide inhibits Tgat GEF activity and transforming potential *in vivo*. (A) TRIP^{E32G} inhibits Tgat GEF activity in cells. RhoA activation in NIH3T3-Tgat cells stably transfected with GFP, GFP-TRIP^{E32G}, or GFP-TRIP$^{T16M/L17S}$ was assayed by the GST-RBD-pull-down assay as described in Fig. 7.3B. The levels of GTP-bound and total RhoA protein are shown in the upper two panels. Expression levels of all

than TRIP^{E32G} at inhibiting Tgat-mediated activation of RhoA in cells (Fig. 7.6A) and was therefore not analyzed further in the study.

Tgat, when stably expressed in NIH3T3 cells, had a strong transformation potential and led to the formation of foci. TRIP^{E32G} was therefore coexpressed in these cells to determine whether it was able to inhibit Tgat-induced transformation. Figure 7.6B shows that the number and size of foci present in Tgat-expressing cells were severely reduced when cells coexpressed TRIP^{E32G}. This reduction was not due to an off-target, general effect of TRIP^{E32G} on cell proliferation or apoptosis (data not shown). Targeting Tgat GEF activity with TRIP^{E32G} peptide was therefore sufficient to impair Tgat-transforming activity.

Finally, since the oncogene Tgat had been shown to induce tumor formation in nude mice, we analyzed the effect of the TRIP^{E32G} peptide on this *in vivo* transforming activity. Balb/c nude mice were subcutaneously inoculated with NIH3T3 cells expressing either Tgat or Tgat and TRIP^{E32G} and tumor formation was scored. When TRIP^{E32G} was coexpressed with Tgat, tumors appeared in fewer mice and with a delay of about 3 weeks, as compared to Tgat-expressing tumors (Fig. 7.6C). In addition, tumor volume and weight were also significantly smaller. Altogether, these data show that expression of TRIP^{E32G} strongly reduced Tgat transformation activity in cells and affected tumor formation in nude mice, most likely by inhibiting Tgat-mediated GTP loading of RhoA.

This second study demonstrates that peptide aptamer optimization is a valid strategy for inhibitor identification, as it allowed us to turn the original TRIPα peptide into an efficient Tgat inhibitor, TRIP^{E32G}. This represented

GFP-tagged proteins are shown in the lower panel. Quantification of the RhoA activity assay from at least three independent experiments is represented in the histogram on the right part of the panel. Error bars represent standard deviation. (B) Focus formation assay of NIH3T3 cells, stably expressing GFP or GFP-Tgat, together with GST or GST-TRIP^{E32G}. Quantification of three independent focus formation assays is represented in the histogram on the right part of the panel. The number of foci formed by Tgat/GST expressing cells was set to 100%. Error bars represent standard deviation. (C) Tumor formation in Balb/c nude mice. NIH3T3 cells stably expressing GFP-Tgat/GST or GFP-Tgat/GST-TRIP^{E32G} were injected subcutaneously into the flanks of Balb/c nude mice and tumor volume was measured every week. The graph is representative of the three independent assays that were performed. Ten weeks post graft, mice were euthanized, tumors were excised and weighed, and the mean tumor weight was plotted on the graph. (*) A paired Student's *t*-test was performed, matching the samples for each mouse, and the *P* value was 0.019. Error bars represent standard deviation in all graphs.

the first example of a peptidic GEF inhibitor that was functional *in vivo* and could therefore serve as a lead for the discovery of new therapeutic drugs.

5. CONCLUSION—PERSPECTIVES

RhoGEFs are directly responsible for Rho GTPase activation following several external stimuli, and in this respect they represent promising targets for inhibition. Indeed, since signaling specificity appears to be mostly determined by the GEFs, RhoGEF inhibitor identification represents an emerging field of investigation. The challenge is that these proteins are not mere enzymes, with a well-defined active site that can be blocked. Rather, protein–protein interactions have to be targeted and therefore an unbiased approach represents an attractive alternative for this challenging issue. This chapter brings the proof-of-concept that peptide aptamer screening represents a valid strategy for inhibitor identification, as illustrated by the discovery of efficient inhibitors targeting the Rho GEF Trio and its oncogenic isoform Tgat. We have discovered the only—peptide—inhibitors so far that efficiently target a RhoGEF. In addition, we have shown that these inhibitors are readily amenable to optimization, by random mutagenesis combined with a selection screen based on interaction strength.

The major advantage of the peptide aptamer screening technology lies in the *in vivo* screening method and the highly combinatorial libraries available, yielding strong affinity inhibitors. The yeast exchange assay [54] is another elegant screening method that has allowed the identification of several chemical RhoGEF inhibitors. This strategy, based on the modification of the yeast two-hybrid screen, has the advantage over virtual screening of identifying molecules directly in cells, and without any bias as to the targeted interaction site [55]. Using this approach, the small-molecule ITX3 targeting Trio GEF1-mediated activation of RhoG/Rac1 has been identified [55,56], as well as the C21 compound targeting DOCK5-mediated activation of Rac1 *in vitro* and *in vivo* [57]. This last molecule is the first example of an inhibitor targeting the recently identified DOCK family of RhoGEFs. This approach will be described in chapter 8 of this volume.

More recently, other strategies have been devised to discover chemical inhibitors of monomeric Rho GTPase/RhoGEF interactions [42]. The focus has been mainly on targeting the GTPase Rac1. The NSC23766 compound was the first inhibitor of Rac1 and has been identified by computer-assisted virtual screening based on structure/function information of the Rac1/Tiam1 complex [58]. This molecule inhibits Rac1-induced cellular processes, but the targeted surface of Rac1 interacts with at least two Rac1-specific GEFs, Tiam1 and Trio GEF1, thus lacking high specificity. Based on the NSC23766 compound used as a lead, EHop-016 was then identified, which inhibits Rac1 two to three times more effectively [59]. An additional chemical Rac inhibitor, EHT1864, has been identified later, which locks the GTPases Rac1, Rac2, and Rac3 in an inert and inactive state and therefore impairs binding to their downstream effectors [60,61]. In the last year, growing interest has been given to this field of investigation as illustrated by the number of publications reporting the identification of novel chemical inhibitors targeting Rho GTPase/RhoGEF tandems. Interestingly, all these small molecules were identified by virtual screening, based on the disruption of the interaction of a specific GEF and its cognate GTPase. One compound, ZCL278, targeted the interaction between intersectin and Cdc42 [62], and two molecules were identified that inhibited RhoA interaction with RhoGEFs of the LARG subfamily, Y16 [63] and G04 [64].

Peptide aptamers have been used to validate a number of therapeutic targets belonging to a variety of protein families (reviewed in Ref. [65]). The advantage of using aptamers is that these are rather small proteins (10–20 kDa), giving rise to reduced immunogenicity, which is an important issue when developing therapeutic molecules. In addition, peptide stability can be optimized by peptide modification such as cyclization or chemical modifications, such as use of D-amino acids (instead of the natural L-aa), or of chemically modified amino acids (phosphomethyl-Phe or 6-chloro-Trp), which are less susceptible to proteolytic degradation. Finally, the exquisite specificity of peptide aptamers due to the targeting of protein–protein interaction reduces the toxicity due to off-targets inhibition. Recently, proofs-of-concept have been obtained and have shown that recombinant peptide aptamers, fused to protein transduction domains, can penetrate cultured cells, inhibit their cognate target protein, and induce the expected phenotype [66]. Recent reports have also shown that peptide aptamers could exert antitumoral activities in animals when expressed in xenografted tumors [67–69]. Developing peptide aptamers *per se* as therapeutic candidates is thus valid and innovative, and it rests on solid bases regarding intracellular delivery and activity in animal models.

The challenge of *in vivo* delivery of peptide aptamers can alternatively be circumvented by the use of an aptamer displacement screen. This represents an elegant method, in which a small-molecule library is screened for compounds that displace the peptide aptamer from its target and reproduce its inhibitory activity [70]. Aptamer displacement is a forward, dual yeast two-hybrid assay that makes use of two luciferase reporter genes (*luc* and *ruc*), whose signals can be measured sequentially in the same wells. Hence, this assay allows screening against two protein interactions at once, each interaction providing a negative control for the other. This scheme enables the elimination of most false positive hits resulting from toxicity in yeast (which produces the inhibition of both reporter genes). This technology has already been used and validated against protein–peptide aptamer interactions [71] and against protein–protein interactions, screening various chemical libraries.

In conclusion, peptide aptamers represent a promising alternative for the discovery of leads for new therapeutic drugs, especially for complex proteins that are not mere enzymes and not easily druggable.

ACKNOWLEDGMENTS

We would like to thank all past and present members of the Debant Lab who participated in the studies that led to the identification of the inhibitory peptides, especially Anne Briançon-Marjollet, Sylvie Fromont, Nathalie Bouquier, Pauline Larrousse, and Camille Auziol. We are grateful to Gudrun Aldrian and Jean Méry for peptide synthesis and to Jacqueline Cherfils, Mahel Zeghouf, and Jean-Christophe Zeeh for their help with solving the enzymatic processes.

This work was supported by grants from the CNRS, the Agence Nationale de la Recherche ANR (PCV) the Association pour la Recherche contre le Cancer, the Fondation de France, and the Cancéropôle Grand Sud-Ouest (GSO).

REFERENCES

[1] S. Etienne-Manneville, A. Hall, Rho GTPases in cell biology, Nature 420 (2002) 629–635.
[2] A. Schmidt, A. Hall, Guanine nucleotide exchange factors for Rho GTPases: turning on the switch, Genes Dev. 16 (2002) 1587–1609.
[3] J.F. Cote, K. Vuori, Identification of an evolutionarily conserved superfamily of DOCK180-related proteins with guanine nucleotide exchange activity, J. Cell Sci. 115 (2002) 4901–4913.
[4] M.K. Chhatriwala, L. Betts, D.K. Worthylake, J. Sondek, The DH and PH domains of Trio coordinately engage Rho GTPases for their efficient activation, J. Mol. Biol. 368 (2007) 1307–1320.
[5] S. Lutz, A. Shankaranarayanan, C. Coco, M. Ridilla, M.R. Nance, C. Vettel, D. Baltus, C.R. Evelyn, R.R. Neubig, T. Wieland, J.J. Tesmer, Structure of Galphaq-p63RhoGEF-RhoA complex reveals a pathway for the activation of RhoA by GPCRs, Science 318 (2007) 1923–1927.
[6] R.J. Rojas, M.E. Yohe, S. Gershburg, T. Kawano, T. Kozasa, J. Sondek, G-alpha-q directly activates p63RhoGEF and Trio via a conserved extension of the DH-associated PH domain, J. Biol. Chem. 282 (2007) 29201–29210.

[7] K.L. Rossman, L. Cheng, G.M. Mahon, R.J. Rojas, J.T. Snyder, I.P. Whitehead, J. Sondek, Multifunctional roles for the PH domain of Dbs in regulating Rho GTPase activation, J. Biol. Chem. 278 (2003) 18393–18400.
[8] K.L. Rossman, C.J. Der, J. Sondek, GEF means go: turning on RHO GTPases with guanine nucleotide-exchange factors, Nat. Rev. Mol. Cell Biol. 6 (2005) 167–180.
[9] E. Sahai, C.J. Marshall, RHO-GTPases and cancer, Nat. Rev. Cancer 2 (2002) 133–142.
[10] D. Toksoz, K.D. Merdek, The Rho small GTPase: functions in health and disease, Histol. Histopathol. 17 (2002) 915–927.
[11] A. Eva, S.A. Aaronson, Isolation of a new human oncogene from a diffuse B-cell lymphoma, Nature 316 (1985) 273–275.
[12] S. Katzav, D. Martin-Zanca, M. Barbacid, Vav, a novel human oncogene derived from a locus ubiquitously expressed in hematopoietic cells, EMBO J. 8 (1989) 2283–2290.
[13] T. Miki, C.L. Smith, J.E. Long, A. Eva, T.P. Fleming, Oncogene ect2 is related to regulators of small GTP-binding proteins, Nature 362 (1993) 462–465.
[14] I. Whitehead, H. Kirk, C. Tognon, G. Trigo-Gonzalez, R. Kay, Expression cloning of lfc, a novel oncogene with structural similarities to guanine nucleotide exchange factors and to the regulatory region of protein kinase C, J. Biol. Chem. 270 (1995) 18388–18395.
[15] I.P. Whitehead, R. Khosravi-Far, H. Kirk, G. Trigo-Gonzalez, C.J. Der, R. Kay, Expression cloning of lsc, a novel oncogene with structural similarities to the Dbl family of guanine nucleotide exchange factors, J. Biol. Chem. 271 (1996) 18643–18650.
[16] O. von Bohlen Und Halbach, Dendritic spine abnormalities in mental retardation, Cell Tissue Res. 342 (2010) 317–323.
[17] A. Debant, C. Serra-Pages, K. Seipel, S. O'Brien, M. Tang, S.H. Park, M. Streuli, The multidomain protein Trio binds the LAR transmembrane tyrosine phosphatase, contains a protein kinase domain, and has separate rac-specific and rho-specific guanine nucleotide exchange factor domains, Proc. Natl. Acad. Sci. U.S.A. 93 (1996) 5466–5471.
[18] J.M. Bellanger, J.B. Lazaro, S. Diriong, A. Fernandez, N. Lamb, A. Debant, The two guanine nucleotide exchange factor domains of Trio link the Rac1 and the RhoA pathways in vivo, Oncogene 16 (1998) 147–152.
[19] A. Blangy, E. Vignal, S. Schmidt, A. Debant, C. Gauthier-Rouviere, P. Fort, TrioGEF1 controls Rac- and Cdc42-dependent cell structures through the direct activation of rhoG, J. Cell Sci. 113 (2000) 729–739.
[20] A. Briancon-Marjollet, A. Ghogha, H. Nawabi, I. Triki, C. Auziol, S. Fromont, C. Piche, H. Enslen, K. Chebli, J.F. Cloutier, V. Castellani, A. Debant, N. Lamarche-Vane, Trio mediates netrin-1-induced Rac1 activation in axon outgrowth and guidance, Mol. Cell. Biol. 28 (2008) 2314–2323.
[21] J. DeGeer, J. Boudeau, S. Schmidt, F. Bedford, N. Lamarche-Vane, A. Debant, Tyrosine phosphorylation of the Rho guanine nucleotide exchange factor Trio regulates netrin-1/DCC-mediated cortical axon outgrowth, Mol. Cell. Biol. 33 (2013) 739–751.
[22] T.P. Newsome, S. Schmidt, G. Dietzl, K. Keleman, B. Asling, A. Debant, B.J. Dickson, Trio combines with dock to regulate Pak activity during photoreceptor axon pathfinding in Drosophila, Cell 101 (2000) 283–294.
[23] R. Steven, T.J. Kubiseski, H. Zheng, S. Kulkarni, J. Mancillas, A. Ruiz Morales, C.W. Hogue, T. Pawson, J. Culotti, UNC-73 activates the Rac GTPase and is required for cell and growth cone migrations in *C. elegans*, Cell 92 (1998) 785–795.
[24] S. Backer, M. Hidalgo-Sanchez, N. Offner, E. Portales-Casamar, A. Debant, P. Fort, C. Gauthier-Rouviere, E. Bloch-Gallego, Trio controls the mature organization of neuronal clusters in the hindbrain, J. Neurosci. 27 (2007) 10323–10332.

[25] S. Estrach, S. Schmidt, S. Diriong, A. Penna, A. Blangy, P. Fort, A. Debant, The human Rho-GEF trio and its target GTPase RhoG are involved in the NGF pathway, leading to neurite outgrowth, Curr. Biol. 12 (2002) 307–312.
[26] S. Schmidt, A. Debant, (2010). TRIO (triple functional domain (PTPRF interacting). Atlas Gene Cytogenet Oncol Haematol. http://AtlasGeneticsOncology.org/Genes/TRIOD43542ch5p15.html.
[27] M. Adamowicz, B. Radlwimmer, R.J. Rieker, D. Mertens, M. Schwarzbach, P. Schraml, A. Benner, P. Lichter, G. Mechtersheimer, S. Joos, Frequent amplifications and abundant expression of TRIO, NKD2, and IRX2 in soft tissue sarcomas, Genes Chromosomes Cancer 45 (2006) 829–838.
[28] C. Baldwin, C. Garnis, L. Zhang, M.P. Rosin, W.L. Lam, Multiple microalterations detected at high frequency in oral cancer, Cancer Res. 65 (2005) 7561–7567.
[29] I. Chattopadhyay, A. Singh, R. Phukan, J. Purkayastha, A. Kataki, J. Mahanta, S. Saxena, S. Kapur, Genome-wide analysis of chromosomal alterations in patients with esophageal squamous cell carcinoma exposed to tobacco and betel quid from high-risk area in India, Mutat. Res. 696 (2010) 130–138.
[30] C. Garnis, J.J. Davies, T.P. Buys, M.S. Tsao, C. MacAulay, S. Lam, W.L. Lam, Chromosome 5p aberrations are early events in lung cancer: implication of glial cell line-derived neurotrophic factor in disease progression, Oncogene 24 (2005) 4806–4812.
[31] B. Salhia, N.L. Tran, A. Chan, A. Wolf, M. Nakada, F. Rutka, M. Ennis, W.S. McDonough, M.E. Berens, M. Symons, J.T. Rutka, The guanine nucleotide exchange factors trio, Ect2, and Vav3 mediate the invasive behavior of glioblastoma, Am. J. Pathol. 173 (2008) 1828–1838.
[32] G.M. Calaf, D. Roy, Gene expression signature of parathion-transformed human breast epithelial cells, Int. J. Mol. Med. 19 (2007) 741–750.
[33] J. Lane, T.A. Martin, R.E. Mansel, W.G. Jiang, The expression and prognostic value of the guanine nucleotide exchange factors (GEFs) Trio, Vav1 and TIAM-1 in human breast cancer, Int Semin Surg Oncol 5 (2008) 23.
[34] B.P. Coe, L.J. Henderson, C. Garnis, M.S. Tsao, A.F. Gazdar, J. Minna, S. Lam, C. Macaulay, W.L. Lam, High-resolution chromosome arm 5p array CGH analysis of small cell lung carcinoma cell lines, Genes Chromosomes Cancer 42 (2005) 308–313.
[35] E. Portales-Casamar, A. Briancon-Marjollet, S. Fromont, R. Triboulet, A. Debant, Identification of novel neuronal isoforms of the Rho-GEF Trio, Biol. Cell 98 (2006) 183–193.
[36] N. Yoshizuka, R. Moriuchi, T. Mori, K. Yamada, S. Hasegawa, T. Maeda, T. Shimada, Y. Yamada, S. Kamihira, M. Tomonaga, S. Katamine, An alternative transcript derived from the trio locus encodes a guanosine nucleotide exchange factor with mouse cell-transforming potential, J. Biol. Chem. 279 (2004) 43998–44004.
[37] T. Mori, R. Moriuchi, E. Okazaki, K. Yamada, S. Katamine, Tgat oncoprotein functions as a inhibitor of RECK by association of the unique C-terminal region, Biochem. Biophys. Res. Commun. 355 (2007) 937–943.
[38] K. Yamada, R. Moriuchi, T. Mori, E. Okazaki, T. Kohno, T. Nagayasu, T. Matsuyama, S. Katamine, Tgat, a Rho-specific guanine nucleotide exchange factor, activates NF-kappaB via physical association with IkappaB kinase complexes, Biochem. Biophys. Res. Commun. 355 (2007) 269–274.
[39] C. Greenman, P. Stephens, R. Smith, G.L. Dalgliesh, C. Hunter, G. Bignell, H. Davies, J. Teague, A. Butler, C. Stevens, S. Edkins, S. O'Meara, I. Vastrik, E.E. Schmidt, T. Avis, S. Barthorpe, G. Bhamra, G. Buck, B. Choudhury, J. Clements, J. Cole, E. Dicks, S. Forbes, K. Gray, K. Halliday, R. Harrison, K. Hills, J. Hinton, A. Jenkinson, D. Jones, A. Menzies, T. Mironenko, J. Perry, K. Raine, D. Richardson, R. Shepherd, A. Small, C. Tofts, J. Varian, et al., Patterns of somatic mutation in human cancer genomes, Nature 446 (2007) 153–158.

[40] J.P. Vaqué, R.T. Dorsam, X. Feng, R. Iglesias-Bartolome, D.J. Forsthoefel, Q. Chen, A. Debant, M.A. Seeger, B.R. Ksander, H. Teramoto, J.S. Gutkind, A genome-wide RNAi screen reveals a Trio-egulated Rho GTPase circuitry transducing mitogenic signals initiated by G protein-coupled receptors, Mol. Cell 49 (2013) 1–15.
[41] J.L. Bos, H. Rehmann, A. Wittinghofer, GEFs and GAPs: critical elements in the control of small G proteins, Cell 129 (2007) 865–877.
[42] D. Vigil, J. Cherfils, K.L. Rossman, C.J. Der, Ras superfamily GEFs and GAPs: validated and tractable targets for cancer therapy? Nat. Rev. Cancer 10 (2010) 842–857.
[43] P. Colas, B. Cohen, T. Jessen, I. Grishina, J. McCoy, R. Brent, Genetic selection of peptide aptamers that recognize and inhibit cyclin-dependent kinase 2, Nature 380 (1996) 548–550.
[44] R.C. Ladner, Constrained peptides as binding entities, Trends Biotechnol. 13 (1995) 426–430.
[45] K. Butz, C. Denk, A. Ullmann, M. Scheffner, F. Hoppe-Seyler, Induction of apoptosis in human papillomaviruspositive cancer cells by peptide aptamers targeting the viral E6 oncoprotein, Proc. Natl. Acad. Sci. U.S.A. 97 (2000) 6693–6697.
[46] M. Crawford, R. Woodman, P. Ko Ferrigno, Peptide aptamers: tools for biology and drug discovery, Brief. Funct. Genomic. Proteomic. 2 (2003) 72–79.
[47] F. Hoppe-Seyler, I. Crnkovic-Mertens, E. Tomai, K. Butz, Peptide aptamers: specific inhibitors of protein function, Curr. Mol. Med. 4 (2004) 529–538.
[48] C. Borghouts, C. Kunz, B. Groner, Peptide aptamers: recent developments for cancer therapy, Expert Opin. Biol. Ther. 5 (2005) 783–797.
[49] S. Schmidt, S. Diriong, J. Mery, E. Fabbrizio, A. Debant, Identification of the first Rho-GEF inhibitor, TRIPalpha, which targets the RhoA-specific GEF domain of Trio, FEBS Lett. 523 (2002) 35–42.
[50] N. Bouquier, S. Fromont, J.C. Zeeh, C. Auziol, P. Larrousse, B. Robert, M. Zeghouf, J. Cherfils, A. Debant, S. Schmidt, Aptamer-derived peptides as potent inhibitors of the oncogenic RhoGEF Tgat, Chem. Biol. 16 (2009) 391–400.
[51] M. Souchet, E. Portales-Casamar, D. Mazurais, S. Schmidt, I. Leger, J.L. Javre, P. Robert, I. Berrebi-Bertrand, A. Bril, B. Gout, A. Debant, T.P. Calmels, Human p63RhoGEF, a novel RhoA-specific guanine nucleotide exchange factor, is localized in cardiac sarcomere, J. Cell Sci. 115 (2002) 629–640.
[52] C.A. Rabiner, R.E. Mains, B.A. Eipper, Kalirin: a dual Rho guanine nucleotide exchange factor that is so much more than the sum of its many parts, Neuroscientist 11 (2005) 148–160.
[53] N. Bouquier, S. Fromont, A. Debant, S. Schmidt, Random mutagenesis of peptide aptamers as an optimization strategy for inhibitor screening, Methods Mol. Biol. 928 (2012) 97–118.
[54] M. De Toledo, K. Colombo, T. Nagase, O. Ohara, P. Fort, A. Blangy, The yeast exchange assay, a new complementary method to screen for Dbl-like protein specificity: identification of a novel RhoA exchange factor, FEBS Lett. 480 (2000) 287–292.
[55] A. Blangy, N. Bouquier, C. Gauthier-Rouviere, S. Schmidt, A. Debant, J.P. Leonetti, P. Fort, Identification of TRIO-GEFD1 chemical inhibitors using the yeast exchange assay, Biol. Cell 98 (2006) 511–522.
[56] N. Bouquier, E. Vignal, S. Charrasse, M. Weill, S. Schmidt, J.P. Leonetti, A. Blangy, P. Fort, A cell active chemical GEF inhibitor selectively targets the Trio/RhoG/Rac1 signaling pathway, Chem. Biol. 16 (2009) 657–666.
[57] V. Vives, M. Laurin, G. Cres, P. Larrousse, Z. Morichaud, D. Noel, J.F. Cote, A. Blangy, The Rac1 exchange factor Dock5 is essential for bone resorption by osteoclasts, J. Bone Miner. Res. 26 (2011) 1099–1110.

[58] Y. Gao, J.B. Dickerson, F. Guo, J. Zheng, Y. Zheng, Rational design and characterization of a Rac GTPase-specific small molecule inhibitor, Proc. Natl. Acad. Sci. U.S.A. 101 (2004) 7618–7623.
[59] B.L. Montalvo-Ortiz, L. Castillo-Pichardo, E. Hernandez, T. Humphries-Bickley, A. De la Mota-Peynado, L.A. Cubano, C.P. Vlaar, S. Dharmawardhane, Characterization of EHop-016, novel small molecule inhibitor of Rac GTPase, J. Biol. Chem. 287 (2012) 13228–13238.
[60] L. Desire, J. Bourdin, N. Loiseau, H. Peillon, V. Picard, C. De Oliveira, F. Bachelot, B. Leblond, T. Taverne, E. Beausoleil, S. Lacombe, D. Drouin, F. Schweighoffer, RAC1 inhibition targets amyloid precursor protein processing by gamma-secretase and decreases Abeta production in vitro and in vivo, J. Biol. Chem. 280 (2005) 37516–37525.
[61] C. Onesto, A. Shutes, V. Picard, F. Schweighoffer, C.J. Der, Characterization of EHT 1864, a novel small molecule inhibitor of Rac family small GTPases, Methods Enzymol. 439 (2008) 111–129.
[62] A. Friesland, Y. Zhao, Y.H. Chen, L. Wang, H. Zhou, Q. Lu, Small molecule targeting Cdc42–intersectin interaction disrupts Golgi organization and suppresses cell motility, Proc. Natl. Acad. Sci. U.S.A. 110 (2013) 1261–1266.
[63] X. Shang, F. Marchioni, C.R. Evelyn, N. Sipes, X. Zhou, W. Seibel, M. Wortman, Y. Zheng, Small-molecule inhibitors targeting G-protein-coupled Rho guanine nucleotide exchange factors, Proc. Natl. Acad. Sci. U.S.A. 110 (2013) 3155–3160.
[64] X. Shang, F. Marchioni, N. Sipes, C.R. Evelyn, M. Jerabek-Willemsen, S. Duhr, W. Seibel, M. Wortman, Y. Zheng, Rational design of small molecule inhibitors targeting RhoA subfamily Rho GTPases, Chem. Biol. 19 (2012) 699–710.
[65] P. Colas, The eleven-year switch of peptide aptamers, J. Biol. 7 (2008) 2.
[66] C. Borghouts, C. Kunz, N. Delis, B. Groner, Monomeric recombinant peptide aptamers are required for efficient intracellular uptake and target inhibition, Mol. Cancer Res. 6 (2008) 267–281.
[67] A. Appert, C.H. Nam, N. Lobato, E. Priego, R.N. Miguel, T. Blundell, L. Drynan, H. Sewell, T. Tanaka, T. Rabbitts, Targeting LMO2 with a peptide aptamer establishes a necessary function in overt T-cell neoplasia, Cancer Res. 69 (2009) 4784–4790.
[68] B. Gibert, E. Hadchity, A. Czekalla, M.T. Aloy, P. Colas, C. Rodriguez-Lafrasse, A.P. Arrigo, C. Diaz-Latoud, Inhibition of heat shock protein 27 (HspB1) tumorigenic functions by peptide aptamers, Oncogene 30 (2011) 3672–3681.
[69] A.L. Rerole, J. Gobbo, A. De Thonel, E. Schmitt, J.P. Pais de Barros, A. Hammann, D. Lanneau, E. Fourmaux, O. Deminov, O. Micheau, L. Lagrost, P. Colas, G. Kroemer, C. Garrido, Peptides and aptamers targeting HSP70: a novel approach for anticancer chemotherapy, Cancer Res. 71 (2011) 484–495.
[70] I.C. Baines, P. Colas, Peptide aptamers as guides for small-molecule drug discovery, Drug Discov. Today 11 (2006) 334–341.
[71] C. Bardou, C. Borie, M. Bickle, B.B. Rudkin, P. Colas, Peptide aptamers for small molecule drug discovery, Methods Mol. Biol. 535 (2009) 373–388.
[72] C. Sardet, M. Vidal, D. Cobrinik, Y. Geng, C. Onufryk, A. Chen, R.A. Weinberg, E2F-4 and E2F-5, two members of the E2F family, are expressed in the early phases of the cell cycle, Proc. Natl. Acad. Sci. U.S.A. 92 (1995) 2403–2407.

CHAPTER EIGHT

Targeting the Dbl and Dock-Family RhoGEFs: A Yeast-Based Assay to Identify Cell-Active Inhibitors of Rho-Controlled Pathways

Anne Blangy[*,†,1], **Philippe Fort**[*,†]
[*]Montpellier University, Montpellier, France
[†]Centre de Recherche de Biochimie Macromoléculaire, CNRS, UMR5237, Montpellier, France
[1]Corresponding author: e-mail address: anne.blangy@crbm.cnrs.fr

Contents

Abstract

The Ras-like superfamily of low molecular weight GTPases is made of five major families (Arf/Sar, Rab, Ran, Ras, and Rho), highly conserved across evolution. This is in keeping with their roles in basic cellular functions (endo/exocytosis, vesicular trafficking, nucleocytoplasmic trafficking, cell signaling, proliferation and apoptosis, gene regulation, F-actin dynamics), whose alterations are associated with various types of diseases, in particular cancer, neurodegenerative, cardiovascular, and infectious diseases. For these reasons, Ras-like pathways are of great potential in therapeutics and identifying inhibitors that decrease signaling activity is under intense research.

Along this line, guanine exchange factors (GEFs) represent attractive targets. GEFs are proteins that promote the active GTP-bound state of GTPases and represent the major entry points whereby extracellular cues are converted into Ras-like signaling.

The Enzymes, Volume 33
ISSN 1874-6047
http://dx.doi.org/10.1016/B978-0-12-416749-0.00008-7

We previously developed the yeast exchange assay (YEA), an experimental setup in the yeast in which activity of a mammalian GEF can be monitored by auxotrophy and color reporter genes. This assay was further engineered for medium-throughput screening of GEF inhibitors, which can readily select for cell-active and specific compounds. We report here on the successful identification of inhibitors against Dbl and CZH/DOCK-family members, GEFs for Rho GTPases, and on the experimental setup to screen for inhibitors of GEFs of the Arf family.

We also discuss on inhibitors developed using virtual screening (VS), which target the GEF/GTPase interface with high efficacy and specificity. We propose that using VS and YEA in combination may represent a method of choice for identifying specific and cell-active GEF inhibitors.

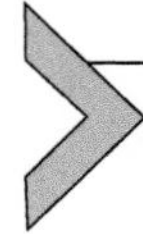

1. INTRODUCTION

1.1. The Ras-like superfamily in metazoans

Proteins that bind to and hydrolyze nucleoside triphosphates are implicated in all aspects of life and have been classified depending on the bound nucleoside (ATP or GTP) and on the structural fold conferring the binding [1]. Among these, the Ras-like GTPase superfamily forms an offshoot of the TRAFAC (translation factor-related) GTPase class in the P-loop GTPase superclass and showed a spectacular expansion in eukaryotes, in particular in metazoans [2].

The Ras-like superfamily is made of five major families (Arf/Sar, Rab, Ran, Ras, and Rho) highly conserved across evolution—ortholog proteins generally sharing 65–85% amino acid sequence similarity, even between distantly related groups such as yeast and mammals—indicative of heavy selective constraints [3]. This is in keeping with their roles in basic cellular functions (endo/exocytosis, vesicular trafficking, nucleocytoplasmic trafficking, cell signaling, proliferation and apoptosis, gene regulation, F-actin dynamics). Since all physiological processes—from fertilization and embryonic development to tissue homeostasis of adult organisms—require the coordinate activity of these basic functions, it was not surprising that alterations in pathways controlled by Ras-like GTPases were associated with developmental defects [4–6] and in many types of diseases, in particular cancer [7–11], neurodegenerative [12–15], cardiovascular [16,17], and infectious [18] diseases.

1.2. Ras-like regulatory modules as therapeutic targets

For these reasons, Ras-like pathways were identified as being of great potential in therapeutics [19]. How can these pathways be targeted? Ras-like GTPases act as binary signaling switches that rely on structural changes between their GDP-bound and GTP-bound conformations—the inactive and active states, respectively—which allow differential binding to a set of downstream "effector" proteins that mediate the cellular effects [20,21]. The canonical regulatory module is submitted to successive activation and inactivation steps, promoted, respectively, by guanine nucleotide exchange factors (GEFs)—many of which are associated with and activated by plasma membrane receptors—and by GTPase activating proteins (GAPs), which stimulate intrinsic GTPase activity. An additional mechanism of inactivation was recently evidenced that relies on proteasome-dependent degradation of the GTP-bound form [22–24].

Most of pathological Ras-like alterations are associated with over-activation of pathways as a consequence either of somatic mutations or over-expression of GEFs or GTPases, eventually enhancing intensity of the signaling. For instance, Ras proteins are found constitutively activated at high frequency in many tumor types by somatic mutations that prevent GTP hydrolysis [25]. In contrast, increased activity of Rho pathways observed in tumors is mostly associated with overexpression of GTPases or RhoGEFs [26], although fast-cycling somatic mutations affecting Rac1 have been identified recently in 8% of UV-light-induced melanoma tumor samples [27,28] and in a few tumor cell lines [29]. Theoretically, drugs may thus either inhibit GEFs or GTPases or act as agonists on GAPs, although this latter step is far more difficult to develop. Choice of the target and screening strategies should be tailored for each Ras-like family, taking into account the diversity of components, how they crosstalk with each other, and which level of specificity is requested. The human genome indeed encodes more than 430 components of Ras-like regulatory modules, and for all families but Ran, which has a single member in each category, components range from 13–15 for Arf/Arl GTPases to 80 for RhoGEFs (Table 8.1).

1.3. Targeting RhoGEFs to inhibit Rho pathways

Over the past two decades, signaling pathways controlled by Rho GTPases emerged as key regulators of eukaryotic basic cell dynamics (cell–cell and cell–ECM adhesion, polarity, migration, contraction), pivotal for normal and pathological processes, from cell migration and differentiation in the developing embryo to common diseases, such as hypertension, cancer, or neurodegenerative diseases [30,31]. Mammalian genomes contain genes for 20 Rho GTPases [32] and 82 RhoGEFs (71 Dbl-related [33] and 11 CZH/DOCK-related proteins [34]). Most RhoGEFs are high molecular

Table 8.1 Complexity of Ras-like regulatory modules in the human genome

	GAP	GEF	GTPase
Ras	24 [2]	37 [2]	38 [2]
Rab	>40 TBC [93]	17 DENN [94] 7 Vps9 [95] 3 Trapp complexes [96] 1 Ric1–Rgp1 [97]	66 [98]
Arf/Arl	3 ELMOD [99] 10 ArfGAP [100]	15 Sec7 [101]	22 [102]
Ran	1 [103]	1 [103]	1 [103]
Rho	59 [104]	71 Dbl-like [33] 11 DOCK [34,105]	22 [32]

weight multidomain proteins; in addition to the catalytic domain that promotes nucleotide exchange (DH for the Dbl family or DHR2 for the DOCK family), other domains mediate association with membrane receptor complexes (e.g., the spectrin, SH3, FYVE, or PH domains, involved in protein or phospholipid interaction) or other enzymatic activities (e.g., kinases, phosphatases, or GEF or GAP domains for other Ras-like families).

Rho pathways can be targeted at various levels to be inhibited: GEFs, GTPases, and downstream effectors. Drugs have been developed already against downstream effectors, such as the ROCK or PAK kinases, inhibited by Y-27632 [35] and IPA3 [36], respectively. However, these proteins have both Rho-dependent and Rho-independent activities [37,38] and in the former case, they can be under the control of multiple Rho members. Consequently, inhibiting at the level of effectors may be of interest when the effector is directly involved in the phenotype—for example, ROCK, whose inhibitor fasudil used against hypertension [39]—but may have too broad an inhibition spectrum when a specific Rho-signaling pathway is targeted.

RhoGEFs and GTPases may both represent potent therapeutic targets. However, a difference between the two protein types is that GTPases can be activated by multiple RhoGEFs whereas most RhoGEFs activate a single GTPase, as illustrated in Fig. 8.1 for 69 Dbl-like proteins of known specificity, compiled from Ref. [33,40]. Furthermore, numerous RhoGEFs get activated upon membrane receptor stimulation and therefore represent major cellular entry points whereby extracellular cues are converted into Rho signaling. Last, although several Rho GTPases exchange their nucleotide spontaneously [40], activity of the most widely expressed members (RhoA, Rac1, and Cdc42) is

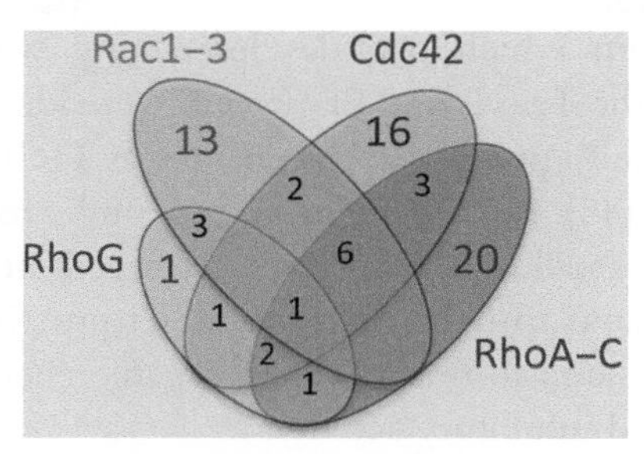

Figure 8.1 Most of RhoGEFs activate a single Rho GTPase. Specificity of 69 RhoGEFs of the Dbl family toward Rac1–3, RhoG, Cdc42, and RhoA–C. *Compiled from Refs. [33,40].* (See color plate.)

positively controlled by RhoGEFs and are overactivated in many cancer types [10]. For these reasons, RhoGEFs emerge as more attractive targets to optimize efficacy and specificity of inhibition [31,41]. However, the interface between the GTPase and the GEF might be even more attractive. Indeed, the exchange reaction involves successive discrete steps [42,43]: first, a low affinity complex is formed between the RhoGEF and the GDP-bound GTPase, which results in the opening of the nucleotide pocket of the GTPase and promotes the loss of the nucleotide; once the nucleotide is lost, the complex thus formed turns into high affinity and maintains the GTPase in an open conformation ready to bind any nucleotide available, GTP being the most abundant one in the cell cytoplasm; once loaded with GTP, the GTPase–GEF complex goes back to a low affinity state and dissociates. The GEF–GTPase complex therefore represents a transient protein–protein intermediate of a regulatory reaction, which may be trapped by drugs in a nonproductive conformation. Inhibiting this complex may be highly specific if the inhibitor interacts with both partners of the complex. A natural example is Brefeldin A, which binds both to Arf1 and to its GEFs and inhibits Arf1 activation by aborting the low- to high-affinity complex transition [44]. Since RhoGEFs outnumber their potential targets in various cellular networks, targeting RhoGEFs/GTPases complexes might be thus more powerful than targeting RhoGEFs on their own, combining efficacy—it targets the very fraction involved in the exchange reaction—and specificity—only a particular GEF–GTPase association may be hit.

1.4. Which types of inhibitors and which screening strategies?

Several Rho pathway inhibitors have been identified already, most of them by screening small molecule libraries. Only a single screen used a library of peptide aptamers from which peptides interacting with the C-terminal exchange domain of the Trio RhoGEF protein were identified. A secondary activity screen identified a unique peptide, named as TRIPalpha, as inhibiting the exchange activity [45]. However, although peptides remain a valid option to identify inhibitors, small molecules probably represent the best druggable type of inhibitors, thanks to their ability to target protein pockets and protein–protein interfaces and to the power of organic chemistry to synthesize analogs with improved inhibitory activity.

Whatever the type of library used, the main issue remains the experimental setup to identify RhoGEF or RhoGEF/GTPase complex inhibitors. The most straightforward approach is to adapt *in vitro* assays for RhoGEF catalytic activity to a suitable screening format. Thanks to fluorescent guanine derivatives, guanine nucleotide exchange assays using purified proteins or domains now allow fast, sensitive, and accurate biochemical analysis of the exchange reaction [46,47]. Such screening strategy was used successfully to select chemical compounds that inhibit *in vitro* the catalytic activity of the LARG RhoGEF toward RhoA [48]. However, such an approach has two major drawbacks: although active RhoA, Rac1, and Cdc42 can be easily purified, it may turn difficult to prepare full-length RhoGEFs or even DH/PH domains because of insolubility and nonproper folding [49]. Furthermore, potent inhibitors selected out of a biochemical activity assay may either not enter living cells thus be inactive or inversely may have nonspecific toxicity.

Another strategy for screening RhoGEF inhibitors may be cell-based assays and microscopy analysis to detect phenotypic changes, for example, cell morphology, adhesion, or cytoskeleton. Recent examples are the identification of compounds that block wound repair and/or cytokinesis in association with a reduced Rho activity [50,51] or that regulate zymosan phagocytosis through deactivation of Rac1–2 and Cdc42 [52]. Pros for this strategy are automation of cell imaging, straightforward identification of cell-active inhibitors, and the possible acquisition of multiple cell parameters on different light channels. Cons are the lack of specificity of the readout—inhibition of multiple pathways may produce identical phenotypes—thus requiring additional steps to identify which target is inhibited.

A last approach to identify inhibitors is virtual screening (VS), based on structural knowledge of targets and libraries of small molecules. VS consists in assessing how a series of small molecules could fit into a specific pocket (or any remarkable substructure) of the target and gives as output a ranking of likelihood to behave as bioactive inhibitors with the desired properties [53].

Compared to other approaches, VS thus targets a particular region of the desired protein and is able to screen libraries whose size is only limited by computing power and parameter settings. In addition to structural data, VS also uses any relevant information from external databases—for example, toxicity of compounds of the library, unwanted binding to other targets—in the final ranking. VS is now acknowledged as a hit-delivery method, classically associated with HTS experiments to optimize compound libraries to be screened or hit follow-up for secondary screens [53]. VS-based screenings have been successful in identifying Rho pathway inhibitors: the largely used Rac inhibitor NSC23766, which targets a region in Rac1 involved in the interaction with its RhoGEFs such as TrioN or Tiam1, was designed by using VS [54–56]. NSC23766 is now classically used for Rac inhibition in living cells and in organisms [57]. More potent Rac inhibitors were since then identified by a similar approach [58]. VS was also successfully used to target a pocket of the Arf1 GTPase, located near the binding interface to its cognate GEF ARNO. Structural data indeed predicted that this pocket undergoes major conformational changes during the exchange reaction, making it a good target for preventing the Arf1/ARNO complex to evolve toward the GTP-loading step. VS led to the identification of LM11, a cell-active noncompetitive inhibitor, which specifically blocked Arf1 but not Arf6 activation by ARNO [59]. Along the same line, the recent use of VS allowed to identify Y16, targeting the LARG–RhoA complex at the binding interface [60]. VS thus seems particularly well adapted to the discovery of Rho inhibitors. However, to get successful hits requires the prior determination of the structure of the targeted protein and the appropriate setting of many parameters for the analysis, without which it may produce low quality output [61].

None of the aforementioned approaches offers the possibility to monitor RhoGEF activity in a living situation. This lacuna makes it necessary to perform additional screening steps for validating the potentiality of the identified hits. For this main reason, we previously developed the yeast exchange assay (YEA), a sensitive test based on a two-hybrid system in which RhoGEF activity is monitored in the yeast through interaction of the activated Rho GTPase with its downstream effector [62]. This assay was next updated and optimized for medium-throughput screens for RhoGEF inhibitors [63]. Theoretically, YEA strains can be engineered for screening inhibitors to any mammalian RhoGEF, even RhoGEFs for RhoA or Cdc42 whose homologues are expressed in the yeast *Saccharomyces cerevisiae* [62], and to any mammalian GEF-activating GTPases of other Ras-like families like Ras, Rab, or Arf.

The experimental procedure described here was established using strains expressing wild-type or V12 mutant Rac1 and the kinectin Rho binding domain as two-hybrid partners [64], and first Dbl-like domain of Trio [65] or the DHR2 domain of DOCK5 [66] as Rac1 RhoGEFs.

2. THE YEA: RATIONALE AND PRINCIPLE

When expressed in the two-hybrid system, the dominant active mutant of mammalian GTPases such as Rac1G12V, RhoGG12V, Cdc42G12V, and RhoAG14V fused to LexA DNA-binding domain can bind to their respective effector proteins fused to the Gal4 activation domain, which is monitored by the expression of the reporter genes [62]. By contrast, wt Rac1 and RhoG do not bind to their effectors, reflecting their inactive conformation state and suggesting that no protein expressed by the yeast is capable of exchange activity on these mammalian GTPases. In comparison, wt RhoA and to a larger extent wt Cdc42 show a higher basal binding activity to their effectors, due to the expression of yeast exchange factors that can activate these GTPases, namely Rom1p and Rom2p for RhoA [67] and Cdc24 for Cdc42 [68].

These observations led us to transform the classical two-hybrid system into a system in which to monitor the activation of mammalian Rho-family GTPases by exchange factors expressed ectopically (Fig. 8.2A) [62]. For instance, in the case of wt Rac1, RhoG, and RhoA expressed with their respective effectors in the two-hybrid system, the expression of β-galactosidase reporter gene increases in yeast upon coexpression of an exchange factor known to activate the GTPase, for example, Tiam-1 [69], TrioD1, and TrioD2, respectively [70,71] (Fig. 8.2B). In the case of Cdc42Hs, the basal activation of the GTPase by Cdc24Sc is very efficient, which precluded the detection of a clear increase in β-galactosidase activity upon expression of a Cdc42Hs GEF such as alpha-PIX [62]. To circumvent this problem, the use of a Cdc24ts yeast strain permits efficient detection of Cdc42 activation by an exogenous exchange factor (Fig. 8.2C). This convenient two-hybrid-derived method which we named the YEA allows easy monitoring of the activation of a GTPase by an exchange factor. This assay proved efficient to analyze various mammalian exchange factors of formerly unknown specificity, including PLEKHG5, PLEKHG6, or ARHGEF17 [62,72,73]. It was also adapted to study the activation of Rho-family GTPases from other species such as *Caenorhabditis elegans* [74]. This system is only based on plasmid constructs; therefore, complex high molecular

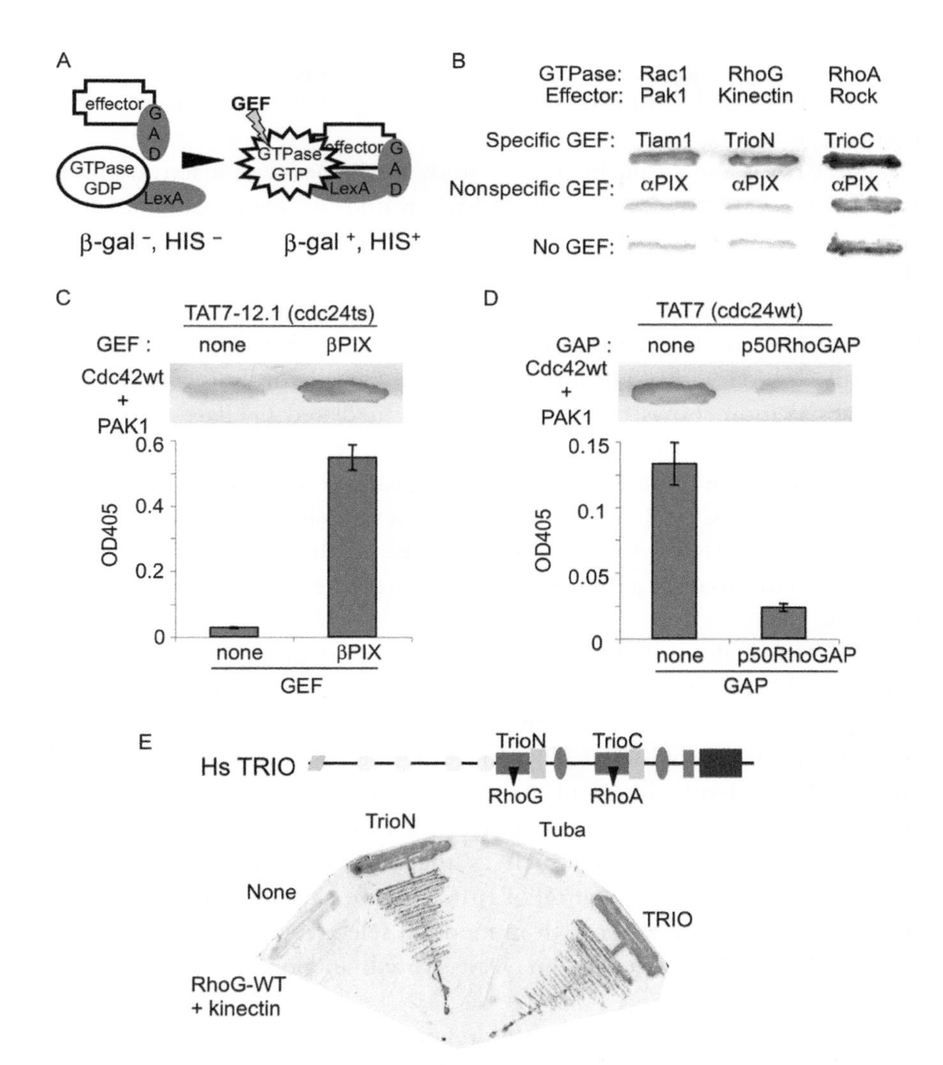

Figure 8.2 The yeast exchange assay (YEA) is suitable for monitoring activities of full-length RhoGEFs and RhoGAPs. (A) Rationale of the YEA: Mammalian wt GTPases, not activated in yeast, can be activated by expression of exogenous GEF. Activation is detected by β-galactosidase activity (β-gal+) and auxotrophy for histidine (HIS+). (B) Filter assay showing β-galactosidase activity in the yeast strain TAT7-12.1, expressing Lex-GTPases and GAD-Effector as indicated, either in the absence of exogenous GEF (No GEF) or in the presence of specific or nonspecific exogenous GEFs as indicated. (C) Filter and ONPG liquid assay showing β-galactosidase activity in the cdc24ts yeast strain TAT7-12.1, expressing Lex-Cdc42Hs-wt, GAD-PAK1 [62], and no GEF or myc-βPIX DH–PH domain. (D) Filter and ONPG liquid assay showing β-galactosidase activity in TAT7 yeast strain, expressing Lex-Cdc42Hs-wt, GAD-PAK1, and no GAP or myc-p50RhoGAP. (E) Filter assay showing β-galactosidase activity in TAT7 yeast strain, expressing Lex-RhoG-wt, GAD-kinectin [62], and no GEF, myc-TrioN [63], myc-Tuba DH–PH domain, or myc-Trio full length. (See color plate.)

weight GEFs can be studied, such as Trio (Fig. 8.2D), as well as many isoforms and mutants.

The YEA is also a valuable tool to study GAP-mediated inactivation of Rho GTPases. In a wild-type Cdc24 background, we expressed Cdc42Hs and its effector PAK1. As expected, Cdc42Hs was efficiently activated (Fig. 8.2E). We further expressed the GAP domain of p50RhoGAP, a GAP active on Cdc42 [75]. The expression of p50RhoGAP led to 80% reduction of β-galactosidase activity, indicating that Cdc42 was preferentially loaded with GDP due to increased GTP hydrolysis (Fig. 8.2E). This observation shows that not only GEF-mediated GTPase activation but also GAP-mediated GTPase inactivation can be studied in the YEA. This can be done either by taking advantage of endogenous GEFs as in the case of Cdc42 and RhoA or by coexpressing a GEF as a fourth partner in the system to activate the GTPase in the case of Rac-related GTPases. This assay could be valuable to study the activity of RhoGAP proteins.

The YEA is also suited to study exchange factors from other families of Ras-related GTPases, for example, Arf family GTPases and their exchange factors. To illustrate this, we expressed in this system the GTPase Arf6Hs with its effector ArhGAP10 [76]. In yeast, the dominant active mutant Arf6Q67L efficiently binds to ArhGAP10, inducing the expression of β-galactosidase, whereas the wild-type GTPase does not, unless an exchange factor is expressed such as the Sec7 domain of ARNO or EFA6 (Fig. 8.3A). The activation of Arf6 by the exchange factor Arno also translates into faster yeast growth in medium selective for histidine auxotrophy while it has no effect on yeast growth in nonselective medium (Fig. 8.3B).

The YEA proved a very versatile reporter system that can be adapted to generate various assays to address the activation or inactivation of Ras-related GTPases in a cellular context. The basic materials required are:

- two-hybrid expression vectors for the GTPase of interest, wild type and dominant active, and for its effector
- yeast plasmid expression vectors for the GEFs or GAPs of interest
- a two-hybrid yeast strain

Background activation of the wt GTPase can be observed in the case of GTPases sensitive to yeast GEFs, which may lead to comparable binding efficiency of the wt GTPAse to its effector as compared to the dominant active mutant. This may lead to similar growth rate in selective medium

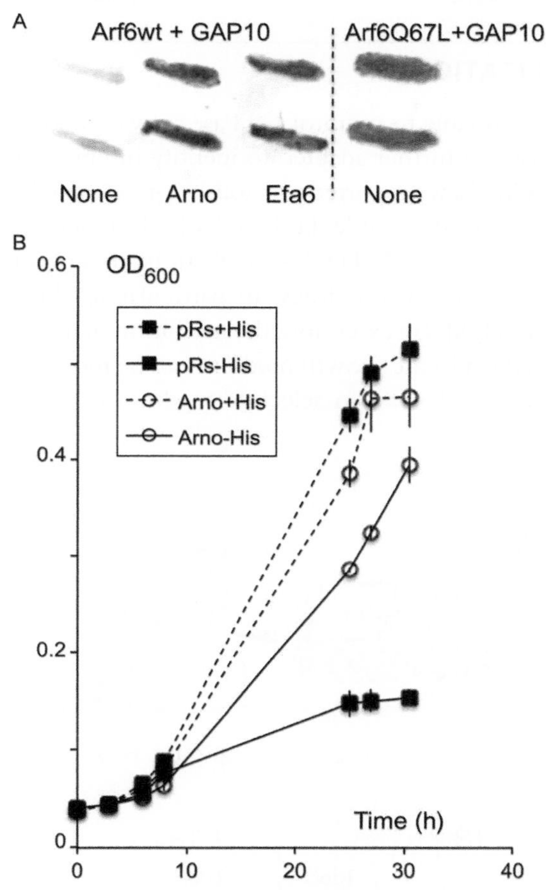

Figure 8.3 The YEA is suitable for monitoring Sec7 activity on Arf GTPases. (A) Filter assay showing β-galactosidase activity in TAT7 yeast strain, expressing LexA–Arf6-wt, GAD-GAP10, and either no Sec7 domain or the Sec7 domain of Arno or Efa6. (B) Growth curves of TAT7 yeasts expressing LexA–Arf6-wt and GAD-GAP10 with (open dots) and without (closed squares) Arno Sec7 domain medium complemented (dotted lines) or not (plain lines) with histidine. (See color plate.)

and reporter gene activity. To overcome this problem and measure the activity of a GEF with the YEA, it can be valuable to use yeast strains bearing viable deletion of the yeast GEF, such as *Rom2* deletion [67] or thermosensitive mutations, as for instance in *Sec7* [77], *Rom1* [67], or *Cdc24* as exemplified here.

3. APPLICATIONS

As an assay suitable to monitor GTPase activation as well as inactivation, the YEA can be further adapted to identify inhibitors for the GEF of interest (Fig. 8.4A). As we reported previously for example [63], the expression of TrioC inhibitory peptide TRIPalpha [45] hinders the activation of RhoC by TrioC in the YEA (Fig. 8.4B). Yeast has long been regarded as a tempting system to screen for drugs, in particular inhibitors of protein–protein interaction [78]. By extension, the inhibition of the GEF by a chemical compound added in the growth medium can be monitored in the YEA through the growth of yeast in selective medium or the activity of the reporter protein.

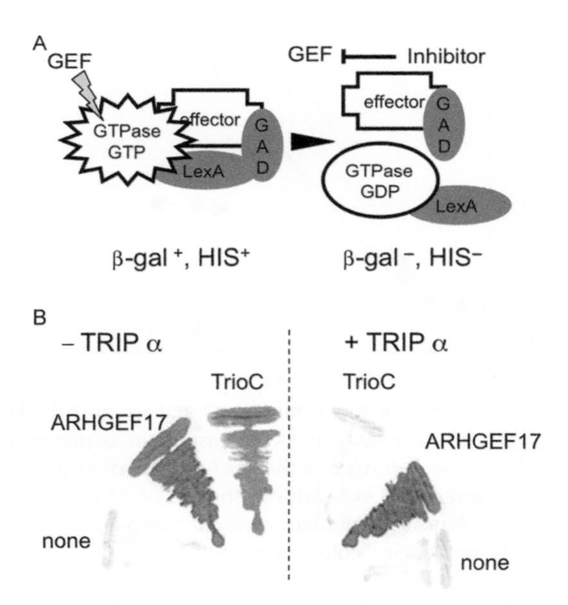

Figure 8.4 The YEA is suitable for detecting GEF inhibitors. (A) Principle of the GEF inhibitor detection system based on the YEA. The wild-type GTPase activated by the GEF binds to its effector and downstream reporter genes are expressed. Addition of a peptide or a chemical compound that inhibits activity of the GEF thus favors the GDP-bound state and hence reduced expression of the reporter genes. (B) Specific inhibition of TrioC by the TRIPα peptide in the YEA. Filter assay showing β-galactosidase activity in TAT7-expressing LexA–RhoC-wt, GAD-ROCK, and no GEF (none), TrioC or ARHGEF17 and expressing or not TRIPα. *This research was originally published in Ref. [63] © Portland Press Limited.* (See color plate.)

To optimize the efficiency of the chemical compounds tested, it can be valuable to take advantage of viable mutations that render yeast more permeable to a wider range of chemical compounds including hydrophilic or charged molecules. For instance, the disruption of *erg6* encodes a methyltransferase in the ergosterol biosynthetic pathway. It is a nonessential gene, the disruption of which affects ergosterol synthesis and membrane function; *erg6*-deficient yeasts as a result are more sensitive to a broad range of compounds including Brefeldin A, the inhibitor of the ARF (ADP-ribosylation factor) GTPase exchange factor Sec7 [79–81]. Targeting genes among the ABC transporter detoxification pathway, which acts as a multidrug exporter, also renders yeast more sensitive to chemical compounds [82] and hyperpermeable two-hybrid yeast strains bearing such mutations have been developed [83]. Using the TAT7 yeast strain disrupted for *erg6*, we screened a library of 2640 heterocyclic commercial chemical compounds to identify inhibitors of two RhoGEFs: TrioN, a Dbl family GEF, and Dock5, a GEF from the Dock/CZH family.

3.1. Inhibitors of the RhoGEF Trio

Trio is a complex exchange factor with two exchange domains (Fig. 8.2E): TrioN/TrioD1 activates RhoG and Rac1 [71,84] and TrioC/TrioD2 activates RhoA and RhoC [63,84]. Trio conserved through evolution is present in Drosophila, *C. elegans*, and vertebrates. It has been involved in a wide range of physiological and pathological processes [85]. Inhibitory aptamers targeting TrioC counteract the transforming activity of Tgat, an oncogenic isoform of Trio bearing only the second exchange domain [86]. To identify inhibitors of TrioN, we used the YEA. Starting with the TAT7 yeast strain disrupted for *erg6*, we engineered a yeast stably expressing LexA–RhoG-wt, GAD-kinectin, and myc-tagged TrioN. We adapted the YEA procedures to 96-well plates [87] and screened a library of 2640 heterocyclic compounds (ChemBridge Corporation) to identify molecules able to slow down yeast growth in selective HIS^- medium. Compounds were tested at a concentration of 200 μM in 2% DMSO, in the presence of 2 mM 3-AT and yeast growth was followed for 24 h. Among 120 compounds that inhibited yeast growth in HIS^- medium, only 64 were not toxic for yeast, as assessed by testing their effect on yeast growth in histidine-complemented (HIS^+) medium. To definitely identify TrioN inhibitors, we turned to *in vitro* exchange assays, using purified RhoG and TrioN. This led us to select two compounds: NPPD [1-(3-nitrophenyl)-1*H*-pyrrole-2,5-dione] and inhibitor of trio exchange (ITX)-1 [2-(5-chloro-2-ethoxybenzylidiene)[1,3]thiazolo[3,2-*a*]benzimidazol-3(2*H*)-one] as inhibitors of TrioN [63,65].

The most efficient compound, NPPD, inhibited TrioN exchange activity *in vitro*, but not that of ARHGEF17/p164RhoGEF, a RhoA exchange factor distantly related to Trio, nor that of ARNO, an exchange factor for Arf1 and Arf6 [63]. Interestingly, pull-down assays from 293T cell extracts showed that NPPD also inhibited the activation of Rac1 by the kalirin GEF1 domain, a very close paralog of TrioN. However, NPPD did not prevent Rac1 activation by DBS, Tiam1, or Vav2. Consequently, NPPD prevented kalirin GEF1-induced secretory granule secretion in AtT-20 corticotrope tumor cells [88]. Unfortunately, NPPD proved toxic for mammalian cells [65] and can therefore only be used for short-term cellular assays in the range of a few hours.

For *in vivo* assays, we turned to the less potent but less toxic molecule ITX1. To identify more potent inhibitors, we performed *in vitro* exchange assays on a series of derivatives and identified ITX3 (2-[(2,5-dimethyl-1-phenyl-1*H*-pyrrol-3-yl)methylene][1,3]thiazolo[3,2-*a*]benzimidazol-3(2*H*)-one) as a more potent inhibitor of TrioN as compared to ITX1 [65]. Similar to NPPD, cellular assays showed that ITX3 inhibits the activation of Rac1 by TrioN but is inactive on Rac1 activation by Tiam1 or Vav2 and on RhoA activation by ArhGEF17 [65]. But contrarily to NPPD, ITX3 can be used in long-term assays, inducing for instance no cell death in PC12 cells after 36 h or in C2C12 cells after 4 days in the presence of 100 μM ITX3. ITX3 also shows specificity for TrioN *in vitro* as it does not interfere with the exchange activity of Dbl toward Cdc42, ArhGEF17 toward RhoA, or ARNO toward Arf1.

As compared to the Rac inhibitor NSC23766, which targets a region in Rac1 involved in the interaction with its RhoGEFs [55], these two molecules inhibit Rac1 and RhoG activation by TrioN and have no effect on the activation of Rac1 by Tiam1, strongly suggesting that these molecules target TrioN. Contrarily to EHT 1864 [89], NPPD [63] and ITX3 [65] did not stimulate the release of nucleotide by the GTPase in the absence of the exchange factor. Therefore, NPPD and ITX3 are inhibitors that target the TrioN RhoGEF and specifically interfere with its ability to promote exchange. They were the first chemical compounds identified to inhibit a Dbl family exchange factor. Availability of structural data of the TrioN/ITX3 complex is necessary to identify the TrioN region involved in the binding and develop a VS approach to identify compounds with better inhibition parameters.

3.2. Inhibitors of the RhoGEF DOCK5

We have identified Dock5 as a Rac1 exchange factor essential for bone resorption by osteoclasts, the deletion of which in the mouse has no major effect apart from inducing an increase in bone mass [66]. As these properties qualify Dock5 as an interesting therapeutic target in the context of osteoporosis, we sought for a chemical inhibitor of Dock5 using the YEA. The TAT7 yeast strain disrupted for *erg6* was transformed with a plasmid expressing LexA–Rac1-wt and GAD-kinectin and another plasmid expressing the DHR2 exchange domain of Dock5. The screening of the library was performed as above, but an extra control was added during the screening procedure, to directly eliminate compounds toxic for yeast growth and supposedly so in mammalian cells. In that purpose, the screening was performed in duplicate: on one hand we compared yeast growth in selective histidine-deprived medium and on the other hand we compared yeast growth in nonselective histidine-complemented medium [87]. Among the compounds inhibiting yeast growth in selective medium we considered only those that did not affect yeast growth in nonselective medium. This procedure led to the identification of 53 potential Dock5 inhibitors.

These molecules were further tested for their effect on osteoclast survival and on the ability of the osteoclasts to solubilize a calcium phosphate matrix, a reporter assay for their bone resorbing activity [66]. Out of the 53 compounds identified in the screening, 6 inhibited mineral resorption by osteoclasts without affecting their survival (Fig. 8.5A), similar to the effect of Dock5 shRNA (Fig. 8.5B) [66]. One of these six compounds, C21: *N*-(3,5-dichlorophenyl)benzenesulfonamide (Fig. 8.5C), was able to inhibit the activation of Rac1 by Dock5 in a cellular assay (Fig. 8.5D) and represents the first and thus far only chemical inhibitor described for a Dock-family GEF. C21 efficiently inhibited bone resorption by osteoclasts in culture (Fig. 8.5E) and had no effect on the activation of Rac1 by TrioN [66]. That Rac1 remains available to other RhoGEFs is of physiological importance given the prominent implication of this GTPase in most basic cellular processes. In this respect, C21 differs from other inhibitors designed from the Rac1 structure [58] and may prove very powerful to inhibit local DOCK5-dependent Rac activity without altering its global activity.

4. CONCLUDING REMARKS

Rho-signaling pathways represent attractive therapeutic targets for cancer and neurologic disorders. Among the different strategies to identify

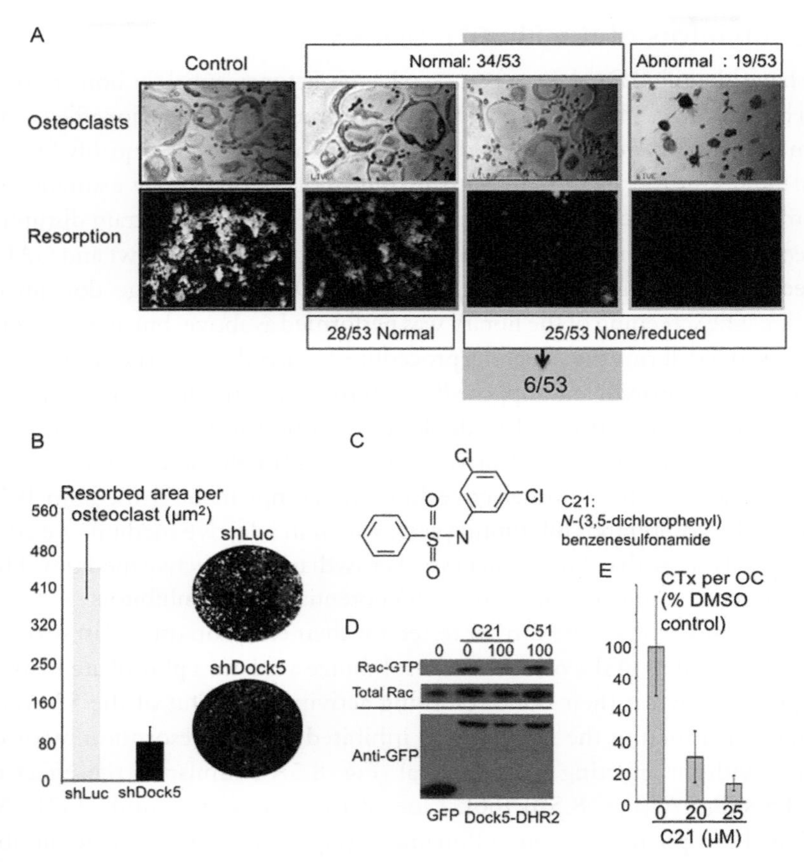

Figure 8.5 Identification of a Dock5 inhibitor that hinders bone resorption by osteoclasts. (A) The 53 compounds isolated in the YEA screen for Dock5 inhibitors were incubated for 2 days with osteoclasts derived from RAW264.7 cells, in 96-well plates or in multiwell calcium phosphate Osteologic Biocoats. Osteoclasts on plastic were stained for TRAP activity (upper panels) and Osteologic Biocoats were submitted to von Kossa staining to reveal minerals (lower panels). Out of the 53 compounds tested, 6 led to normal osteoclasts that did not resorb the mineral. (B) Osteoclasts derived from RAW264.7 cells expressing shRNA against luciferase or Dock5 were differentiated on plastic or on calcium phosphate Osteologic Biocoats for 2 days. Bar graph shows the average area resorbed per osteoclast from three wells. (C) Structure of *N*-(3,5-dichlorophenyl) benzenesulfonamide. (D) Total and GTP-bound Rac1 (upper panels) in 293T cells expressing GFP and GFP-fused Dock5-DHR2 (lower panels) and treated for 1 h with 0 or 100 μM of C21 or of the control negative compound C51. (E) Collagen telopeptide production (Ctx) per osteoclast in the presence of the indicated concentration of C21. Graph shows average and SD CTx concentration per osteoclast in one experiment performed in triplicate, expressed as % of the DMSO control. *Panels B and E were originally published in Ref. [66] © John Wiley and Sons.* (See color plate.)

inhibitors of Rho pathways, VS is probably one of the best promising approach, in particular for targeting the GEF/GTPase interface, which appears as the most efficient and specific target. However, VS requires accurate structural data, which are available hitherto for a restricted number of protein complexes and for only catalytic GEF domains. VS thus requires secondary screens to validate the target and the bioavailability of the candidate inhibitors. The YEA method we describe here, derived from the two-hybrid screen, offers the possibility to perform medium-throughput screens for identifying in a single step cell-active compounds, while requiring a moderate investment in terms of biological material and expertise for the experimental setup. Combined with VS to limit the number of candidate compounds to be screened, this approach thus represents a method of choice for finding specific inhibitors of nucleotide exchange catalytic activity of full-length RhoGEFs.

5. MATERIALS AND METHODS

The GAP domain of p50RhoGAP containing amino acid residues 205–439 [75], nicely provided by Yi Zheng, Philadelphia, USA, and full-length Trio [70], kindly provided by Anne Debant, Montpellier, France, were fused to the myc tag and inserted into pRs426Met yeast expression vector [62]. The Cdc42 GEFs Tuba (aa 470–1279) and βPIX were fused to the myc tag and inserted into pRs426Met yeast expression vector. Two-hybrid constructs expressing the Arf-binding domain of ArhGAP10 fused to Gal4-activation domain and wild type and Q67L dominant active human Arf6 fused to LexA DNA-binding domain [76] were a kind gift from Philippe Chavrier, Paris, France. The Sec7 domains of ARNO (residues 52–242) [44] and of EFA6 (residues 51–252) [90], respectively and kindly provided by Jacqueline Cherfils, Gif sur Yvette, France and Michel Franco, Sofia Antipolis, France, were fused to the myc tag and inserted into pRs426Met yeast expression vector. Other expression vectors, TAT7, TAT7-12.1 strains, yeast growth media, and protocols were described earlier, as mentioned in figure legends [62,63,65,66,87].

To obtain osteoclasts, RAW264.7 cells were grown for 4 days in the presence of 50 ng/ml RANKL (Peprotech, Neuilly sur Seine, France) as described previously [91]. For resorption assays, osteoclasts were differentiated in multiwell chambers coated with calcium phosphate (Osteologic Biocoat; BD Biosciences, Le Pont de Claix, France) or in 96-well plates containing a bovine bone slice (IDS Nordic Bioscience, Paris, France).

Osteoclasts were stained for TRAP and multiwell chambers stained with von Kossa as described previously [92]. The entire well surfaces were imaged using a Zeiss AxioimagerZ1. Osteoclasts were counted and resorbed areas measured using ImageJ. CTx levels in culture medium were determined using Crosslap kit (IDS Nordic Bioscience) as described [66]. GTP-bound Rac1 was pulled down from extracts of 293T cell expressing GFP or Dock5 DHR2 domain fused to GFP [66] using GST-fused PAK1 CRIB domain as described previously [65].

REFERENCES

[1] D.D. Leipe, Y.I. Wolf, E.V. Koonin, L. Aravind, Classification and evolution of P-loop GTPases and related ATPases, J. Mol. Biol. 317 (2002) 41–72.
[2] D. Díez, F. Sánchez-Jiménez, J.A.G. Ranea, Evolutionary expansion of the Ras switch regulatory module in eukaryotes, Nucleic Acids Res. 39 (2011) 5526–5537.
[3] A.M. Rojas, G. Fuentes, A. Rausell, A. Valencia, The Ras protein superfamily: evolutionary tree and role of conserved amino acids, J. Cell Biol. 196 (2012) 189–201.
[4] W.E. Tidyman, K.A. Rauen, The RASopathies: developmental syndromes of Ras/MAPK pathway dysregulation, Curr. Opin. Genet. Dev. 19 (2009) 230–236.
[5] L. Larizza, C. Gervasini, F. Natacci, P. Riva, Developmental abnormalities and cancer predisposition in neurofibromatosis type 1, Curr. Mol. Med. 9 (2009) 634–653.
[6] J.E. Johndrow, C.R. Magie, S.M. Parkhurst, Rho GTPase function in flies: insights from a developmental and organismal perspective, Biochem. Cell Biol. 82 (2004) 643–657.
[7] J.H. Camonis, M.A. White, Ral GTPases: corrupting the exocyst in cancer cells, Trends Cell Biol. 15 (2005) 327–332.
[8] K. Mardilovich, M.F. Olson, M. Baugh, Targeting Rho GTPase signaling for cancer therapy, Future Oncol. (London, England) 8 (2012) 165–177.
[9] Y. Pylayeva-Gupta, E. Grabocka, D. Bar-Sagi, RAS oncogenes: weaving a tumorigenic web, Nat. Rev. Cancer 11 (2011) 761–774.
[10] R. Rathinam, A. Berrier, S.K. Alahari, Role of Rho GTPases and their regulators in cancer progression, Front. Biosci. 16 (2011) 2561–2571.
[11] S.C. Smith, D. Theodorescu, The Ral GTPase pathway in metastatic bladder cancer: key mediator and therapeutic target, Urol. Oncol. 27 (2009) 42–47.
[12] J. Antoine-Bertrand, J.-F. Villemure, N. Lamarche-Vane, Implication of rho GTPases in neurodegenerative diseases, Curr. Drug Targets 12 (2011) 1202–1215.
[13] L.M. Harrison, Rhes: a GTP-binding protein integral to striatal physiology and pathology, Cell. Mol. Neurobiol. 32 (2012) 907–918.
[14] A.-C. Laurent, M. Breckler, M. Berthouze, F. Lezoualc'h, Role of Epac in brain and heart, Biochem. Soc. Trans. 40 (2012) 51–57.
[15] R.L. Stornetta, J.J. Zhu, Ras and Rap signaling in synaptic plasticity and mental disorders, Neuroscientist 17 (2011) 54–78.
[16] T. Hirase, K. Node, Endothelial dysfunction as a cellular mechanism for vascular failure, Am. J. Physiol. Heart Circ. Physiol. 302 (2012) H499–H505.
[17] K.E. Porter, N.A. Turner, Statins and myocardial remodelling: cell and molecular pathways, Expert Rev. Mol. Med. 13 (2011) e22.
[18] R.D. Hayward, V. Koronakis, Pathogens reWritE Rho's rules, Cell 124 (2006) 15–17.
[19] D. Vigil, J. Cherfils, K.L. Rossman, C.J. Der, Ras superfamily GEFs and GAPs: validated and tractable targets for cancer therapy? Nat. Rev. Cancer 10 (2010) 842–857.

[20] E.E. Bosco, J.C. Mulloy, Y. Zheng, Rac1 GTPase: a "Rac" of all trades, Cell. Mol. Life Sci. 66 (2009) 370–374.
[21] F. Raimondi, G. Portella, M. Orozco, F. Fanelli, Nucleotide binding switches the information flow in ras GTPases, PLoS Comput. Biol. 7 (2011) e1001098.
[22] J.C. Schwamborn, M. Müller, A.H. Becker, A.W. Püschel, Ubiquitination of the GTPase Rap1B by the ubiquitin ligase Smurf2 is required for the establishment of neuronal polarity, EMBO J. 26 (2007) 1410–1422.
[23] S. Torrino, O. Visvikis, A. Doye, et al., The E3 ubiquitin-ligase HACE1 catalyzes the ubiquitylation of active Rac1, Dev. Cell 21 (2011) 959–965.
[24] H.-R. Wang, Y. Zhang, B. Ozdamar, et al., Regulation of cell polarity and protrusion formation by targeting RhoA for degradation, Science (New York, NY) 302 (2003) 1775–1779.
[25] E. Diaz-Flores, K. Shannon, Targeting oncogenic Ras, Genes Dev. 21 (2007) 1989–1992.
[26] F.M. Vega, A.J. Ridley, Rho GTPases in cancer cell biology, FEBS Lett. 582 (2008) 2093–2101.
[27] M.J. Davis, B.H. Ha, E.C. Holman, et al., RAC1P29S is a spontaneously activating cancer-associated GTPase, Proc. Natl. Acad. Sci. U.S.A. 110 (2013) 912–917.
[28] E. Hodis, I.R. Watson, G.V. Kryukov, et al., A landscape of driver mutations in melanoma, Cell 150 (2012) 251–263.
[29] M. Kawazu, T. Ueno, K. Kontani, et al., Transforming mutations of RAC guanosine triphosphatases in human cancers, Proc. Natl. Acad. Sci. U.S.A. 110 (2013) 3029–3034.
[30] H. Shirai, M. Autieri, S. Eguchi, Small GTP-binding proteins and mitogen-activated protein kinases as promising therapeutic targets of vascular remodeling, Curr. Opin. Nephrol. Hypertens. 16 (2007) 111–115.
[31] G. Fritz, I. Just, B. Kaina, Rho GTPases are over-expressed in human tumors, Int. J. Cancer 81 (1999) 682–687.
[32] A. Boureux, E. Vignal, S. Faure, P. Fort, Evolution of the Rho family of ras-like GTPases in eukaryotes, Mol. Biol. Evol. 24 (2007) 203–216.
[33] K.L. Rossman, C.J. Der, J. Sondek, GEF means go: turning on RHO GTPases with guanine nucleotide-exchange factors, Nat. Rev. Mol. Cell Biol. 6 (2005) 167–180.
[34] N. Meller, S. Merlot, C. Guda, CZH proteins: a new family of Rho-GEFs, J. Cell Sci. 118 (2005) 4937–4946.
[35] M. Uehata, T. Ishizaki, H. Satoh, et al., Calcium sensitization of smooth muscle mediated by a Rho-associated protein kinase in hypertension, Nature 389 (1997) 990–994.
[36] S.W. Deacon, A. Beeser, J.A. Fukui, et al., An isoform-selective, small-molecule inhibitor targets the autoregulatory mechanism of p21-activated kinase, Chem. Biol. 15 (2008) 322–331.
[37] H. Ueda, R. Morishita, H. Itoh, et al., Galpha11 induces caspase-mediated proteolytic activation of Rho-associated kinase, ROCK-I, in HeLa cells, J. Biol. Chem. 276 (2001) 42527–42533.
[38] T.H. Loo, Y.W. Ng, L. Lim, E. Manser, GIT1 activates p21-activated kinase through a mechanism independent of p21 binding, Mol. Cell Biol. 24 (2004) 3849–3859.
[39] S.G. Raja, Evaluation of clinical efficacy of fasudil for the treatment of pulmonary arterial hypertension, Recent Pat. Cardiovasc. Drug Discov. 7 (2012) 100–104.
[40] M. Jaiswal, R. Dvorsky, M.R. Ahmadian, Deciphering the molecular and functional basis of Dbl Family proteins: a novel systematic approach toward classification of elective activation of the Rho family proteins, J. Biol. Chem. 288 (2013) 4486–4500.
[41] G. Lazer, S. Katzav, Guanine nucleotide exchange factors for RhoGTPases: good therapeutic targets for cancer therapy? Cell. Signal. 23 (2010) 969–979.
[42] J. Cherfils, P. Chardin, GEFs: structural basis for their activation of small GTP-binding proteins, Trends Biochem. Sci. 24 (1999) 306–311.

[43] F. Wittinghofer, Ras signalling. Caught in the act of the switch-on, Nature 394 (1998) 317, 319–320.
[44] J.C. Zeeh, M. Zeghouf, C. Grauffel, et al., Dual specificity of the interfacial inhibitor brefeldin a for arf proteins and sec7 domains, J. Biol. Chem. 281 (2006) 11805–11814.
[45] S. Schmidt, S. Diriong, J. Mery, et al., Identification of the first Rho-GEF inhibitor, TRIPalpha, which targets the RhoA-specific GEF domain of Trio, FEBS Lett. 523 (2002) 35–42.
[46] D.A. Leonard, T. Evans, M. Hart, et al., Investigation of the GTP-binding/GTPase cycle of Cdc42Hs using fluorescence spectroscopy, Biochemistry 33 (1994) 12323–12328.
[47] Z. Surviladze, S.M. Young, L.A. Sklar, High-throughput flow cytometry bead-based multiplex assay for identification of Rho GTPase inhibitors, Method Mol. Biol. (Clifton, NJ) 827 (2012) 253–270.
[48] C.R. Evelyn, T. Ferng, R.J. Rojas, et al., High-throughput screening for small-molecule inhibitors of LARG-stimulated RhoA nucleotide binding via a novel fluorescence polarization assay, J. Biomol. Screen. 14 (2009) 161–172.
[49] Y. Zheng, M.J. Hart, R.A. Cerione, Guanine nucleotide exchange catalyzed by dbl oncogene product, Methods Enzymol. 256 (1995) 77–84.
[50] A.B. Castoreno, Y. Smurnyy, A.D. Torres, et al., Small molecules discovered in a pathway screen target the Rho pathway in cytokinesis, Nat. Chem. Biol. 6 (2010) 457–463.
[51] A.G. Clark, J.R. Sider, K. Verbrugghe, et al., Identification of small molecule inhibitors of cytokinesis and single cell wound repair, Cytoskeleton (Hoboken, N.J.) 69 (2012) 1010–1020.
[52] J.S. Bang, Y.J. Kim, J. Song, et al., Small molecules that regulate zymosan phagocytosis of macrophage through deactivation of Rho GTPases, Bioorg. Med. Chem. 20 (2012) 5262–5268.
[53] Y. Tanrikulu, B. Kruger, E. Proschak, The holistic integration of virtual screening in drug discovery, Drug Discov. Today 18 (2013) 358–364.
[54] H. Akbar, J. Cancelas, D.A. Williams, et al., Rational design and applications of a Rac GTPase-specific small molecule inhibitor, Methods Enzymol. 406 (2006) 554–565.
[55] Y. Gao, J.B. Dickerson, F. Guo, et al., Rational design and characterization of a Rac GTPase-specific small molecule inhibitor, Proc. Natl. Acad. Sci. U.S.A. 101 (2004) 7618–7623.
[56] N. Nassar, J. Cancelas, J. Zheng, et al., Structure–function based design of small molecule inhibitors targeting Rho family GTPases, Curr. Top. Med. Chem. 6 (2006) 1109–1116.
[57] H. Akbar, J. Kim, K. Funk, et al., Genetic and pharmacologic evidence that Rac1 GTPase is involved in regulation of platelet secretion and aggregation, J. Thromb. Haemostasis 5 (2007) 1747–1755.
[58] N. Ferri, A. Corsini, P. Bottino, et al., Virtual screening approach for the identification of new Rac1 inhibitors, J. Med. Chem. 52 (2009) 4087–4090.
[59] J. Viaud, M. Zeghouf, H. Barelli, et al., Structure-based discovery of an inhibitor of Arf activation by Sec7 domains through targeting of protein–protein complexes, Proc. Natl. Acad. Sci. U.S.A. 104 (2007) 10370–10375.
[60] X. Shang, F. Marchioni, C.R. Evelyn, et al., Small-molecule inhibitors targeting G-protein-coupled Rho guanine nucleotide exchange factors, Proc. Natl. Acad. Sci. U.S.A. 110 (2013) 3155–3160.
[61] T. Scior, A. Bender, G. Tresadern, et al., Recognizing pitfalls in virtual screening: a critical review, J. Chem. Inf. Model. 52 (2012) 867–881.
[62] M. De Toledo, K. Colombo, T. Nagase, et al., The yeast exchange assay, a new complementary method to screen for Dbl-like protein specificity: identification of a novel RhoA exchange factor, FEBS Lett. 480 (2000) 287–292.

[63] A. Blangy, N. Bouquier, C. Gauthier-Rouviere, et al., Identification of TRIO-GEFD1 chemical inhibitors using the yeast exchange assay, Biol. Cell 98 (2006) 511–522.
[64] E. Vignal, A. Blangy, M. Martin, et al., Kinectin is a key effector of RhoG microtubule-dependent cellular activity, Mol. Cell Biol. 21 (2001) 8022–8034.
[65] N. Bouquier, E. Vignal, S. Charrasse, et al., A cell active chemical GEF inhibitor selectively targets the Trio/RhoG/Rac1 signaling pathway, Chem. Biol. 16 (2009) 657–666.
[66] V. Vives, M. Laurin, G. Cres, et al., The Rac1 exchange factor Dock5 is essential for bone resorption by osteoclasts, J. Bone Miner. Res. 26 (2011) 1099–1110.
[67] K. Ozaki, K. Tanaka, H. Imamura, et al., Rom1p and Rom2p are GDP/GTP exchange proteins (GEPs) for the Rho1p small GTP binding protein in *Saccharomyces cerevisiae*, EMBO J. 15 (1996) 2196–2207.
[68] R. Li, Y. Zheng, Residues of the Rho family GTPases Rho and Cdc42 that specify sensitivity to Dbl-like guanine nucleotide exchange factors, J. Biol. Chem. 272 (1997) 4671–4679.
[69] F.N. van Leeuwen, R.A. van der Kammen, G.G. Habets, J.G. Collard, Oncogenic activity of Tiam1 and Rac1 in NIH3T3 cells, Oncogene 11 (1995) 2215–2221.
[70] J.M. Bellanger, J.B. Lazaro, S. Diriong, et al., The two guanine nucleotide exchange factor domains of Trio link the Rac1 and the RhoA pathways in vivo, Oncogene 16 (1998) 147–152.
[71] A. Blangy, E. Vignal, S. Schmidt, et al., TrioGEF1 controls Rac- and Cdc42-dependent cell structures through the direct activation of rhoG, J. Cell Sci. 113 (Pt 4) (2000) 729–739.
[72] R. D'Angelo, S. Aresta, A. Blangy, et al., Interaction of ezrin with the novel guanine nucleotide exchange factor PLEKHG6 promotes RhoG-dependent apical cytoskeleton rearrangements in epithelial cells, Mol. Biol. Cell 18 (2007) 4780–4793.
[73] M. De Toledo, V. Coulon, S. Schmidt, et al., The gene for a new brain specific RhoA exchange factor maps to the highly unstable chromosomal region 1p36.2–1p36.3, Oncogene 20 (2001) 7307–7317.
[74] H. Qadota, A. Blangy, G. Xiong, G.M. Benian, The DH–PH region of the giant protein UNC-89 activates RHO-1 GTPase in *Caenorhabditis elegans* body wall muscle, J. Mol. Biol. 383 (2008) 747–752.
[75] B. Zhang, Y. Zheng, Regulation of RhoA GTP hydrolysis by the GTPase-activating proteins p190, p50RhoGAP, Bcr, and 3BP-1, Biochemistry 37 (1998) 5249–5257.
[76] T. Dubois, O. Paleotti, A.A. Mironov, et al., Golgi-localized GAP for Cdc42 functions downstream of ARF1 to control Arp2/3 complex and F-actin dynamics, Nat. Cell Biol. 7 (2005) 353–364.
[77] P. Novick, C. Field, R. Schekman, Identification of 23 complementation groups required for post-translational events in the yeast secretory pathway, Cell 21 (1980) 205–215.
[78] M. Vidal, H. Endoh, Prospects for drug screening using the reverse two-hybrid system, Trends Biotechnol. 17 (1999) 374–381.
[79] N. Shah, R.D. Klausner, Brefeldin A reversibly inhibits secretion in *Saccharomyces cerevisiae*, J. Biol. Chem. 268 (1993) 5345–5348.
[80] J.P. Vogel, J.N. Lee, D.R. Kirsch, et al., Brefeldin A causes a defect in secretion in *Saccharomyces cerevisiae*, J. Biol. Chem. 268 (1993) 3040–3043.
[81] P. Chardin, F. McCormick, Brefeldin A: the advantage of being uncompetitive, Cell 97 (1999) 153–155.
[82] B. Rogers, A. Decottignies, M. Kolaczkowski, et al., The pleitropic drug ABC transporters from *Saccharomyces cerevisiae*, J. Mol. Microbiol. Biotechnol. 3 (2001) 207–214.

[83] V. Khazac, E. Golemis, L. Weber, Development of a yeast two-hybrid screen for selection of Ras–Raf protein interaction inhibitors, Methods Mol. Biol. 310 (2005) 253–271.

[84] A. Debant, C. Serra-Pages, K. Seipel, et al., The multidomain protein Trio binds the LAR transmembrane tyrosine phosphatase, contains a protein kinase domain, and has separate rac-specific and rho-specific guanine nucleotide exchange factor domains, Proc. Natl. Acad. Sci. U.S.A. 93 (1996) 5466–5471.

[85] J. van Rijssel, J.D. van Buul, The many faces of the guanine-nucleotide exchange factor trio, Cell Adh. Migr. 6 (2012) 482–487.

[86] N. Bouquier, S. Fromont, J.C. Zeeh, et al., Aptamer-derived peptides as potent inhibitors of the oncogenic RhoGEF Tgat, Chem. Biol. 16 (2009) 391–400.

[87] A. Blangy, P. Fort, Using a modified yeast two-hybrid system to screen for chemical GEF inhibitors, Methods Mol. Biol. (Clifton, NJ) 928 (2012) 81–95.

[88] F. Ferraro, X.M. Ma, J.A. Sobota, et al., Kalirin/Trio Rho guanine nucleotide exchange factors regulate a novel step in secretory granule maturation, Mol. Biol. Cell 18 (2007) 4813–4825.

[89] A. Shutes, C. Onesto, V. Picard, et al., Specificity and mechanism of action of EHT 1864, a novel small molecule inhibitor of Rac family small GTPases, J. Biol. Chem. 282 (2007) 35666–35678.

[90] P. Chavrier, M. Franco, Expression, purification, and biochemical properties of EFA6, a Sec7 domain-containing guanine exchange factor for ADP-ribosylation factor 6 (ARF6), Methods Enzymol. 329 (2001) 272–279.

[91] H. Brazier, S. Stephens, S. Ory, et al., Expression profile of RhoGTPases and RhoGEFs during RANKL-stimulated osteoclastogenesis: identification of essential genes in osteoclasts, J. Bone Miner. Res. 21 (2006) 1387–1398.

[92] H. Brazier, G. Pawlak, V. Vives, A. Blangy, The Rho GTPase Wrch1 regulates osteoclast precursor adhesion and migration, Int. J. Biochem. Cell Biol. 41 (2009) 1391–1401.

[93] M. Fukuda, TBC proteins: GAPs for mammalian small GTPase Rab? Biosci. Rep. 31 (2011) 159–168.

[94] S.-i. Yoshimura, A. Gerondopoulos, A. Linford, et al., Family-wide characterization of the DENN domain Rab GDP–GTP exchange factors, J. Cell Biol. 191 (2010) 367–381.

[95] A. Delprato, D.G. Lambright, Structural basis for Rab GTPase activation by VPS9 domain exchange factors, Nat. Struct. Mol. Biol. 14 (2007) 406–412.

[96] S. Chen, H. Cai, S.-K. Park, et al., Trs65p, a subunit of the Ypt1p GEF TRAPPII, interacts with the Arf1p exchange factor Gea2p to facilitate COPI-mediated vesicle traffic, Mol. Biol. Cell 22 (2011) 3634–3644.

[97] G.V. Pusapati, G. Luchetti, S.R. Pfeffer, Ric1–Rgp1 complex is a guanine nucleotide exchange factor for the late Golgi Rab6A GTPase and an effector of the medial Golgi Rab33B GTPase, J. Biol. Chem. 287 (2012) 42129–42137.

[98] T.H. Klöpper, N. Kienle, D. Fasshauer, S. Munro, Untangling the evolution of Rab G proteins: implications of a comprehensive genomic analysis, BMC Biol. 10 (2012) 71.

[99] M.P. East, J.B. Bowzard, J.B. Dacks, R.A. Kahn, ELMO domains, evolutionary and functional characterization of a novel GTPase-activating protein (GAP) domain for Arf protein family GTPases, J. Biol. Chem. 287 (2012) 39538–39553.

[100] A. Spang, Y. Shiba, P.A. Randazzo, Arf GAPs: gatekeepers of vesicle generation, FEBS Lett. 584 (2010) 2646–2651.

[101] R. Cox, R.J. Mason-Gamer, C.L. Jackson, N. Segev, Phylogenetic analysis of Sec7-domain-containing Arf nucleotide exchangers, Mol. Biol. Cell 15 (2004) 1487–1505.

[102] R.A. Kahn, L. Volpicelli-Daley, B. Bowzard, et al., Arf family GTPases: roles in membrane traffic and microtubule dynamics, Biochem. Soc. Trans. 33 (2005) 1269–1272.
[103] D. Yudin, M. Fainzilber, Ran on tracks—cytoplasmic roles for a nuclear regulator, J. Cell Sci. 122 (2009) 587–593.
[104] M.M. Brandão, K.L. Silva-Brandão, F.F. Costa, S.T.O. Saad, Phylogenetic analysis of RhoGAP domain-containing proteins, Genomics Proteomics Bioinformatics 4 (2006) 182–188.
[105] J.F. Cote, K. Vuori, Identification of an evolutionarily conserved superfamily of DOCK180-related proteins with guanine nucleotide exchange activity, J. Cell Sci. 115 (2002) 4901–4913.

CHAPTER NINE

Inhibitors of the ROCK Serine/ Threonine Kinases: Key Effectors of the RhoA Small GTPase

Dominico Vigil, Channing J. Der[1]
University of North Carolina at Chapel Hill, Lineberger Comprehensive Cancer Center, Chapel Hill, North Carolina, USA
[1]Corresponding author: e-mail address: cjder@med.unc.edu

Contents

Abstract

Aberrant activation of the RhoA small GTPase has been implicated in cancer and other human diseases. Therefore, inhibitors of RhoA may have important therapeutic value. However, similar to the Ras small GTPases, RhoA itself is not considered a tractable target and is currently considered to be "undruggable." While recent efforts suggest that direct inhibitors of the Ras oncoprotein may yet be developed, the most promising directions for anti-Ras inhibitors involve inhibitors of protein kinases that are activated downstream of Ras. By analogy, protein kinases activated downstream of RhoA may provide more attractive directions for the development of anti-RhoA inhibitors. Among the multitude of RhoA effectors, the ROCK serine/threonine kinases have emerged as attractive targets for anti-RhoA drug discovery. In this review, we summarize the current status of the development of small molecule inhibitors of ROCK.

1. INTRODUCTION

The Rho-associated, coiled-coil domain containing protein kinases (ROCK1 and ROCK2; subsequently referred to collectively as ROCK

The Enzymes, Volume 33
ISSN 1874-6047
http://dx.doi.org/10.1016/B978-0-12-416749-0.00009-9

unless a specific isoform is discussed) were originally discovered by screening for RhoA effectors [1] and subsequently demonstrated to be major mediators of the functions of RhoA in regulation of actin and myosin cytoskeletal organization [2–4] (Fig. 9.1A). Phosphorylation of myosin light chain (MLC) phosphatase target subunit 1 (MYPT1) on T696 and T853 by ROCK inactivates its phosphatase activity toward MLC and thus results in increased phosphorylation of MLC on S19 and S20, thus increasing actomyosin contractility [5] (Fig. 9.1B). Additionally, ROCK can also directly phosphorylate MLC, although the full physiological relevance of direct phosphorylation versus indirect increase in phosphorylation by inactivation

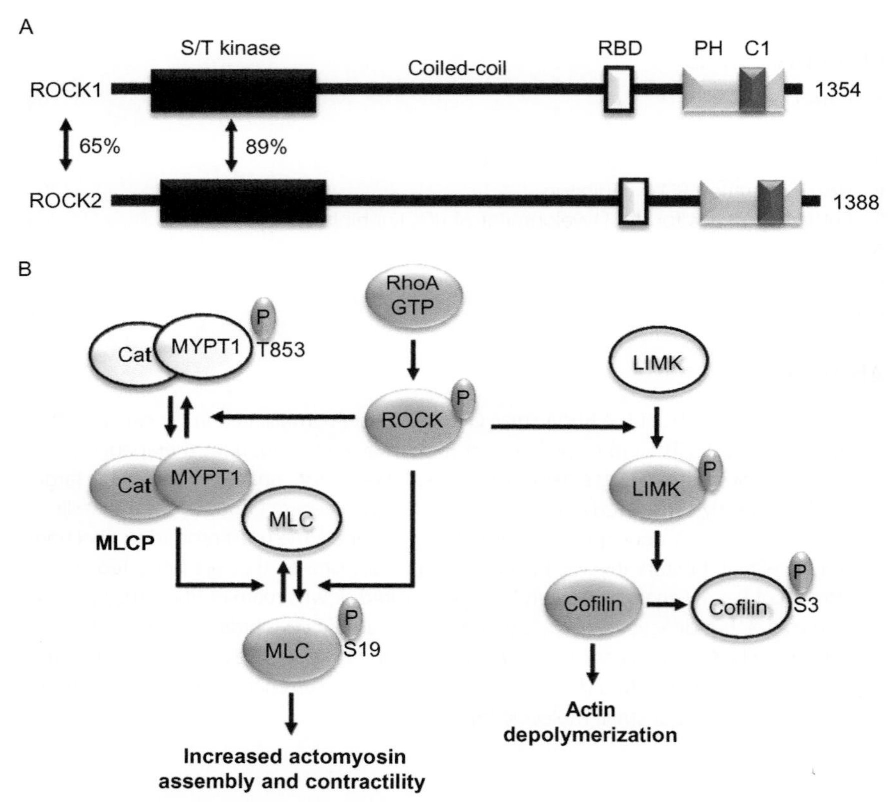

Figure 9.1 ROCK structure and signaling. (A) Human ROCK serine/threonine kinases. Domain structure was determined, in part, by SMART (http://smart.embl-heidelberg.de/). % values indicate percentage amino acid identity. S/T, serine/threonine; RBD, Rho-GTP-binding domain; PH, pleckstrin homology domain; C1, phorbol esters/diacylglycerol-binding domain. (B) ROCK signaling and key substrates. Open circles indicate inactive forms.

of MYPT1 is not clear [6]. In parallel, direct phosphorylation of the LIM kinases (LIMK1 and LIMK2) by ROCK causes its activation and phosphorylation of cofilin, which inactivates its actin disassembly function [7]. Together, these two parallel pathways combine to mediate many of the effects of ROCK on the cytoskeleton [8,9], although several other ROCK substrates including Ezrin and myristoylated alanine-rich C kinase substrate have been described that also mediate some of the functions of ROCK [10,11]. Because of its important functions on the cytoskeleton, cellular motility, and cell growth and apoptosis, many studies have shown the importance of ROCK activity in diseases where improper regulation of these processes is crucial for disease pathology, including growth/invasion/metastasis in cancer, high blood pressure, glaucoma, cerebral cavernous malformation, and a large host of other diseases that are reviewed extensively elsewhere [8,12–15]. Our goal in this chapter is to analyze the current state of small molecule ROCK inhibitors as tools and therapeutic leads, and strategies that we believe are necessary to further develop these next-generation inhibitors.

2. FASUDIL AND Y-27632

Importantly, pharmacological inhibitors of ROCK have been and continue to be instrumental in understanding the many functions of ROCK and upstream Rho GTPases in many diverse cellular processes and diseases. In this chapter, we discuss several different classes of such ROCK inhibitors, paying extra attention to their relative potency, selectivity, and their on-target cellular or *in vivo* properties and thus their potential use as tools to study ROCK function in cancer or other diseases. We also discuss how their further development and proper use as tools can be used to validate ROCK as a therapeutic target in multiple disease classes.

Fasudil (HA-1077) was originally discovered as a small molecule that blocked cerebral vasospasm in animal models [16] and was used to demonstrate that ROCK phosphorylates MYPT to regulate myosin contractility [17]. Importantly, fasudil has been shown to be safe and effective in clinical trials for cerebral vasospasm [12,18–20] and has been evaluated in clinical trials for pulmonary hypertension and Raynaud's phenomenon, which, like cerebral vasospasm, have abnormal muscle contractility at the heart of their pathophysiology [21,22]. In addition, fasudil has also been shown to block cancer invasion, metastasis, or angiogenesis in animal models of small-cell lung, glioblastoma, ovarian, and breast cancer [23–27].

However, much care must be taken in interpreting any results with fasudil, since an early study showed significant inhibition of several other kinases [28]. Furthermore, in a more comprehensive study of a panel of 300 kinases, fasudil was shown to inhibit MAP4K4, MAP4K5, PRK1, and PRKX more potently than ROCK, and ARK5, CDK3, CDK5, CK1a1, DAPK1, EPHA1, FRK/PTK5, MAP4K2, MAP4K4, MAPKAPK5, MEKK2, MELK, MNK2, MSK1, MSK2, MST3, p70S6K, PAK5, PIM1, protein kinase C (PKC) epsilon, PRKD1, PRKD3, PKA, PKG1A, and RSK3 only slightly less potently than ROCK [29]. Especially as several of the abovementioned kinases have well-documented roles in cancer and cytoskeletal organization, extreme caution must thus be used to interpret any results obtained with fasudil, as there is virtually no way to know whether the phenotype is solely dependent on its inhibition of ROCK.

In addition to fasudil, the structurally unrelated Y-27632 was originally discovered as a small molecule that inhibits smooth muscle contraction by acting as an ATP-competitive inhibitor of ROCK [30] (Fig. 9.2A). Shortly thereafter, it was demonstrated that Y-27632 and ROCK dominant negatives blocked actomyosin contractility, invasion, Ras-dependent transformation, and anchorage-independent growth of cancer cell lines, strongly implicating ROCK in cancer initiation and progression [31,32]. Y-27632, often alone, has been used to implicate ROCK in an astounding array of biological processes too numerous to discuss here. However, as with fasudil, strong caution must be used to interpret results obtained with this inhibitor alone, as two independent studies profiling Y-27632 against significant portions of the kinome each demonstrated that it significantly inhibits 15 additional kinases besides ROCK including multiple PKC family members, which are themselves involved in a vast array of cellular processes [29,33]. In addition to not being very selective inhibitors of ROCK, both fasudil and Y-27632 do not show very strong potency *in vitro* with purified proteins or in cells (most commonly micromolar for both; Table 9.1 and [34]) compared to other important kinase inhibitor tool compounds and clinically approved kinase inhibitors, many of which are in the single- or double-digit nanomolar range, meaning that high amounts of these inhibitors are needed to inhibit ROCK activity in cells and *in vivo*, increasing the chance of off-target effects on other cellular targets besides kinases. These known issues with fasudil and Y-27632 have prompted a flurry of recent research into the discovery and use of more potent and selective ROCK inhibitors as tools and leads for therapeutics.

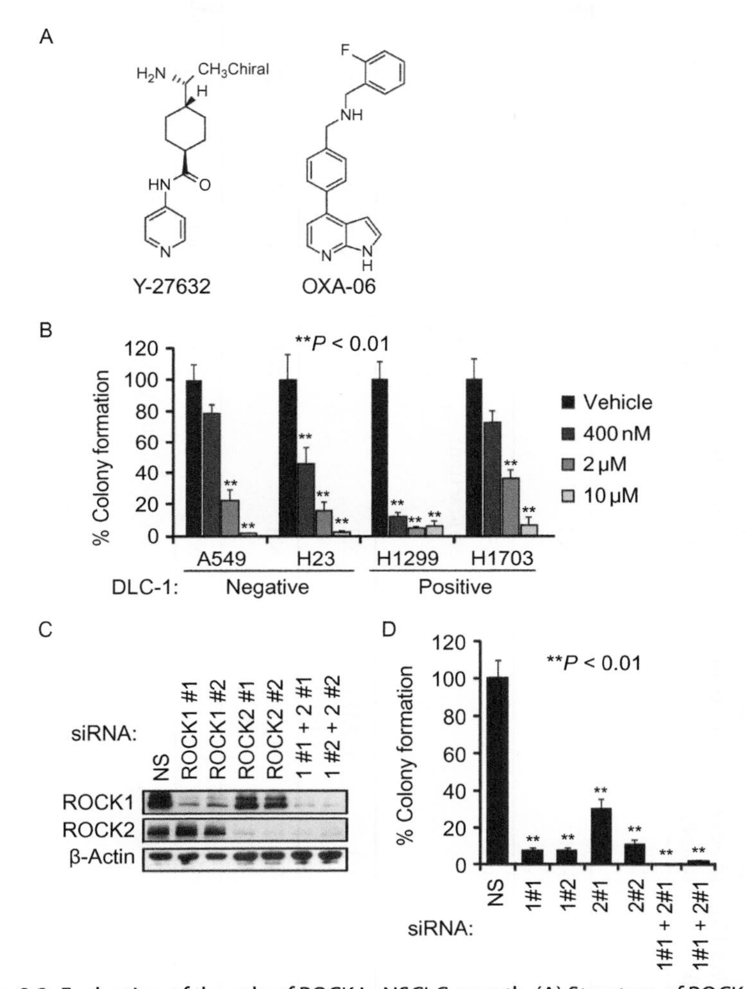

Figure 9.2 Evaluation of the role of ROCK in NSCLC growth. (A) Structure of ROCK inhibitors. (B) Colony formation of NSCLC cell lines in soft agar in growth medium supplemented with vehicle (DMSO) or the indicated concentrations of OXA-06. The number of viable proliferating colonies was stained by MTT and counted after 14 (A549 and H1299) or 30 (H23 and H1703) days. Data shown are the percentage colonies relative to vehicle ± the average standard deviation of triplicate wells. (C) siRNA suppression of ROCK expression. H1299 NSCLC cells were transfected with nonspecific (NS) siRNA or two different siRNA sequences targeting ROCK1 (1#1 and 1#2) or ROCK2 (2#1 and 2#2). After 48 h, lysates were harvested and western blot analysis was done using the indicated antibodies. (D) Suppression of ROCK1 or ROCK2 expression alone is sufficient to impair NSCLC anchorage-independent growth. Forty-eight-hour post-transfection cells were suspended in soft agar. The number of viable proliferating colonies was stained by MTT (3-(4,5-dimethylthiazol-2-yl)-2,5-diphenyltetrazolium bromide) and counted after 14 days. Data shown are the percentage soft agar colonies relative to vehicle ± the average standard deviation of triplicate wells.

Table 9.1 ROCK inhibitors

Organization	Compound name	*In vitro* IC_{50} (nM)	*In vitro* selectivity (# active/# tested)	Cellular IC_{50} (phosphorylation) (nM)	Cellular IC_{50} (phenotype) (nM)	*In vivo* effects	Target indication	References
Scripps	SR-6494	0.4	0/5	<6; MLC (T18/S19)	NT	NT	General	[39]
Moffitt Cancer Center	5g-Mes	0.45	5/20	NT	NT	NT	Cancer	[57]
Bayer	Azaindole 1	0.6	0/112	NT	65 (muscle cell contraction)	B.p. decrease in rats/dogs	Hypertension	[41]
Scripps	Compound 2	<1	0/1	4; MLC (T18/S19)	NT	NT	General	[40]
Scripps	Compound 35	<1	7/442	51; MLC (T18/S19)	NT	NT	General	[42]
Scripps	Compound 47	1	0/1	NT	NT	NT	General	[58]
GSK	GSK269962A	1.6	1/10	NT	35 (vasodilation)	B.p. decrease in rats	Hypertension	[43]
Kitasato University	HA-1152P	1.6	0/2	NT	NT	NT	Vasospasm	[59]
Scripps	Compound 30	1.7	0/1	85; MLC (T18/S19)	NT	NT	General	[60]

GSK	Compound 6n	1.8	3/30	NT	35 (vasodilation)	B.p. decrease in rats	Hypertension	[61]
Scripps	SR-3850	<2	0/6	< 6; MLC (T18/S19)	NT	NT	General	[62]
Scripps	Compound 43	2	0/1	12; MLC (T18/S19)	NT	NT	General	[63]
Scripps	Compound 22	2	0/1	46; MLC (T18/S19)	NT	NT	General	[64]
Scripps	SR-3677	3	5/353	4; MLC (T18/S19)	NT	NT	General	[65]
Kirin Brewery	Compound 24v	3	NT	NT	1000 (lymphoma chemotaxis)	NT	General	[66]
Nagoya University	Y-39983	4	0/2	NT	NT	Ocular pressure decrease in rabbits	Intraoccular pressure	[67]
GSK	Compound 21	5	0/31	NT	256 (vasodilation)	B.p. decrease in rats	Hypertension	[68]

Continued

Table 9.1 ROCK inhibitors—cont'd

Organization	Compound name	*In vitro* IC_{50} (nM)	*In vitro* selectivity (# active/# tested)	Cellular IC_{50} (phosphorylation) (nM)	Cellular IC_{50} (phenotype) (nM)	*In vivo* effects	Target indication	References
GSK	SB-7720770-B	5.6	2/10	NT	39 (vasodilation)	B.p. decrease in rats	Hypertension	[43]
Moffitt Cancer Center	RKI-1447	6	NT	~300; MCL2 (S19)	710 (soft agar growth)	Reduced breast tumor volume	Breast cancer	[69]
Pfizer	PF-4950834	8	4/228	76; MLC (T18/S19)	69 (T cell migration)	NT	Inflammation	[70]
GSK	Compound 13	9	0/30	NT	130 (vasodilation)	B.p. decrease in rats	Hypertension	[71]
OSI (Astellas)	OXA-06	10	6/167	300; MYPT1 (T696)	~100–3000 (soft agar growth)	NT	Lung cancer	[33]
Kirin Brewery	Compound 30-S	25	NT	1000–10,000; MYPT1 (S853)	1000 (lymphoma chemotaxis)	NT	General	[72]
Sanofi-Aventis	SAR407899	36	0/78	~2000; MYPT (T696)	122–280 (vasodilation)	B.p. decrease in rats	Hypertension	[73]

Moffitt Cancer Center	RKI-18	349	NT	~2000; MCL2 (S19)	8000 (soft agar growth)	NT	Breast cancer	[74]
Yoshitomi Pharmaceutical	Y-27632	800	15/167	Wide range from many studies	Wide range	Wide range	Wide range	[30]
Asahi Chemical	Fasudil (HA-1077)	1900	29/300	Wide range from many studies	Wide range	Wide range	Vasospasm	[75]
Harvard University	Rockout	25,000	15/300	NT	12,000 (cell blebbing)	NT	General	[76]
Moffitt Cancer Center	RKI-24	100 (ROCK2), 1700 (ROCK1)	NT	<1000; MCL2 (S19)	NT	NT	Breast cancer	[54]
Surface Logix	SLx-2119	105 (ROCK2), 24,000 (ROCK1)	NT	NT	NT	NT	Fibrosis	[55]

NT, not tested.

3. PROPER USE OF ROCK INHIBITORS AS TOOLS

Before we discuss specific examples of some of these next-generation ROCK inhibitors, it is helpful to set the stage by discussing theoretically both what properties a good ROCK inhibitor should have and what experimental components an ideal study using inhibitors to define the involvement of ROCK in biological process or as a therapeutic disease target should be. First, for a tool compound, the ideal properties should of course include excellent potency (ideally IC_{50} values in the picomolar or low nanomolar range against ROCK1 and ROCK2), selectivity (ideally 100-fold or more selective for ROCK against a very large panel of kinases), and cell-based inhibition of at least two ROCK downstream phosphorylation pathways (ideally IC_{50} values in the nanomolar range for both). This last requirement, although unfortunately not performed in many publications using ROCK inhibitors, is essential to provide evidence of cellular activity that should correlate well to cellular phenotype if the phenotype is on-target for ROCK inhibition. Although there are many described substrates of ROCK, the two most commonly utilized and well-characterized pathways to measure ROCK activity and inhibition are the direct ROCK target MYPT1 and its downstream substrate MLC, and the direct ROCK target LIMK and its downstream target cofilin, as described earlier and shown in Fig. 9.1B.

For these cellular on-target ROCK inhibition assays, there are a multitude of phospho-specific antibodies for the ROCK-dependent phosphorylation sites on all four of these proteins, and all have been used extensively in the literature to detect inhibition of ROCK activity by inhibitors (Fig. 9.1B). In our experience in multiple pancreatic and non-small-cell lung cancer (NSCLC) cell lines, phosphorylation of both MYPT1 T853 and cofilin S3 were more sensitive to ROCK inhibition than MYPT1 T696 [33,35], but cell type differences are likely and thus multiple or all of these phosphorylation sites should be evaluated for different cell lines to determine which will most accurately report ROCK inhibition in various cell types or tissues. Most studies have used western blots with the above antibodies to determine on-target cell activity for ROCK inhibitors, but the use of quantitative assays such as enzyme-linked immunosorbent assays has been used to determine the cell-based effects of ROCK inhibitors in a much higher throughput and more quantitative manner than western blots, enabling their use to both characterize individual compounds in a large number of cell lines and promote compound screening funnel progression for ROCK inhibitors [34,35]. Further advances in cell-based techniques for detection of downstream substrate phosphorylation will increasingly allow even more straightforward and precise detection of ROCK activity in response to inhibitors. However, there are several caveats to using downstream ROCK substrate phosphorylation as a surrogate for ROCK inhibition in cells. Firstly, other kinases may also be responsible for the given phosphorylation sites and thus ROCK inhibition may not fully or even substantially block phosphorylation of these sites. For example, Integrin-linked Kinase can also phosphorylate S696 on MYPT1 [36]. Thus, it is always prudent to characterize multiple downstream targets to find the phosphorylation activities that give the best dynamic range in response to ROCK inhibition. Secondly, the particular downstream ROCK phosphorylation pathways measured may not necessarily be the ones important for the activity of ROCK for a given phenotype, another good reason for measurement of multiple downstream substrates. Even considering these caveats, measurement of the above-described phosphorylation events in response to ROCK inhibitors is essential to interpreting any phenotypes.

After determining that ROCK downstream phosphorylation is potently inhibited by the ROCK inhibitors, cellular phenotypes (such as migration, contractility, and anchorage-independent growth that are hypothesized to be dependent on ROCK) should be quantified in a concentration response manner with the same concentrations used to quantify inhibition of cellular target phosphorylation. If truly on-target, the IC_{50}s for compound inhibition of downstream phosphorylation should be comparable to IC_{50}s for ROCK-dependent cellular phenotype. Importantly, to more strongly validate ROCK's involvement in a biological process, at least two structurally unrelated ROCK inhibitors with different kinase selectivity profiles as well as siRNA or shRNA targeting both ROCK1 and ROCK2 should be utilized. Even if a particular inhibitor is shown to not inhibit any other kinase, there is still the possibility of non-kinase off-target effects that could produce the desired phenotype. If each of the different inhibitors as well as genetic knockdown leads to the same phenotypes, the chances of off-target effects leading to the observed phenotype are dramatically reduced.

While it is beyond the scope of this chapter to discuss all of the studies using ROCK inhibitors as cellular tools to investigate ROCK as a therapeutic disease target, we describe our study with a novel ROCK inhibitor (OXA-06) that fulfills most of our suggested requirements for properly using ROCK inhibitors to study ROCK in disease [33]. In our study, we hypothesized that ROCK may be a therapeutic target in the subset of NSCLC cell lines that are characterized by the loss of expression of DLC1 (deleted in liver cancer 1), a Rho GTPase-selective GTPase activating protein (RhoGAP) and tumor suppressor. Loss or decreased *DLC1* RNA expression was found in 95% NSCLC tumor tissue [37] and we found that a majority of NSCLC cell lines lacked DLC-1 protein expression [38]. Since DLC-1 is a RhoGAP and negative regulator of RhoA activity, the loss of DLC-1 expression should lead to increased formation of RhoA-GTP and consequently should cause increased ROCK activation. Thus, the goal of our study was to assess the loss of DLC-1 protein expression as a biomarker for NSCLC cell lines that would be sensitive to ROCK inhibition.

To this end, we discovered a novel ROCK inhibitor (OXA-06) that is significantly more potent and somewhat more selective than Y-27632 (OXA-06 was active against only 6 of the 167 kinases tested) (Fig. 9.2A). We then utilized OXA-06 to demonstrate dose-dependent inhibition of phosphorylation of MYPT1 and cofilin in a variety of NSCLC cell lines, as well as a blockade of anchorage-independent (Fig. 9.2B) but not dependent growth and invasion with IC_{50} values comparable to IC_{50}s of inhibition of MYPT1 and cofilin inhibition. Furthermore, the structurally unrelated Y-27632 as well as siRNA against ROCK1 and ROCK2 in the same NSCLC cell lines also decreased downstream target phosphorylation, reduced anchorage-independent (Fig. 9.2C) but not dependent growth, and caused the same cell cycle defects as OXA-06 (not shown). Taken together, this study provides a comprehensive approach to assess the role of ROCK in a specific biological process. Additionally, our body of evidence strongly implicates ROCK as a therapeutic target in NSCLC.

For *in vivo* tools to investigate biology as well as lead molecules for therapeutics, ROCK inhibitors should fulfill all the required criteria discussed above for cellular tools and additionally of course have favorable PK/PD and toxicity profiles in animal models. Additionally, a suitable biomarker such as downstream target phosphorylation reduction in plasma should always be used to ensure that the inhibitor is reducing ROCK activity *in vivo* that correlates with physiological activity of the inhibitor. More work needs to be done to before suitable biomarker assays are developed that could be measured in humans for clinical trial purposes, but eventually this would be the only sure way to monitor *in vivo* inhibitor activity.

4. NEXT-GENERATION ROCK INHIBITORS

Table 9.1 summarizes published ROCK inhibitors, ordered by their *in vitro* potency against ROCK. In many cases, only inhibition against ROCK1 or ROCK2 was measured, or which ROCK isoform was measured was not indicated, so for simplicity we just report one measured IC_{50} for each. In cases in which inhibition of both ROCK1 and ROCK2 was measured, for simplicity, we simply report the most potent IC_{50}. In all cases except the two isoform-specific inhibitors, inhibition of ROCK1 and ROCK2 was within a few fold. While it is important to keep in mind that these inhibition constants were all measured under different experimental conditions using a wide variety of assays, our table nonetheless provides an important starting point to roughly compare the different published inhibitors in terms of their *in vitro* and cell-based potencies as well as cell-based or *in vivo* phenotypic functions. A quick glance at the table strongly shows that many ROCK inhibitors are much more potent (some more than 1000-fold) and selective than fasudil and Y-27632, highlighting that these next-generation ROCK inhibitors should be used in place of or at least in combination with fasudil and Y-27632. Importantly, there are five ROCK inhibitors with picomolar potency *in vitro*, and two of these demonstrate single-digit nanomolar inhibition of cellular MLC phosphorylation [39,40] although in these two cases only very limited kinase selectivity profiles were determined, so inhibition of off-target kinases cannot be ruled out. Although most of these inhibitors have not been screened for activity against large panels of kinases, azaindole 1 (Bayer) is a picomolar inhibitor that showed more than 100-fold selectivity against the 112 kinases profiled [41]. In addition, Compound 35 (Scripps) is a picomolar inhibitor that showed activity against only 7 of 442 tested when used at 3 μM (more than 3000 times its IC_{50} against ROCK), the highest off-target activity being 91% at 3 μM, suggesting at least 100-fold selectivity for ROCK against the rest of the kinome [42]. In addition to being potent and selective, many of these inhibitors also demonstrated favorable bioavailability and toxicity profiles in mice or rats, suggesting that they may be used for testing against diverse *in vivo* disease models [40,41,43]. Additionally, the vast majority of these inhibitors do not report both cellular phosphorylation IC_{50} values and cellular phenotypic IC_{50} values, as recommended above. Nevertheless, it is obvious that there are many good leads for both tools and potential therapeutics for ROCK inhibitors based on the excellent potencies, kinase selectivity profiles, and *in vivo* properties.

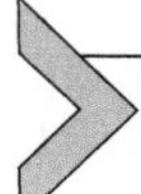

5. FUTURE DIRECTIONS FOR THE DEVELOPMENT OF ROCK INHIBITORS

While Y-27632 and fasudil have shown benefit in mouse models of diverse diseases such as cancer, glaucoma, hypertension, stroke, and other diseases [44–50], their relatively poor potency and selectivity against other kinases strongly cautions that better tool compounds and therapeutic leads are sorely needed. The treasure trove of next-generation ROCK inhibitors discussed in this chapter can importantly serve as new tools and leads to investigate disease biology in the same animals used to test Y-27632 and fasudil, as long as the guidelines described above are used to demonstrate on-target mechanism. In addition, novel mouse models such as a genetically engineered mouse models for cancer could be used to test the safety and efficacy of these inhibitors in more physiologically relevant settings. Additionally, in cancer for instance, large panels of cell lines could be used with structurally diverse next-generation ROCK inhibitors to investigate the genetic underpinnings for drug sensitivity such that rational patient selection could be accomplished for clinical trials. In addition to the investigation of next-generation ROCK inhibitors as therapeutics, such molecules could be used as molecular tools to investigate the proteomics of Rho signaling and cytoskeletal processes in a time-dependent and reversible manner that genetic means cannot. Although the ROCK inhibitors discussed here could be used as described above, novel types of ROCK inhibitors would also be important to increase the size and diversity of the ROCK inhibitor toolkit.

While there are a large number of potent ATP-competitive ROCK inhibitors, selectivity will always be a challenging problem due to the high conservation of the ATP-binding pocket between kinases, especially within the AGC family. One approach to circumvent this could be the development of allosteric ROCK inhibitors. Besides being activated by Rho-GTP, ROCK also has a pleckstrin homology (PH) domain that binds lipids and ROCK2 but not ROCK1 can be activated in the absence of RhoA-GTP by arachidonic acid and phosphatidylinositol phosphates [51]. Such inhibitors have the potential for exquisite selectivity against other kinases, in analogy to the Merck allosteric Akt inhibitor MK-2206 [52], as well having completely different mechanisms of action such as inhibiting localization that could also target nonkinase functions of ROCK.

While the vast majority of the inhibitors listed have roughly equal potency against ROCK1 and ROCK2 due to the extremely high sequence conservation between the two, it is clear that the two isoforms have some overlapping but in many cases nonredundant functions in many different aspects of biology that are just beginning to become understood [8,44]. For example, ROCK2 is important in migration polarity while ROCK1 is crucial for proper cell elongation during migration [53]. Potent and isoform-specific inhibitors could be extremely useful in the understanding of these different isoforms in various diseases and could potentially have dramatically lower on-target side effects than pan-inhibitors. Two inhibitors have demonstrated selectivity for ROCK2 over ROCK1 [54,55]. In particular, SLx-2119 from Surface Logix shows more than 200-fold selectivity, demonstrating that despite their high sequence conservation, it is possible to develop isoform-specific molecules.

6. CONCLUSIONS

In this chapter we describe the use of ROCK inhibitors as a therapeutic strategy for diseases where RhoA activation has a driver role. Since ROCK activity does promote key functions associated with RhoA activation, in particular those relating to cytoskeletal organization and cell motility, this strategy will likely be useful for many RhoA-driven pathologic states. However, RhoA does utilize additional effectors with functions also likely to be important in human disease. Therefore, additional strategies may be needed for effective inhibition of RhoA in some settings. RhoA does utilize additional protein kinase substrates [56], such as the PKN and citron protein kinase, that may serve as additional targets for RhoA inhibition in some settings.

ACKNOWLEDGMENTS

Due to space limitations, we apologize to colleagues whose work we could not include. We thank Jenni Sells for assistance with manuscript preparation. Our research was supported in part by grants from the US National Institutes of Health to C. J. Der and a fellowship from the American Cancer Society to D. Vigil.

REFERENCES

[1] T. Ishizaki, M. Maekawa, K. Fujisawa, K. Okawa, A. Iwamatsu, A. Fujita, N. Watanabe, Y. Saito, A. Kakizuka, N. Morii, S. Narumiya, The small GTP-binding protein Rho binds to and activates a 160 kDa Ser/Thr protein kinase homologous to myotonic dystrophy kinase, EMBO J. 15 (8) (1996) 1885–1893.
[2] T. Ishizaki, M. Naito, K. Fujisawa, M. Maekawa, N. Watanabe, Y. Saito, S. Narumiya, p160ROCK, a Rho-associated coiled-coil forming protein kinase, works downstream of Rho and induces focal adhesions, FEBS Lett. 404 (2–3) (1997) 118–124.
[3] K. Kimura, M. Ito, M. Amano, K. Chihara, Y. Fukata, M. Nakafuku, B. Yamamori, J. Feng, T. Nakano, K. Okawa, A. Iwamatsu, K. Kaibuchi, Regulation of myosin phosphatase by Rho and Rho-associated kinase (Rho-kinase), Science 273 (5272) (1996) 245–248.
[4] M. Amano, M. Ito, K. Kimura, Y. Fukata, K. Chihara, T. Nakano, Y. Matsuura, K. Kaibuchi, Phosphorylation and activation of myosin by Rho-associated kinase (Rho-kinase), J. Biol. Chem. 271 (34) (1996) 20246–20249.

[5] M. Ito, T. Nakano, F. Erdodi, D.J. Hartshorne, Myosin phosphatase: structure, regulation and function, Mol. Cell. Biochem. 259 (1–2) (2004) 197–209.
[6] G. Totsukawa, Y. Yamakita, S. Yamashiro, D.J. Hartshorne, Y. Sasaki, F. Matsumura, Distinct roles of ROCK (Rho-kinase) and MLCK in spatial regulation of MLC phosphorylation for assembly of stress fibers and focal adhesions in 3T3 fibroblasts, J. Cell Biol. 150 (4) (2000) 797–806.
[7] M. Maekawa, T. Ishizaki, S. Boku, N. Watanabe, A. Fujita, A. Iwamatsu, T. Obinata, K. Ohashi, K. Mizuno, S. Narumiya, Signaling from Rho to the actin cytoskeleton through protein kinases ROCK and LIM-kinase, Science 285 (5429) (1999) 895–898.
[8] N. Rath, M.F. Olson, Rho-associated kinases in tumorigenesis: re-considering ROCK inhibition for cancer therapy, EMBO Rep. 13 (10) (2012) 900–908.
[9] M. Amano, Y. Fukata, K. Kaibuchi, Regulation and functions of Rho-associated kinase, Exp. Cell Res. 261 (1) (2000) 44–51.
[10] T. Matsui, M. Maeda, Y. Doi, S. Yonemura, M. Amano, K. Kaibuchi, S. Tsukita, Rho-kinase phosphorylates COOH-terminal threonines of ezrin/radixin/moesin (ERM) proteins and regulates their head-to-tail association, J. Cell Biol. 140 (3) (1998) 647–657.
[11] H. Nagumo, M. Ikenoya, K. Sakurada, K. Furuya, T. Ikuhara, H. Hiraoka, Y. Sasaki, Rho-associated kinase phosphorylates MARCKS in human neuronal cells, Biochem. Biophys. Res. Commun. 280 (3) (2001) 605–609.
[12] M.F. Olson, Applications for ROCK kinase inhibition, Curr. Opin. Cell Biol. 20 (2) (2008) 242–248.
[13] L. Yao, M.J. Romero, H.A. Toque, G. Yang, R.B. Caldwell, R.W. Caldwell, The role of RhoA/Rho kinase pathway in endothelial dysfunction, J. Cardiovasc. Dis. Res. 1 (4) (2010) 165–170.
[14] M. Amano, M. Nakayama, K. Kaibuchi, Rho-kinase/ROCK: a key regulator of the cytoskeleton and cell polarity, Cytoskeleton (Hoboken) 67 (9) (2010) 545–554.
[15] B.T. Richardson, C.F. Dibble, A.L. Borikova, G.L. Johnson, Cerebral cavernous malformation is a vascular disease associated with activated RhoA signaling, Biol. Chem. 394 (1) (2013) 35–42.
[16] M. Takayasu, Y. Suzuki, M. Shibuya, T. Asano, M. Kanamori, T. Okada, N. Kageyama, H. Hidaka, The effects of HA compound calcium antagonists on delayed cerebral vasospasm in dogs, J. Neurosurg. 65 (1) (1986) 80–85.
[17] Y. Suzuki, M. Yamamoto, H. Wada, M. Ito, T. Nakano, Y. Sasaki, S. Narumiya, H. Shiku, M. Nishikawa, Agonist-induced regulation of myosin phosphatase activity in human platelets through activation of Rho-kinase, Blood 93 (10) (1999) 3408–3417.
[18] Y. Suzuki, M. Shibuya, S. Satoh, H. Sugiyama, M. Seto, K. Takakura, Safety and efficacy of fasudil monotherapy and fasudil–ozagrel combination therapy in patients with subarachnoid hemorrhage: sub-analysis of the post-marketing surveillance study, Neurol. Med. Chir. (Tokyo) 48 (6) (2008) 241–247, discussion 7–8.
[19] A. Masumoto, M. Mohri, H. Shimokawa, L. Urakami, M. Usui, A. Takeshita, Suppression of coronary artery spasm by the Rho-kinase inhibitor fasudil in patients with vasospastic angina, Circulation 105 (13) (2002) 1545–1547.
[20] G.J. Velat, M.M. Kimball, J.D. Mocco, B.L. Hoh, Vasospasm after aneurysmal subarachnoid hemorrhage: review of randomized controlled trials and meta-analyses in the literature, World Neurosurg. 76 (5) (2011) 446–454.
[21] B. Kojonazarov, A. Myrzaakhmatova, T. Sooronbaev, T. Ishizaki, A. Aldashev, Effects of fasudil in patients with high-altitude pulmonary hypertension, Eur. Respir. J. 39 (2) (2012) 496–498.
[22] A. Fava, P.K. Wung, F.M. Wigley, L.K. Hummers, N.R. Daya, S.R. Ghazarian, F. Boin, Efficacy of Rho kinase inhibitor fasudil in secondary Raynaud's phenomenon, Arthritis Care Res. (Hoboken) 64 (6) (2012) 925–929.

[23] X. Yang, J. Di, Y. Zhang, S. Zhang, J. Lu, J. Liu, W. Shi, The Rho-kinase inhibitor inhibits proliferation and metastasis of small cell lung cancer, Biomed. Pharmacother. 66 (3) (2012) 221–227.
[24] H. Nakabayashi, K. Shimizu, HA1077, a Rho kinase inhibitor, suppresses glioma-induced angiogenesis by targeting the Rho-ROCK and the mitogen-activated protein kinase kinase/extracellular signal-regulated kinase (MEK/ERK) signal pathways, Cancer Sci. 102 (2) (2011) 393–399.
[25] L. Deng, G. Li, R. Li, Q. Liu, Q. He, J. Zhang, Rho-kinase inhibitor, fasudil, suppresses glioblastoma cell line progression in vitro and in vivo, Cancer Biol. Ther. 9 (11) (2010) 875–884.
[26] S. Ogata, K. Morishige, K. Sawada, K. Hashimoto, S. Mabuchi, C. Kawase, C. Ooyagi, M. Sakata, T. Kimura, Fasudil inhibits lysophosphatidic acid-induced invasiveness of human ovarian cancer cells, Int. J. Gynecol. Cancer 19 (9) (2009) 1473–1480.
[27] H. Ying, S.L. Biroc, W.W. Li, B. Alicke, J.A. Xuan, R. Pagila, Y. Ohashi, T. Okada, Y. Kamata, H. Dinter, The Rho kinase inhibitor fasudil inhibits tumor progression in human and rat tumor models, Mol. Cancer Ther. 5 (9) (2006) 2158–2164.
[28] J. Bain, H. McLauchlan, M. Elliott, P. Cohen, The specificities of protein kinase inhibitors: an update, Biochem. J. 371 (Pt 1) (2003) 199–204.
[29] T. Anastassiadis, S.W. Deacon, K. Devarajan, H. Ma, J.R. Peterson, Comprehensive assay of kinase catalytic activity reveals features of kinase inhibitor selectivity, Nat. Biotechnol. 29 (11) (2011) 1039–1045.
[30] M. Uehata, T. Ishizaki, H. Satoh, T. Ono, T. Kawahara, T. Morishita, H. Tamakawa, K. Yamagami, J. Inui, M. Maekawa, S. Narumiya, Calcium sensitization of smooth muscle mediated by a Rho-associated protein kinase in hypertension, Nature 389 (6654) (1997) 990–994.
[31] K. Itoh, K. Yoshioka, H. Akedo, M. Uehata, T. Ishizaki, S. Narumiya, An essential part for Rho-associated kinase in the transcellular invasion of tumor cells, Nat. Med. 5 (2) (1999) 221–225.
[32] E. Sahai, T. Ishizaki, S. Narumiya, R. Treisman, Transformation mediated by RhoA requires activity of ROCK kinases, Curr. Biol. 9 (3) (1999) 136–145.
[33] D. Vigil, T.Y. Kim, A. Plachco, A.J. Garton, L. Castaldo, J.A. Pachter, H. Dong, X. Chen, B. Tokar, S.L. Campbell, C.J. Der, ROCK1 and ROCK2 are required for non-small cell lung cancer anchorage-independent growth and invasion, Cancer Res. 72 (20) (2012) 5338–5347.
[34] T. Schroter, E. Griffin, A. Weiser, Y. Feng, P. LoGrasso, Detection of myosin light chain phosphorylation – a cell-based assay for screening Rho-kinase inhibitors, Biochem. Biophys. Res. Commun. 374 (2) (2008) 356–360.
[35] A.J. Garton, L. Castaldo, J.A. Pachter, Quantitative high-throughput cell-based assays for inhibitors of ROCK kinases, Methods Enzymol. 439 (2008) 491–500.
[36] A. Muranyi, J.A. MacDonald, J.T. Deng, D.P. Wilson, T.A. Haystead, M.P. Walsh, F. Erdodi, E. Kiss, Y. Wu, D.J. Hartshorne, Phosphorylation of the myosin phosphatase target subunit by integrin-linked kinase, Biochem. J. 366 (Pt 1) (2002) 211–216.
[37] B.Z. Yuan, A.M. Jefferson, K.T. Baldwin, S.S. Thorgeirsson, N.C. Popescu, S.H. Reynolds, DLC-1 operates as a tumor suppressor gene in human non-small cell lung carcinomas, Oncogene 23 (7) (2004) 1405–1411.
[38] K.D. Healy, L. Hodgson, T.Y. Kim, A. Shutes, S. Maddileti, R.L. Juliano, K.M. Hahn, T.K. Harden, Y.J. Bang, C.J. Der, DLC-1 suppresses non-small cell lung cancer growth and invasion by RhoGAP-dependent and independent mechanisms, Mol. Carcinog. 47 (5) (2008) 326–337.
[39] Y. Yin, L. Lin, C. Ruiz, M.D. Cameron, J. Pocas, W. Grant, T. Schroter, W. Chen, D. Duckett, S. Schurer, P. Lograsso, Y. Feng, Benzothiazoles as Rho-associated kinase (ROCK-II) inhibitors, Bioorg. Med. Chem. Lett. 19 (23) (2009) 6686–6690.

[40] E.H. Sessions, S. Chowdhury, Y. Yin, J.R. Pocas, W. Grant, T. Schroter, L. Lin, C. Ruiz, M.D. Cameron, P. LoGrasso, T.D. Bannister, Y. Feng, Discovery and optimization of indole and 7-azaindoles as Rho kinase (ROCK) inhibitors (part-II), Bioorg. Med. Chem. Lett. 21 (23) (2011) 7113–7118.
[41] R. Kast, H. Schirok, S. Figueroa-Perez, J. Mittendorf, M.J. Gnoth, H. Apeler, J. Lenz, J.K. Franz, A. Knorr, J. Hutter, M. Lobell, K. Zimmermann, K. Munter, K.H. Augstein, H. Ehmke, J.P. Stasch, Cardiovascular effects of a novel potent and highly selective azaindole-based inhibitor of Rho-kinase, Br. J. Pharmacol. 152 (7) (2007) 1070–1080.
[42] X. Fang, Y. Yin, Y.T. Chen, L. Yao, B. Wang, M.D. Cameron, L. Lin, S. Khan, C. Ruiz, T. Schroter, W. Grant, A. Weiser, J. Pocas, A. Pachori, S. Schurer, P. Lograsso, Y. Feng, Tetrahydroisoquinoline derivatives as highly selective and potent Rho kinase inhibitors, J. Med. Chem. 53 (15) (2010) 5727–5737.
[43] C. Doe, R. Bentley, D.J. Behm, R. Lafferty, R. Stavenger, D. Jung, M. Bamford, T. Panchal, E. Grygielko, L.L. Wright, G.K. Smith, Z. Chen, C. Webb, S. Khandekar, T. Yi, R. Kirkpatrick, E. Dul, L. Jolivette, J.P. Marino Jr., R. Willette, D. Lee, E. Hu, Novel Rho kinase inhibitors with anti-inflammatory and vasodilatory activities, J. Pharmacol. Exp. Ther. 320 (1) (2007) 89–98.
[44] C. Hahmann, T. Schroeter, Rho-kinase inhibitors as therapeutics: from pan inhibition to isoform selectivity, Cell. Mol. Life Sci. 67 (2) (2010) 171–177.
[45] M. Surma, L. Wei, J. Shi, Rho kinase as a therapeutic target in cardiovascular disease, Future Cardiol. 7 (5) (2011) 657–671.
[46] M. Dong, B.P. Yan, J.K. Liao, Y.Y. Lam, G.W. Yip, C.M. Yu, Rho-kinase inhibition: a novel therapeutic target for the treatment of cardiovascular diseases, Drug Discov. Today 15 (15–16) (2010) 622–629.
[47] T.A. McKinsey, D.A. Kass, Small-molecule therapies for cardiac hypertrophy: moving beneath the cell surface, Nat. Rev. Drug Discov. 6 (8) (2007) 617–635.
[48] A. Lai, W.H. Frishman, Rho-kinase inhibition in the therapy of cardiovascular disease, Cardiol. Rev. 13 (6) (2005) 285–292.
[49] S.A. Doggrell, Rho-kinase inhibitors show promise in pulmonary hypertension, Expert Opin. Investig. Drugs 14 (9) (2005) 1157–1159.
[50] E. Hu, D. Lee, Rho kinase as potential therapeutic target for cardiovascular diseases: opportunities and challenges, Expert Opin. Ther. Targets 9 (4) (2005) 715–736.
[51] A. Yoneda, H.A. Multhaupt, J.R. Couchman, The Rho kinases I and II regulate different aspects of myosin II activity, J. Cell Biol. 170 (3) (2005) 443–453.
[52] H. Hirai, H. Sootome, Y. Nakatsuru, K. Miyama, S. Taguchi, K. Tsujioka, Y. Ueno, H. Hatch, P.K. Majumder, B.S. Pan, H. Kotani, MK-2206, an allosteric Akt inhibitor, enhances antitumor efficacy by standard chemotherapeutic agents or molecular targeted drugs in vitro and in vivo, Mol. Cancer Ther. 9 (7) (2010) 1956–1967.
[53] F.M. Vega, G. Fruhwirth, T. Ng, A.J. Ridley, RhoA and RhoC have distinct roles in migration and invasion by acting through different targets, J. Cell Biol. 193 (4) (2011) 655–665.
[54] R. Li, M.P. Martin, Y. Liu, B. Wang, R.A. Patel, J.Y. Zhu, N. Sun, R. Pireddu, N.J. Lawrence, J. Li, E.B. Haura, S.S. Sung, W.C. Guida, E. Schonbrunn, S.M. Sebti, Fragment-based and structure-guided discovery and optimization of Rho kinase inhibitors, J. Med. Chem. 55 (5) (2012) 2474–2478.
[55] M. Boerma, Q. Fu, J. Wang, D.S. Loose, A. Bartolozzi, J.L. Ellis, S. McGonigle, E. Paradise, P. Sweetnam, L.M. Fink, M.C. Vozenin-Brotons, M. Hauer-Jensen, Comparative gene expression profiling in three primary human cell lines after treatment with a novel inhibitor of Rho kinase or atorvastatin, Blood Coagul. Fibrinolysis 19 (7) (2008) 709–718.

[56] X.R. Bustelo, V. Sauzeau, I.M. Berenjeno, GTP-binding proteins of the Rho/Rac family: regulation, effectors and functions in vivo, Bioessays 29 (4) (2007) 356–370.
[57] R. Pireddu, K.D. Forinash, N.N. Sun, M.P. Martin, S.S. Sung, B. Alexander, J.Y. Zhu, W.C. Guida, E. Schonbrunn, S.M. Sebti, N.J. Lawrence, Pyridylthiazole-based ureas as inhibitors of Rho associated protein kinases (ROCK1 and 2), MedChemComm 3 (6) (2012) 699–709.
[58] S. Chowdhury, E.H. Sessions, J.R. Pocas, W. Grant, T. Schroter, L. Lin, C. Ruiz, M.D. Cameron, S. Schurer, P. LoGrasso, T.D. Bannister, Y. Feng, Discovery and optimization of indoles and 7-azaindoles as Rho kinase (ROCK) inhibitors (part-I), Bioorg. Med. Chem. Lett. 21 (23) (2011) 7107–7112.
[59] Y. Sasaki, M. Suzuki, H. Hidaka, The novel and specific Rho-kinase inhibitor (*S*)-(+)-2-methyl-1-[(4-methyl-5-isoquinoline)sulfonyl]-homopiperazine as a probing molecule for Rho-kinase-involved pathway, Pharmacol. Ther. 93 (2–3) (2002) 225–232.
[60] E.H. Sessions, M. Smolinski, B. Wang, B. Frackowiak, S. Chowdhury, Y. Yin, Y.T. Chen, C. Ruiz, L. Lin, J. Pocas, T. Schroter, M.D. Cameron, P. LoGrasso, Y. Feng, T.D. Bannister, The development of benzimidazoles as selective rho kinase inhibitors, Bioorg. Med. Chem. Lett. 20 (6) (2010) 1939–1943.
[61] R.A. Stavenger, H. Cui, S.E. Dowdell, R.G. Franz, D.E. Gaitanopoulos, K.B. Goodman, M.A. Hilfiker, R.L. Ivy, J.D. Leber, J.P. Marino Jr., H.J. Oh, A.Q. Viet, W. Xu, G. Ye, D. Zhang, Y. Zhao, L.J. Jolivette, M.S. Head, S.F. Semus, P.A. Elkins, R.B. Kirkpatrick, E. Dul, S.S. Khandekar, T. Yi, D.K. Jung, L.L. Wright, G.K. Smith, D.J. Behm, C.P. Doe, R. Bentley, Z.X. Chen, E. Hu, D. Lee, Discovery of aminofurazan-azabenzimidazoles as inhibitors of Rho-kinase with high kinase selectivity and antihypertensive activity, J. Med. Chem. 50 (1) (2007) 2–5.
[62] Y.T. Chen, T.D. Bannister, A. Weiser, E. Griffin, L. Lin, C. Ruiz, M.D. Cameron, S. Schurer, D. Duckett, T. Schroter, P. LoGrasso, Y. Feng, Chroman-3-amides as potent Rho kinase inhibitors, Bioorg. Med. Chem. Lett. 18 (24) (2008) 6406–6409.
[63] X. Fang, Y.T. Chen, E.H. Sessions, S. Chowdhury, T. Vojkovsky, Y. Yin, J.R. Pocas, W. Grant, T. Schroter, L. Lin, C. Ruiz, M.D. Cameron, P. LoGrasso, T.D. Bannister, Y. Feng, Synthesis and biological evaluation of 4-quinazolinones as Rho kinase inhibitors, Bioorg. Med. Chem. Lett. 21 (6) (2011) 1844–1848.
[64] E.H. Sessions, Y. Yin, T.D. Bannister, A. Weiser, E. Griffin, J. Pocas, M.D. Cameron, C. Ruiz, L. Lin, S.C. Schurer, T. Schroter, P. LoGrasso, Y. Feng, Benzimidazole- and benzoxazole-based inhibitors of Rho kinase, Bioorg. Med. Chem. Lett. 18 (24) (2008) 6390–6393.
[65] Y. Feng, Y. Yin, A. Weiser, E. Griffin, M.D. Cameron, L. Lin, C. Ruiz, S.C. Schurer, T. Inoue, P.V. Rao, T. Schroter, P. Lograsso, Discovery of substituted 4-(pyrazol-4-yl)-phenylbenzodioxane-2-carboxamides as potent and highly selective Rho kinase (ROCK-II) inhibitors, J. Med. Chem. 51 (21) (2008) 6642–6645.
[66] M. Iwakubo, A. Takami, Y. Okada, T. Kawata, Y. Tagami, H. Ohashi, M. Sato, T. Sugiyama, K. Fukushima, H. Iijima, Design and synthesis of Rho kinase inhibitors (II), Bioorg. Med. Chem. 15 (1) (2007) 350–364.
[67] H. Tokushige, M. Inatani, S. Nemoto, H. Sakaki, K. Katayama, M. Uehata, H. Tanihara, Effects of topical administration of y-39983, a selective rho-associated protein kinase inhibitor, on ocular tissues in rabbits and monkeys, Invest. Ophthalmol. Vis. Sci. 48 (7) (2007) 3216–3222.
[68] C.A. Sehon, G.Z. Wang, A.Q. Viet, K.B. Goodman, S.E. Dowdell, P.A. Elkins, S.F. Semus, C. Evans, L.J. Jolivette, R.B. Kirkpatrick, E. Dul, S.S. Khandekar, T. Yi, L.L. Wright, G.K. Smith, D.J. Behm, R. Bentley, C.P. Doe, E. Hu, D. Lee, Potent, selective and orally bioavailable dihydropyrimidine inhibitors of Rho kinase

(ROCK1) as potential therapeutic agents for cardiovascular diseases, J. Med. Chem. 51 (21) (2008) 6631–6634.
[69] R.A. Patel, K.D. Forinash, R. Pireddu, Y. Sun, N. Sun, M.P. Martin, E. Schonbrunn, N.J. Lawrence, S.M. Sebti, RKI-1447 is a potent inhibitor of the Rho-associated ROCK kinases with anti-invasive and antitumor activities in breast cancer, Cancer Res. 72 (19) (2012) 5025–5034.
[70] L.E. Rajagopalan, M.S. Davies, L.E. Kahn, C.M. Kornmeier, H. Shimada, T.A. Steiner, B.S. Zweifel, J.M. Wendling, M.A. Payne, R.F. Loeffler, B.L. Case, M.B. Norton, M.D. Parikh, O.V. Nemirovskiy, R.J. Mourey, J.L. Masferrer, T.P. Misko, S.A. Kolodziej, Biochemical, cellular, and anti-inflammatory properties of a potent, selective, orally bioavailable benzamide inhibitor of Rho kinase activity, J. Pharmacol. Exp. Ther. 333 (3) (2010) 707–716.
[71] K.B. Goodman, H. Cui, S.E. Dowdell, D.E. Gaitanopoulos, R.L. Ivy, C.A. Sehon, R.A. Stavenger, G.Z. Wang, A.Q. Viet, W. Xu, G. Ye, S.F. Semus, C. Evans, H.E. Fries, L.J. Jolivette, R.B. Kirkpatrick, E. Dul, S.S. Khandekar, T. Yi, D.K. Jung, L.L. Wright, G.K. Smith, D.J. Behm, R. Bentley, C.P. Doe, E. Hu, D. Lee, Development of dihydropyridone indazole amides as selective Rho-kinase inhibitors, J. Med. Chem. 50 (1) (2007) 6–9.
[72] M. Iwakubo, A. Takami, Y. Okada, T. Kawata, Y. Tagami, M. Sato, T. Sugiyama, K. Fukushima, S. Taya, M. Amano, K. Kaibuchi, H. Iijima, Design and synthesis of rho kinase inhibitors (III), Bioorg. Med. Chem. 15 (2) (2007) 1022–1033.
[73] M. Lohn, O. Plettenburg, Y. Ivashchenko, A. Kannt, A. Hofmeister, D. Kadereit, M. Schaefer, W. Linz, M. Kohlmann, J.M. Herbert, P. Janiak, S.E. O'Connor, H. Ruetten, Pharmacological characterization of SAR407899, a novel rho-kinase inhibitor, Hypertension 54 (3) (2009) 676–683.
[74] R.A. Patel, Y. Liu, B. Wang, R. Li, S.M. Sebti, Identification of novel ROCK inhibitors with anti-migratory and anti-invasive activities, Oncogene (2013).
[75] T. Asano, T. Suzuki, M. Tsuchiya, S. Satoh, I. Ikegaki, M. Shibuya, Y. Suzuki, H. Hidaka, Vasodilator actions of HA1077 in vitro and in vivo putatively mediated by the inhibition of protein kinase, Br. J. Pharmacol. 98 (4) (1989) 1091–1100.
[76] J.C. Yarrow, G. Totsukawa, G.T. Charras, T.J. Mitchison, Screening for cell migration inhibitors via automated microscopy reveals a Rho-kinase inhibitor, Chem. Biol. 12 (3) (2005) 385–395.

A Two-Hybrid Approach to Identify Inhibitors of the RAS–RAF Interaction

Vladimir Khazak*[,2], Susanne Eyrisch[†], Juran Kato[‡,1], Fuyuhiko Tamanoi[‡], Erica A. Golemis*[,§,2]

*Program in Biology, Priaxon Inc., Philadelphia, Pennsylvania, USA
[†]Program in Chemoinformatics, Priaxon AG, Munich, Germany
[‡]Department of Microbiology, Immunology, and Molecular Genetics, University of California, Los Angeles, Los Angeles, California, USA
[§]Program in Developmental Therapeutics, Fox Chase Cancer Center, Philadelphia, Pennsylvania, USA
[1]Current address: Pharmaceutical Research Division, Takeda Pharmaceutical Company Limited, Kanagawa 251-8555, Japan.
[2]Corresponding authors: e-mail address: khazak@priaxon.com; erica.golemis@fccc.edu

Contents

The Enzymes, Volume 33
ISSN 1874-6047
http://dx.doi.org/10.1016/B978-0-12-416749-0.00010-5

Abstract

MCP compounds were developed with the idea to inhibit RAS/RAF interaction. They were identified by carrying out high-throughput screens of chemical compounds for their ability to inhibit RAS/RAF interaction in the yeast two-hybrid assay. A number of compounds including MCP1, MCP53, and MCP110 were identified as active compounds. Their inhibition of the RAS signaling was demonstrated by examining RAF and MEK activities, phosphorylation of ERK as well as characterizing their effects on events downstream of RAF. Direct evidence for the inhibition of RAS/RAF interaction was obtained by carrying out co-IP experiments. MCP compounds inhibit proliferation of a wide range of human cancer cell lines. Combination studies with other drugs showed that MCP compounds synergize with MAPK pathway inhibitors as well as with microtubule-targeting chemotherapeutics. In particular, a strong synergy with paclitaxel was observed. Efficacy to inhibit tumor formation was demonstrated using mouse xenograft models. Combination of MCP110 and paclitaxel was particularly effective in inhibiting tumor growth in a mouse xenograft model of colorectal carcinoma.

1. INTRODUCTION

At present, the great expansion of genome-level information provided by application of next-generation sequencing, gene expression profiling, and proteomics is transforming our understanding of signaling pathway dysregulation in cancer. To best exploit this information therapeutically, it will be important to possess an arsenal of small molecules that can disrupt protein targets nominated as critical to the disease process. The yeast two-hybrid (Y2H) system is one of the first tractable approaches used for systematic query of protein–protein interactions (PPIs) in a cellular context [1,2]. Reengineering core Y2H reagents to increase capacity led to demonstration that this system could be used to inhibit protein interactions with peptides [3], RNA [4], protein complexes [5], and even small molecular weight compounds [6–8]. PPIs have long been considered to be promising but challenging drug targets [9]. Although a limited number of attempts have been made to use Y2H approaches in drug discovery research [7,10], this strategy never became a major avenue for discovering novel PPI inhibitors because of several factors. Among these factors, the reported low permeability of yeast cell wall membrane to organic small molecular weight compounds was a major consideration [11]. As we describe in this chapter, we successfully screened for compounds disrupting PPIs in yeast using specially engineered strains that overcame this limiting factor.

In our screening efforts, we sought to identify small molecules that disrupted the interaction between the oncogenes RAS and RAF [12,13]. The RAS GTPases (H, N, and K) and RAS-regulated signal transduction pathways are among the most validated targets for pharmacological interventions in multiple cancers. Mutational activation of RAS proteins occurs in about 30% of all cancers [13]. Almost 90% of pancreatic cancers, 40% of colorectal and 30% of lung carcinomas carry mutational activated RAS proteins [14]. RAS was also found frequently mutated in malignant melanomas (~20%), acute myeloid leukemia (17%), and several other malignancies [15,16]. In their active GTP-bound state, interactions between the RAS effector domain and the RAS-binding domains (RBD) of multiple effector proteins allow RAS to transiently transmit signals affecting cell growth, survival, and migration [17,18]. Mutationally activated RAS proteins lose their ability to hydrolyze GTP into GDP, and thus constantly induce signaling through their effector partners [19], leading to uncontrolled cell proliferation, increased survival, deregulation of normal cell adhesion and cell migration, and eventually to malignant transformation, invasion, and metastasis [20]. Beyond mutational activation of RAS, wild-type RAS proteins are essential in transmitting responses to extracellular stimuli that include growth factors, receptor tyrosine kinases (RTKs), integrins, and G protein-coupled receptors [21]; hence, inhibition of RAS can be imagined to have therapeutic benefit in limiting the activity of mutationally activated upstream oncoproteins, which include many RTKs.

Despite extended efforts to find effective and selective inhibitors of oncogenic RAS, no progress has been achieved in more than two decades, and RAS is now widely considered to be essentially "undruggable" [22]. In contrast, the pharmacological inhibition of RAS activators and effectors (especially the three major RAF–MEK–ERK, PI3K–AKT, and RALGEF effector pathways) has become a mainstream therapeutic strategy [20]. Kinase inhibitors targeting RAF1, BRAF, MEK, ERK, PI3K, and AKT have been extensively evaluated in clinical trials, with some showing considerable efficacy in subsets of patients [23,24]. For example, the multi-kinase inhibitor sorafenib that inhibits RAF1 as well as the RTKs VEGFR-2, -3, and additional RTKs has been recently approved for the treatment of renal and hepatocellular carcinomas [25,26]. A recently discovered inhibitor of mutationally activated BRAFV600, Zelboraf, became the first drug approved for the treatment of malignant melanoma in the last 30 years [27]. Unfortunately, neither of these drugs is effective against cancers with oncogenically mutated RAS, either as mono- or combination therapy [28,29], and both drugs lose efficacy over time as tumor mutations lead to acquired resistance [30]. In this context, the ability to target RAS–RAF signaling by an alternative mechanism becomes highly desirable.

In this chapter, we will summarize our development, optimization, and detailed characterization of a group of MCP compounds that represent first successful PPI inhibitors of the interaction between the human HRAS and RAF1 proteins. We will describe our approach to compound selection, which involved development of Y2H strains with increased permeability, and provide clear evidence of the ability of MCP compounds to inhibit oncogenic RAS-induced malignant transformation. Moreover, we will demonstrate the ability of MCP compounds to synergize with MAPK pathway targeted tyrosine kinase inhibitors and approved chemotherapeutic agents both in cell lines and *in vivo*, using *Caenorhabditis elegans* and in mouse xenograft models. Finally, we will share the lessons learned in the process of developing MCP compounds and provide robust decision tree applicable for future development of PPI inhibitors using a Y2H approach.

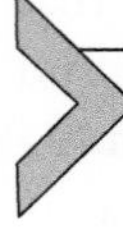

2. OUTLINE OF DISCOVERY PROCESS FOR MCP COMPOUNDS

To efficiently select specific and potent PPI inhibitors and thus direct synthetic chemical optimization efforts, we developed a robust decision tree (Fig. 10.1). We began with a primary compound screen using an optimized Y2H system including a *Saccharomyces cerevisiae* strain with significantly improved permeability to small organic molecules. In parallel, a compound library was assessed for ability to inhibit the interaction of RAS and RAF, versus an unrelated PPI, between two subunits of human RNA polymerase II, hsRPB7 and hsRPB4 [31,32]. Compounds that disrupted RAS–RAF but not hsRPB7–hsRPB4 interactions were selected for future analysis.

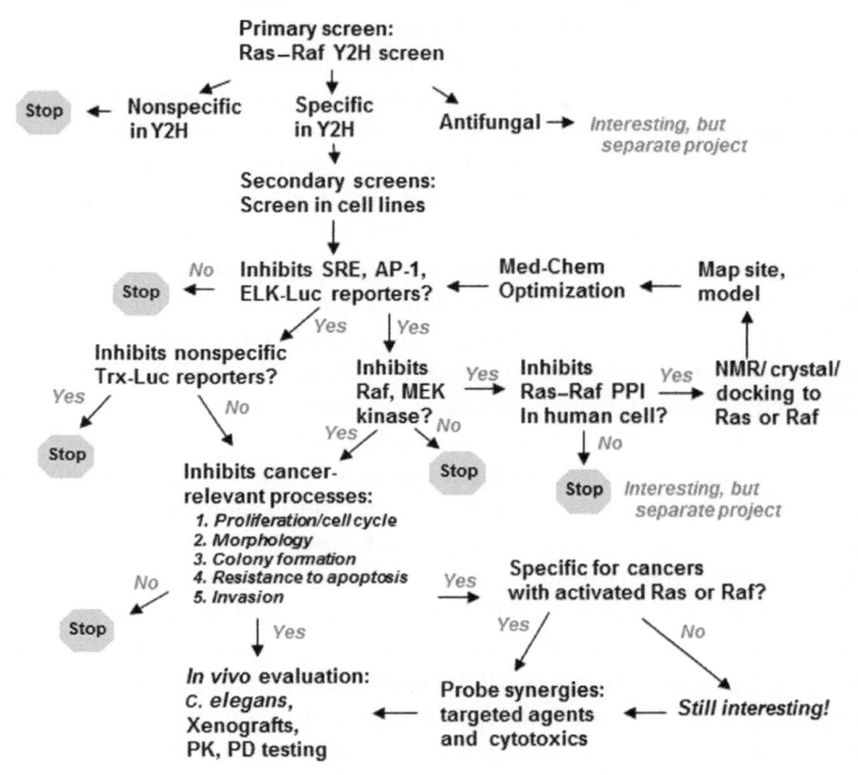

Figure 10.1 Decision tree for selection and characterization of RAS–RAF interaction inhibitors.

As a side note, this screen inevitably identified a subset of compounds that were generally cytotoxic to *S. cerevisiae*, but not necessarily cytotoxic to human cells. These compounds were later tested as promising antifungal agents in two strains of clinically relevant yeast *Candida albicans* and became an integral part of a separated drug discovery project [33].

To assess relatively large number of compounds identified in the primary screen and synthesized in SAR optimization efforts, we used a secondary screen to identify compound activity against critical RAS–RAF effectors. RAS activation of RAF induces signaling through MEK1/2 and ERK1/2 [34], resulting in the activation of ERK-regulated transcription factors AP1 and ELK1. We used mammalian cell lines stably transfected with luciferase reporter genes under the control of AP1- or ELK1-responsive promoters [35]. Compounds that induced a dose-dependent inhibition of cFos-SRE-Luc, AP-1-Luc, or the Elk1-Luc reporter genes in Chinese hamster ovary (CHO) and HeLa cell lines were next evaluated for fit to the "drug-like" parameters summarized by Lipinski [36]. Drug-like compounds were selected for further rounds of validation and chemical optimization. An important next step was to use an *in vitro* kinase assay involving Ras-induced activation of RAF1 protein [35,37]. We then asked if compounds inhibited phosphorylation of RAF1, MEK1, and ERK1 kinases in cells transiently transfected with mutationally activated HRAS versus RAF proteins or in a human fibrosarcoma cell line expressing activated NRAS [35]. MCP110 passed all of these tests. Next, to gain insight into the binding target of MCP110, we assessed whether this and related compounds could block the *in vitro* interaction between oncogenic RAS and the RAF-RBD fused to GST [38]. Positive activity in this assay was further supported by the demonstration that MCP110 reversed the transformed morphology of NIH3T3 fibroblasts stably transfected with oncogenic RAS, but not with a truncated version of RAF1 protein that lacked the RBD domain (RAF22W). Together, these results strongly supported activity of MCP110 targeting the RAS–RAF interface.

In subsequent analysis of MCP110, we have used molecular docking experiments to drive the SAR optimization process. We more broadly characterized biological activity of MCP compounds, measuring curtailment of anchorage-dependent and -independent growth, induction of apoptosis, and modulation of cell cycle progression. For these purposes we used a panel of human cancer cell lines with oncogenic K-, H-, or NRas proteins or mutationally activated BRAF [39–41]. We then gaged the ability of MCP compounds to synergize with several targeted therapeutics and cytotoxic drugs in various human cancer cell lines [39]. Finally, we assessed MCP bioavailability, pharmacokinetics and pharmacodynamics properties, and ability to inhibit cancer tumor growth *in vivo*. MCP compounds were shown to block mutated RAS/*let-60*-associated multivulva (Muv) phenotypes in *C. elegans* [38] and to inhibit tumor xenograft growth in mouse models [39].

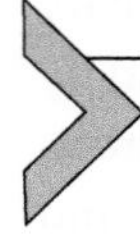

3. CREATION OF A Y2H STRAIN WITH IMPROVED PERMEABILITY

One of the major limitations of using the Y2H system for drug screening is the relatively low permeability of yeast cells to many low molecular weight compounds [42]. Upregulation of hexose transporters HXT9 or HXT11 significantly improves uptake of a number of pharmacophores by *S. cerevisiae* [43,44]. In addition, highly active ABC transporter proteins, members of pleotropic drug resistance (PDR) network, allow yeast to efficiently export xenobiotic compounds [45]. Many of the *S. cerevisiae* genes that regulate cell membrane permeability and the PDR network have been identified [46] and represent excellent targets to increase yeast cell penetration by organic molecules. The Pdr1p and Pdr3p proteins are members of a C6 zinc family of transcription factors that are necessary for transcription of the ABC drug efflux [47] transporters Pdr5p and Snq2p [48–50]. Inactivation of the PDR1 and PDR3 genes significantly compromises expression of Pdr5p and Snq2p and thus strongly decreases the efflux of xenobiotic compounds [50].

We exploited these properties of PDR and HXT mutants to create hyperpermeable yeast strains for Y2H high-throughput screening (HTS) of compound libraries. Through a series of successive genomic integrations, two cassettes containing galactose-inducible copies of yeast HXT9 and HXT11 genes (Fig. 10.2A) were inserted in the PDR1 and PDR3 chromosomal loci of the *S. cerevisiae* Y2H strains SKY48 and SKY191 [2,51]. The resulting strains SKY54 (*MATα ura3 his3 trp1 3LexA-operator—Leu2 λcI-operator-Lys2 pdr1::GAL1pro-HXT9 pdr3::GAL1pro-HXT11*) and SKY197 (*MATα ura3 his3 trp1 1LexA-operator—Leu2 λcI-operator-Lys2 pdr1:: GAL1pro-HXT9 pdr3::GAL1pro-HXT11*) were challenged with a panel of known antifungal agents to probe their permeability, using as controls the parental strains SKY48 and SKY191, respectively. Typical results showing discrete sensitivity of dose-dependent growth inhibition of parental versus hyperpermeable strains challenged with cyclohemide (CYH), 4-nitroquinoline-oxide (4-NQO), sulfomethuron methyl (SMM), and zeocin (Zeo) presented in Fig. 10.2B. Using an agar-plate compound permeability assay described in detail in Ref. [33], we showed that inactivation of PDR1 and PDR3 genes in SKY197 resulted in dramatic increase in permeability to CYH and 4-NQO compounds and to a lesser extent increased sensitivity to Zeo relative to the parental SKY191 line, even when grown on glucose (i.e., as a result of loss of PDR1 and PDR3). Importantly, when both strains were incubated on media containing galactose and raffinose as sugar sources, and thus inducing expression of the HXT9 and HXT11 transporters, the sensitivity to CYH and 4-NQO increased even further. The diameters of the death zones around 4-NQO application sites in the modified SKY197 strain increased an average on 32–40% in comparison with parental SKY191 cells, and the minimal inhibitory concentration of CYH was reduced from 0.5 mg/ml in the SKY191 parental strain to 0.01 mg/ml in the hyperpermeable SKY197 strain. However, permeability was not

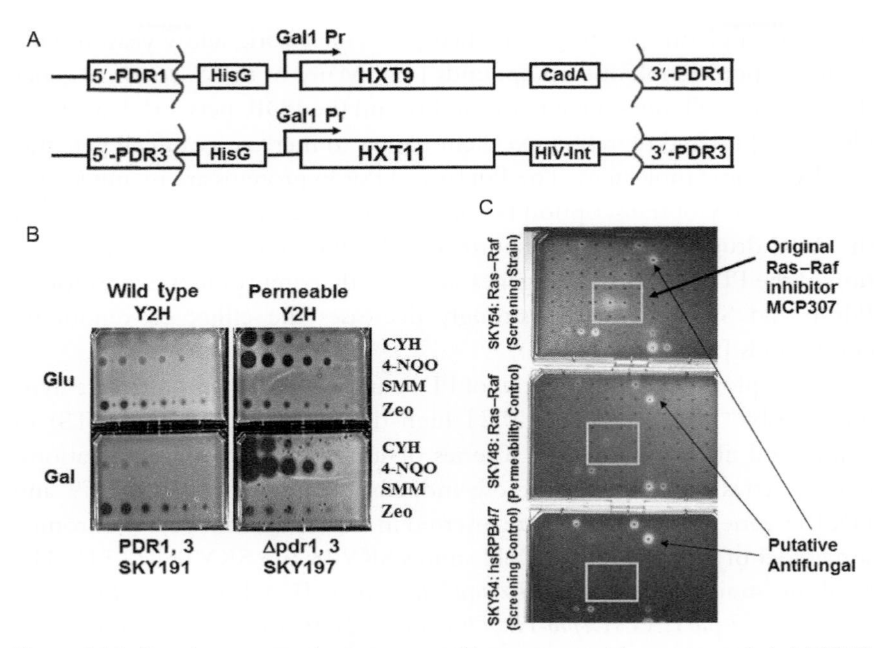

Figure 10.2 Development and assessment of hyperpermeable yeast two-hybrid (Y2H) strains for screening RAS–RAF interaction inhibitors. (A) Schematic shows integrative cassettes for creating the hyperpermeable Y2H strains SKY54 and SKY197. The *pdr1* and *pdr3* genes in the yeast chromosomes VII and II (respectively) were inactivated by integration of galactose-inducible copies of hexose transporters *HXT9* and *HXT11*. DNA fragments of bacterial and viral origins (HisG, CadA, and HIV-Int) with no homology to yeast DNA were used for stepwise homologous integrations. (B) Sensitivity of hyperpermeable yeast strain SKY197 and parental SKY191 to selected antifungal compounds. Yeast cells growing as a monolayer in soft agar supplemented with YPD (glucose) or YPG/R (Gal/Raf) were treated with serial dilutions of cyclohemide (CYH), 4-nitroquinoline-oxide (4-NQO), sulfomethurone methyl (SMM), and zeocin (Zeo). (C) Representative HTS screening plates with hyperpermeable yeast SKY54–HRAS–RAF1, parental SKY48–HRAS–RAF1, and control strain SKY54–hsRPB7–hsRPB4. Yeasts were treated with 96 compounds from a proprietary compound library. A candidate RAS–RAF interaction disrupting hit compound (boxed) and putative antifungal compounds (arrows) are indicated.

increased to all compounds: for instance, neither the parental SKY191 nor the optimized SKY197 yeast exhibited sensitivity toward SMM.

The SKY54 strain showed equal sensitivity to SKY197 for these selected antifungal agents and was desirable for further development, given its greater sensitivity in Y2H applications [35]. Subsequent challenge of this strain versus its parent, SKY48, with a diverse chemical library containing 73,400 small molecular weight compounds, revealed that 3011 library compounds produced at least some lethality, versus 1959 compounds for SKY48. These results suggested that the modified SKY54 strain achieved a 35% improvement in permeability, with the number of death-inducing hits increasing from 2.7% to 4.1%, and was probably permeable to a significant majority of compounds (as most would not be expected to be cytotoxic). This level of permeability was hence similar to that achieved in screening the same compound library in *Escherichia coli* strains specifically created for HTS applications (data not shown). These results confirmed that these Y2H strains were excellent platform to search for small-molecule PPI inhibitors.

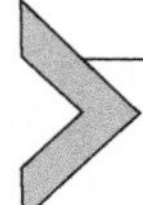

4. SELECTION OF NOVEL RAS–RAF PROTEIN INTERACTION INHIBITORS IN HTS IN HYPERPERMEABLE YEAST SKY54

To adapt the RAS–RAF interaction as a target for HTS in SKY54 yeast, we performed several optimization steps. First, we chose to use the HRAS protein for the assay rather than the more frequently mutated N- and KRas proteins, as the latter do not express as efficiently in yeast cells [52]. In addition, KRAS has several membrane-targeting sequences on its C-terminus, which makes it more problematic to get KRAS transported to the cell nucleus, required for efficient interaction detection by Y2H [53]; HRAS only has a CAAX box. Second, we used a mutated derivative of HRAS, C186G, with the cysteine to glycine substitution disrupting the membrane-targeting CAAX box [54], thus increasing the pool of nuclear (nonmembrane-tethered) HRAS. Third, we used HRAS–C186G as the "Bait" fusion with the phage λ cI DNA-binding domain (DBD) and full length human RAF1 as a "Prey" fusion with an activation domain (AD). We used full length RAF1 rather than just the RBD of RAF, to allow potential selection of allosteric inhibitors targeting other regions of RAF. Fourth, we used the SKY54 strain, based on its possession of multiple copies of the cI-operator bound by the λ DBD, leading to a higher level of sensitivity to the Bait/Prey combination than the SKY197 strain, which had fewer operators [55]. Fifth, we created and screened a parallel control SKY54 yeast strain, expressing λ-cI-DBD–hsRPB7 and AD–hsRPB4 [31], allowing us to counterselect nonspecific antifungal compounds. Sixth, we also expressed the DBD–HRAS–C186G and AD–RAF1 plasmids in the parental SKY48 background and used this for confirmatory screening, on the theory that compounds penetrating this strain as well as SKY54 to inhibit

RAS–RAF interactions might have exceptional penetrating capacity and/or potency.

For screening, all three yeast strains were plated in an agarose monolayer supplemented with X-Gal, a sterile BU-salt solution and UHW media with galactose and raffinose, following standard procedures [53]. The constitutive RAS–RAF, Bait–Prey interaction triggered activation of the yeast cI-operator-LacZ reporter gene and consequent development of blue-colored yeast lawn. A chemical library of 73,400 compounds was applied as 1 μl of a 2.5 mM compound stock solution using a microarray format. A representative panel of screening plates in Fig. 10.2C demonstrates putative PPI inhibitors that were able to inhibit color development selectively with the specific RAS–RAF combination, as well as putative antifungal compounds with the clear death zones around compound application sites regardless of Bait–Prey combination [35]. We identified 3011 molecules that impaired growth of yeast SKY54 and 708 compounds that inhibited β-galactosidase activity of this strain with various level of specificity toward DBD–HRAS/AD–RAF1 versus the DBD–hsRPB7/AD–hsRPB4 pairs of proteins. In contrast, use of SKY48 identified only 1959 growth inhibitory compounds or about 65% of the hits selected in SKY54. These results confirmed our observations of a significant increase in permeability of SKY54 yeast, making this strain extremely effective as a screening tool. As confirmation, the SKY54–HRAS–RAF1 and SKY54–hsRPB7–hsRPB4 strains were then treated in a microtiter plate format with the 708 compounds identified in the initial HTS screen, using compounds at 30 μM concentration for 19 h, and then performing a liquid β-galactosidase assay as described in Ref. [53] with modifications detailed in Ref. [33]. This identified 38 compounds that selectively inhibited activation of the cI-operator-LacZ reporter only in the SKY54–HRAS–RAF1 strain, reducing detectable reporter activity to 3–45% of that observed following treatment of the strain with a DMSO vehicle control. These 38 compounds represented approximately 0.05% of initial combinatorial compound library and were selected for additional characterization.

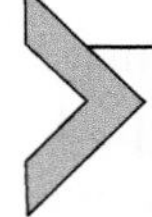

5. EVALUATION OF Y2H-SELECTED COMPOUNDS IN SRE- AND AP-1-LUCIFERASE SECONDARY SCREENS IN MAMMALIAN CELL LINES

To create a robust screening tool for secondary evaluation of putative RAS–RAF interaction inhibitors and to support downstream med-chemical

SAR efforts, we created CHO cell line with a SRE-luciferase reporter gene [55] (containing the ERK1/2-responsive serum response element (SRE) from the *c-fos* promoter) stably integrated [56]. Since compounds that block signaling in the RAF/MEK/ERMK pathway usually also inhibit cell proliferation, we integrated use of a WST-1 proliferation assay into our protocol, allowing us to normalize luciferase activity of tested compounds versus the number of cells after 2 days of treatment. In addition, we used the NIH3T3 and HEK293 cell lines transiently cotransfected with pAP1-luciferase and pcDNA3-HRASV12.

From the 38 initial hit compounds, 13 reduced reporter-dependent luciferase activity by two- to fourfold in comparison with DMSO-treated control cells. Two of the initial hits, MCP1 and MCP307, showed the best dose-dependent inhibition of luciferase activity in both the CHO (Fig. 10.3A) and HEK293 (data not shown) model cell lines [35]. We considered the possibility that the drugs might be acting by binding and destabilizing RAF, which is known to be a fragile protein [57]. However, Western analysis confirmed that none of the active molecules destabilized RAF1 protein, in contrast to the control agent radicicol (Fig. 10.3B). These and additional data led us to select MCP1 as a primary target for additional characterization and SAR optimization [58]. Based on these data, we

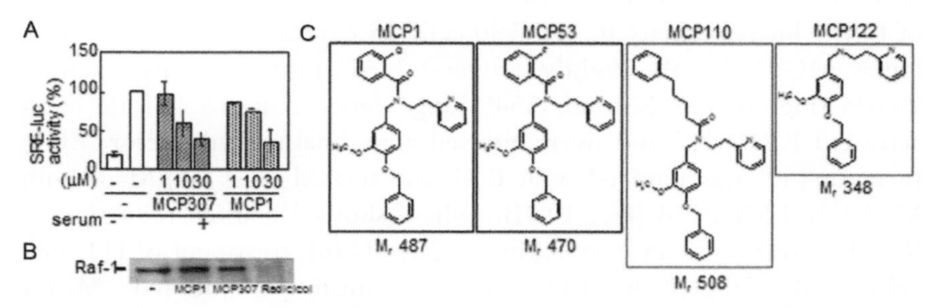

Figure 10.3 MCP compounds inhibit the transcriptional activation of RAS effector-responsive luciferase reporter genes. (A) CHO cells stably transfected with cFOS-SRE-Luc reporter gene were serum starved, and treated with DMSO (–) vehicle control, or MCP307 and MCP1 compounds, and then stimulated with 10% fetal bovine serum. After 5-h stimulation the luciferase assay was performed. (B) MCP compounds do not destabilize RAF1 protein. HT1080 cells were treated with DMSO (–), 40 µM of MCP1 and MCP307, or 20 µM of radicicol, and the level of RAF1 was determined by Western blotting with anti-RAF1 monoclonal antibody. (C) The chemical structures and molecular weights of optimized and negative control MCP compounds.

selected the luciferase assay in the CHO-cFos-SRE-Luc cell line with follow-up evaluation in a WST-1 proliferation assay in the same cell line and in HCT116 cells to assess chemically optimized analogues in SAR optimization efforts. The chemical structures and molecular weights of MCP1, two optimized analogues MCP53 and MCP110, and a weakly active analogue MCP122 are presented in Fig. 10.3C: all of these compounds were used in follow-up experiments.

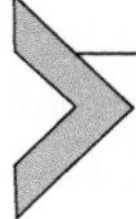

6. MCP COMPOUNDS INHIBIT RAS-INDUCED PHOSPHORYLATION OF THE MAPK PATHWAY KINASES

If MCP compounds are specifically targeting the RAS–RAF interaction, they would be predicted to block activation of RAF dependent on RAS but have no activity in blocking catalytically activated RAF and downstream RAF effectors MEK and ERK1/2. We measured the ability of MCP compounds to inhibit the RAS-induced activation of RAF1 kinase in cells with oncogenically mutated RAS proteins [35]. First, the human fibrosarcoma cell line HT1080, which expresses activated NRASK61, was treated with MCP compounds or DMSO vehicle control; then RAF1 kinase was immunoprecipitated and its activity evaluated in an *in vitro* kinase assay. MCP110 caused a dramatic dose-dependent inhibition of RAF1 kinase activity in HT1080 cells. In contrast, its less active analogue MCP122 only slightly reduced RAF kinase activity even at 20 μM (Fig. 10.4A). Second, A549 lung adenocarcinoma cell line with activated KRASS12 and overexpressed epidermal growth factor (EGF) receptor [59] was induced with EGF and treated with 20 μM of both MCP1 and MCP110 (Fig. 10.4B), which showed marked reduction of RAF1 activation in an *in vitro* kinase assay. Third, treatment of HT1080 cells with 20 μM MCP110 also significantly inhibited MEK1 (Fig. 10.4C) and ERK1/2 (Fig. 10.4D) activity, at effective concentrations similar to the MEK inhibitors PD98035 and U0126. Fourth, in contrast to these results, MCP110 did not inhibit kinase activity of preactivated, purified RAF1 protein, in contrast with effective inhibition of this RAF1 by the RAF1 kinase inhibitor ZM336372 (Calbiochem) (data not shown). Together, these data place the action of MCPs between RAS and before RAF1 in the MAPK signaling cascade.

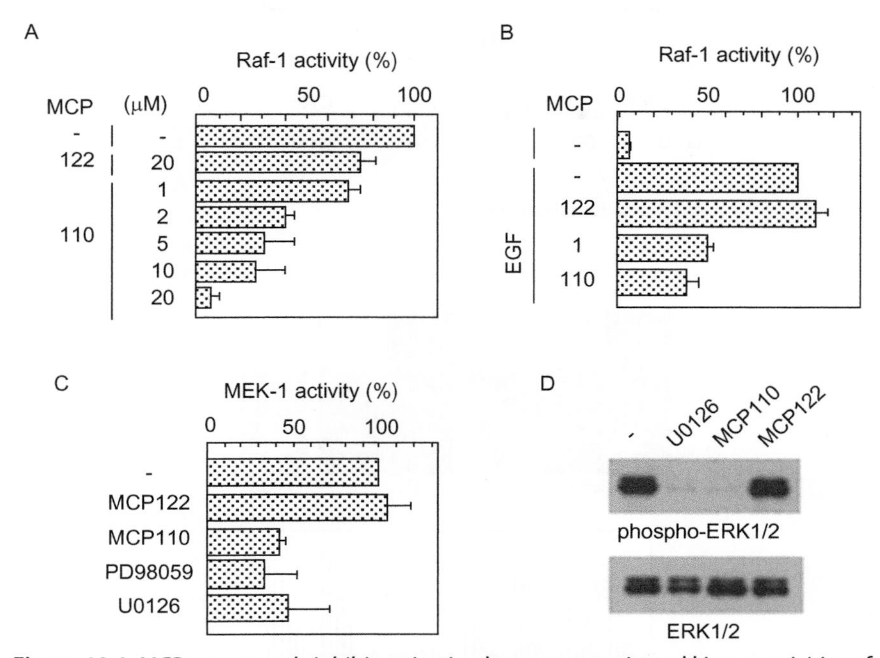

Figure 10.4 MCP compounds inhibit activation but not preactivated kinase activities of RAF1, MEK1/2, and ERK1/2. (A) HT1080 cells were incubated with DMSO (−), negative control MCP122, or optimized MCP110 at indicated concentrations and endogenous RAF1 activity was elucidated. (B) A549 cells were serum starved in the presence of DMSO (−), or 20 μM of MCP compounds and then stimulated with epidermal growth factor (EGF). RAF1 protein was immunoprecipitated and endogenous RAF1 activity was calculated. (C) HT1080 cells were incubated with DMSO (−), 20 μM of MCP compounds or control MEK inhibitors PD98058 (20 μM), and U0126 (10 μM). MEK1 protein was immunoprecipitated and endogenous MEK-1 activity was measured. (D) HT1080 cells were serum starved when treated with DMSO (−), MCP compounds (20 μM), or U0126 inhibitor (10 μM) and the levels of phosphorylated and total ERK1/2 proteins were determined by Western blotting assay.

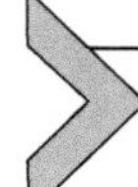

7. MCP COMPOUNDS BLOCK THE INTERACTION BETWEEN RAS AND THE RAF-RBD

In Y2H screening, the MCP compounds inhibited the interaction between full length human HRAS and RAF1. The schematic structure of RAF1 is presented in Fig. 10.5A. A RBD (amino acids 51–131) and a *c*ysteine *r*ich *d*omain (CRD) (amino acids 139–184) are located in the regulatory N-terminal domain of RAF proteins (reviewed in Ref. [60–62]).

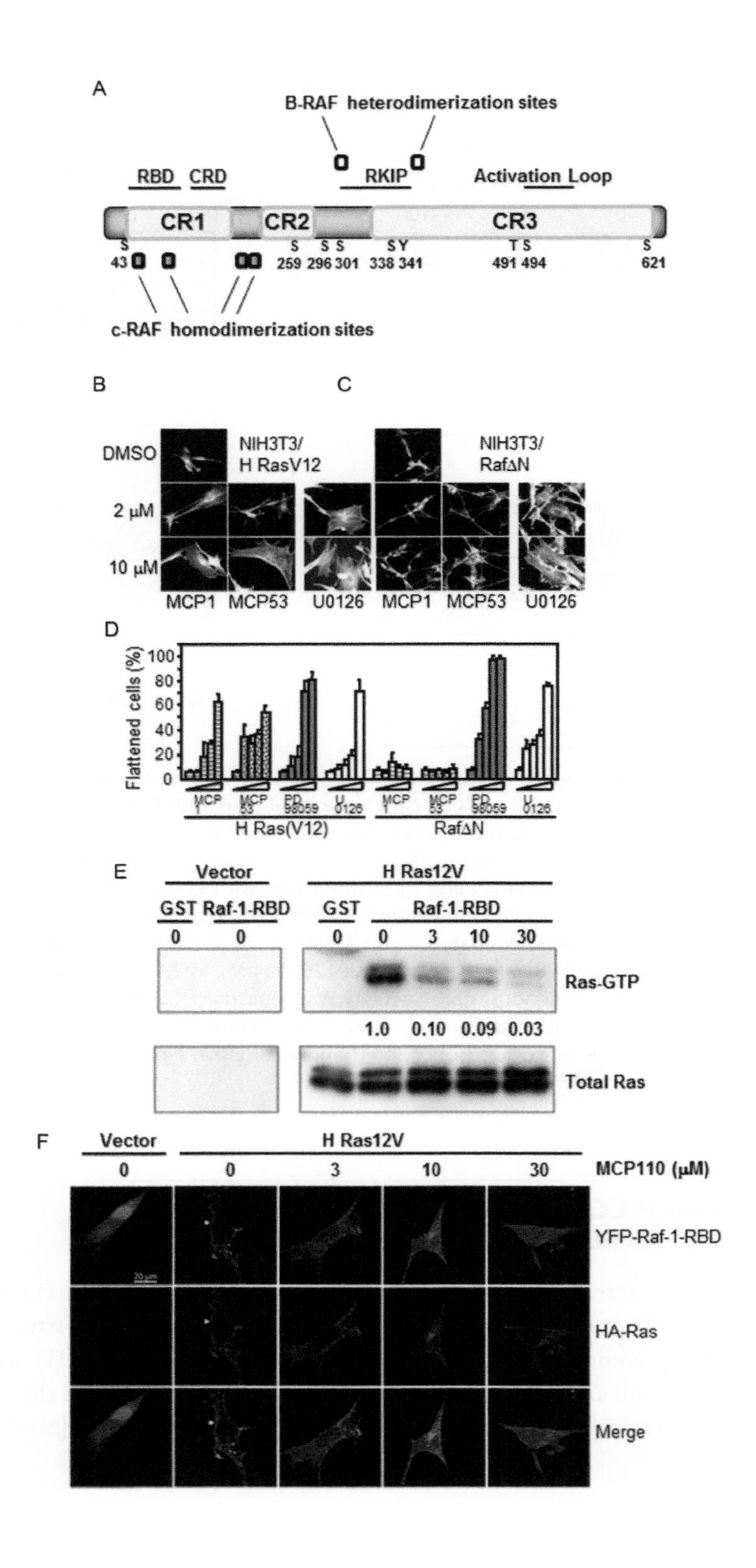
A
B-RAF heterodimerization sites
RBD CRD
RKIP
Activation Loop
CR1
CR2
CR3
S 43
S 259
S 296
S 301
S 338
Y 341
T 491
S 494
S 621
c-RAF homodimerization sites
B
C
DMSO
2 μM
10 μM
NIH3T3/ H RasV12
NIH3T3/ RafΔN
MCP1
MCP53
U0126
D
Flattened cells (%)
H Ras(V12)
RafΔN
E
Vector
H Ras12V
GST
Raf-1-RBD
Ras-GTP
1.0 0.10 0.09 0.03
Total Ras
F
MCP110 (μM)
YFP-Raf-1-RBD
HA-Ras
Merge

The effector interaction region of RAS encompasses residues 21–60, with residues 31–40 particularly important for contacts with RAF [63]. Although the RBD is the main site of interaction between RAS and RAF, the process of RAS activation of RAF involves converting RAF1 from a homodimeric configuration dependent on energetic contributions from both RBD and CRD to a RAF1/BRAF heterodimer [64]. In concept, MCP compounds might bind to the RAS effector domain, or the RAF-RBD or CRD, directly blocking interactions. Alternatively, MCP compounds may work as allosteric inhibitors of RAS–RAF interaction by binding outside of interaction surfaces of one of these proteins and indirectly modulating their folding.

To assess whether the N-terminus of RAF1 mediated MCP activity, we assessed the relative activity of MCP1, MCP53, and the MEK1 inhibitor U0126 in reversing the transformed morphology of NIH3T3 fibroblasts transformed with the HRASV12 or RAF22w (RAFΔN) oncogenes [35]. MCP1, MCP53, and U0126 restore actin filaments and cell spreading in HRASV12-transformed cells in a dose-dependent manner (Fig. 10.5B and D). MCP compounds failed to produce similar changes in cells transformed with oncogenic RAF22w, while U0126 was equally active in this cell line (Fig. 10.5C and D). These results implied that the presence of the RBD, the CRD, or both was absolutely critical for the ability of MCPs to inhibit RAS–RAF binding.

Figure 10.5 MCP compounds inhibit interaction between RAS and RAF proteins in mammalian cells. (A) Schematic representation of the human RAF1 protein structure: CR1–CR3 conserved regions; RBD, RAS-binding domain; CRD, cysteine rich domain; RKIP, minimal RKIP-binding domain. The activation loop and domains involved in homodimerization and heterodimerization with BRAF are indicated, with regulatory phosphorylation sites shown below the structure. (B) MCP compounds reverse spindle-like morphology of HRASV12-transformed NIH3T3 cells. (C) MCP compounds do not reverse morphology of oncogenic RAF22W (RAFΔN)-transformed NIH3T3 cells. (D) Percentage of flattened cells with fibroblast-like morphology in both RAS and RAF22W-transformed NIH3T3 cells treated with increasing concentrations (0, 1, 2, 5, and 10 μM) of indicated compounds is shown. (E) MCP110 inhibits interaction between HRASV12 and RAF1-RBD domain in mammalian cells in a dose-dependent manner. HRASV12–GTP was immunoprecipitated by GST–RBD from NIH3T3 cells treated with indicated concentrations of MCP110 or vehicle solvent, with pulled down RAS–GTP and total RAS visualized by Western blotting, and quantified by densitometry: numbers shown between the panels represent ratio of RAS–GTP to total RAS. (F) MCP110 disrupts RAF-RBD translocation to the plasma membrane in NIH3T3 fibroblasts transformed with HRASV12. (See color plate.)

To further refine the MCP-binding domain, we next asked if MCP110 could block the ability of a GST–RAF-RBD fusion protein to immunoprecipitate GTP-bound RAS from mammalian cells [65]. Indeed, treatment of NIH3T3 cells transiently transfected with HRASV12 and GST–RAF-RBD by MCP110 showed dose-dependent inhibition of the amount of activated HRASV12–GTP immunoprecipitated with GST–RAF-RBD (Fig. 10.5E, upper panel) [38]. Importantly, MCP110 did not cause destabilization of RAS itself (Fig. 10.5E, lower panel). We also determined that the localization of YFP-tagged RAF-RBD [66] in NIH3T3 cells was affected by coexpression of HA-HRASV12, with MCP110 blocking HA-HRASV12-induced relocalization. In the absence of HA-HRASV12, YFP-tagged RAF-RBD fluorescence staining was evenly distributed throughout cytoplasm and nucleus of transiently transfected NIH3T3 cells (Fig. 10.5F, upper panel, left). Coexpression of HA-HRASV12 causes a dramatic rearrangement, with most of the YFP–RBD excluded from the nucleus and redistributed to the RAS localization sites on the plasma membrane and in the Golgi apparatus. Treatment with MCP110 produced a dose-dependent relocalization of YFP–RBD back to the cytoplasm and the nucleus (Fig. 10.5F, right panels), again implying that the RBD domain of RAF1 is sufficient for MCP110 to successfully modulate RAS–RAF interaction in living cells [38].

8. COMPUTATIONAL DOCKING OF MCP110 TO RAF

None of the results discussed to this point addressed the question of whether MCP compounds bound predominantly to RAS or to the RAF-RBD. Disappointingly, extensive efforts to directly measure binding using bacterially expressed HRAS and RAF-RBD by isothermal titration calorimetry, analytical ultracentrifugation, or NMR were unsuccessful, implying the possible importance of posttranslational modifications unique to eukaryotic cells as important to the interaction.

As an alternative approach, we have used computational docking (Fig. 10.6). Small-molecule inhibitors of the KRAS–SOS interaction have been identified that bind into a hydrophobic-binding pocket close to the RAF-binding interface [67]. As this pocket is also accessible in HRAS [67], we hypothesized that MCP110 may bind to the same pocket and either protrude into the RAF-binding interface or inhibit RAF binding allosterically.

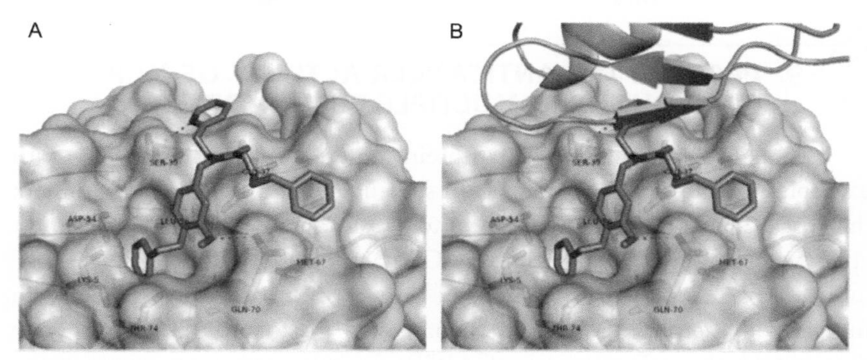

Figure 10.6 Molecular docking of MCP110 on human RAS protein. (A) MCP110 binds to the region of HRAS protein that involved in interaction with SOS. (B) Binding between HRAS and RAF1 protein is modulated by MCP110. RAF1 is shown as a green ribbon diagram. To develop models, docking experiments were performed with AutoDock4.2 [107]. Human HRAS was extracted from crystal structures representing different conformations of the RAF1 bound (PDB entry 3KUD [108]) and active state (PDB entries 3TGP [109] and 4DLU [110]), superposed onto the SOS-inhibitor bound KRAS structure (PDB entry 4EPY [67]) and energy optimized hydrogen atoms were added using BallVIEW [111]. The ligand MCP110 was sketched and energy optimized in Marvin 3.5.6 (ChemAxon). The ligand and the protein structures were loaded into AutoDockTools (ADT 1.5.6) and prepared for docking by adding Gasteiger atomic charges. The SOS-inhibitor was used to define the center of the grid maps for the docking runs. The grid dimensions were set to 22.5 Å × 22.5 Å × 22.5 Å and the default value of 0.375 Å was used as grid spacing. Grid maps were calculated with AutoGrid4 for each protein structure. For the docking procedure the standard Lamarckian Genetic Algorithm protocol was used with an initial population of 150 randomly placed individuals, a maximum number of 2,500,000 energy evaluations, a mutation rate of 0.02, a crossover rate of 0.80, and an elitism value of 1. 30 independent docking runs were carried out for each protein structure. The selection of the most reasonable docking pose was based on visual inspection. (For interpretation of the references to color in this figure legend, the reader is referred to the online version of this chapter.)

Docking to different HRAS conformations supports this hypothesis (Fig. 10.6). In several reasonable binding modes, the benzyloxy-methoxybenzyl moiety sits in the hydrophobic pocket while the 2-ethylpyridine moiety reaches the RAF-binding interface. Further, the less active analogue MCP122 lacks the 2-ethylpyridine moiety and hence is predicted to bind less well to this pocket. Our working model is that MCP110 binds to this pocket on RAS family members and blocks interaction with RAF.

9. PROBING THE ANTICANCER ACTIVITY OF MCP COMPOUNDS IN MULTIPLE TUMOR TYPES

The literature describing protein-targeted inhibitors is rich with examples of activity in the context of specific genetic or epigenetic milieu: the entire goal of personal medicine is to understand which drugs work most effectively in which tumors. Since MCP compounds modulate RAS–RAF interactions, we first broadly assessed activity in human cancer cells with mutationally activated N-, KRAS, or BRAF oncogenes [13,21,68]. We determined the effectiveness of MCPs in reversing oncogenic reprogramming of cancer cell morphology, anchorage-dependent and -independent growth or colony formation ability, cell cycle progression, resistance to apoptosis, motility, and invasion.

9.1. Activity of MCP compounds in an NRAS-mutated fibrosarcoma cell line

Oncogenic mutations of NRAS are frequently identified in malignant melanomas, liver and thyroid carcinomas, sarcomas, myeloid leukemias, and malignancies of lymphoid origins [69]. The HT1080 human fibrosarcoma cell line expresses NRAS61K [70–72]; a rounded cell morphology depends on activated MAPK pathway [72] and the activity of this mutated NRAS oncogene [73]. MCP1 and MCP53 were as effective as control U0126 inhibitor in reverting HT1080 cells to flat untransformed phenotype (Fig. 10.7A) [35]. Human sarcomas in general and HT1080 cells in particular are highly invasive, reflecting the high frequency of metastases associated with these malignancies [74]. MCP1, MCP110, and MCP53 as well as the MEK1/2 inhibitors PD98059 and U0126 were all active in reducing invasion of HT1080 cells in a transwell assay with extracellular matrix coated membranes, causing an approximately 40% reduction in the number of invading cells [35]. The reduced activity MCP analogue MCP122 produced only minor inhibition of invasion in this assay (Fig. 10.7B). Finally, as presented in Fig. 10.7C, 10 μM of MCP1 completely abolished any colony formation in comparison with DMSO-treated HT1080 cells in a soft-agar assay of anchorage-independent growth [35].

9.2. Activity of MCP compounds in human cancer cell lines with mutated KRAS and BRAF genes

The highest incidence of KRAS mutations is found in pancreatic ductal adenocarcinomas, lung and colon carcinomas [75,76], with these mutations

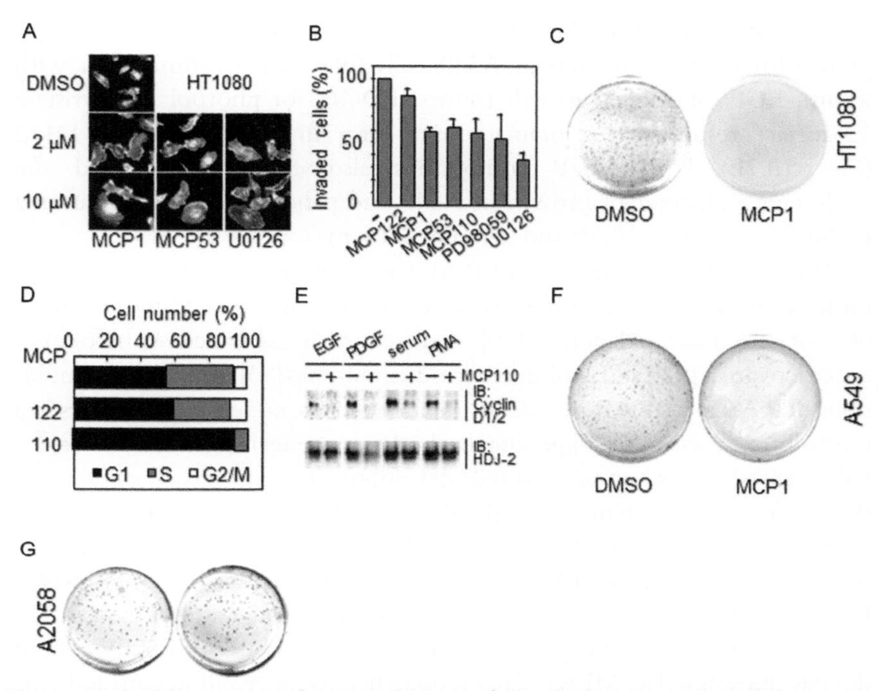

Figure 10.7 MCP compounds revert RAS-transformed phenotypes in cell lines with oncogenically mutated N- and KRAS genes. (A) MCP compounds induce morphological reversion to flat cells in NRAS61K-mutated HT1080 cells. (B) 10 μM MCP1, MCP53, and MCP110 inhibited HT1080 invasion properties of HT1080 cells through extracellular matrix-coated membranes, while MCP122 did not. (C) 10 μM MCP1 abolished anchorage-independent growth of HT1080 cells in a soft-agar colony formation assay. (D) 20 μM MCP110 induced G_1–S cell cycle arrest in KRASV12-mutated A549 cell line. (E) 20 μM MCP110 resulted in loss of cyclin D1/2 proteins in A549 cells induced by EGF and PDGF, fetal bovine serum, or phorbol 12-myristate 13-acetate (PMA). (F) 10 μM MCP1 strongly inhibited anchorage-independent growth of A549 cells in soft-agar colony formation assay. (G) 10 μM MCP1 failed to reduce anchorage-independent growth of A2058 melanoma cells with mutated BRAFV600 in a soft-agar colony formation assay.

occurring at early stages of tumor progression, emphasizing their critical importance in the process of tumorigenesis. The lung adenocarcinoma cell line A549 harbors activated KRASS12. As shown in Fig. 10.7D, treatment of A549 cells with 20 μM of MCP110 led to cell cycle arrest in G_1, while treatment with MCP122 did not [35]. These changes were not associated with increase of cell population in sub-G_1 phase of the cell cycle, suggesting no significant apoptosis induction. G_1 cell cycle arrest is typically associated with degradation of cyclin D1/2 [77]. We determined that MCP110 pretreatment of serum-starved A549 cells followed by stimulation with serum, EGF or platelet growth factors (PDGF), or phorbol 12-myristate 13-acetate resulted in a dramatic reduction in levels of cyclin D1/2 (Fig. 10.7E) [35]. MCP compounds also efficiently reduced the anchorage-independent growth of A549 and pancreatic ductal carcinoma PANC-1 cells (Fig. 10.7F and data not shown).

BRAF is the only member of RAF family of proteins that is frequently mutated in human cancers [78,79], with somatic mutations found in almost 60% of malignant melanomas [80], and frequently associated with colorectal [81], ovarian [82], and papillary thyroid carcinomas [83]. Importantly, mutations of RAS and BRAF are usually identified in the same type of cancers in a mutually exclusive basis, suggesting that both oncogenes trigger the same essential signaling events critical for tumor development and progression. Of more than 30 missense mutations identified in BRAF, almost 90% of them were E600V substitutions in the activation loop of the RAF kinase catalytic domain [84]. BRAFV600 kinase activity exceeds the activity of wild-type BRAF by more than 400-fold, causing constitutive activation of the ERK1/2 pathway [84]. Interestingly, several BRAF mutants but not E600V cannot directly phosphorylate MEK1/2 but require heterodimerization with and activation of RAF1, which then phosphorylates MEK1/2. These findings suggest that in some cases, RAS activation of RAF1 may remain necessary even in cells with activated BRAF.

We evaluated anchorage-independent growth in three human cell lines with BRAFV600 mutations: the malignant melanomas A2508 and SK-MEL-28, and the colorectal carcinoma COLO 205. MCP110 treatment completely abrogated colony formation in the SK-MEL-28 and COLO 205 cell lines [40] but had no appreciable activity in A2508 cells (Fig. 10.7G and data not shown). Based on the COSMIC database resource [76], all three cell lines harbor similar mutations in BRAF and TP53 genes. However, A2508 additionally has inactivating mutations in CDKN2a and PTEN. Potentially, mutation of CDKN2a, a negative regulator of G_1–S cell cycle transition [85], may help cells avoid MCP110-induced G_1 cell cycle arrest. In addition, the mutation of PTEN tumor suppressor gene activates the PI3K signaling cascade constitutively activated [86], which also may provide a route to escape the inhibition of the RAF–MEK–ERK pathway induced by MCP110. Together, these results indicated that MCP110 compounds were potentially useful in tumors bearing BRAF mutations.

9.3. MCP compounds sensitize hematopoietic cancer cells harboring RAS mutations to apoptosis induction

Many cancer cell lines of the T cell and B cell lineages possess activating mutations in RAS family genes. The acute lymphoblastic leukemia cell line CEM (KRASD12), the acute promyelocytic leukemia cell line HL-60 (NRASL61), and the multiple myeloma cell line MM1.S (KRASA12) were incubated with MCP1, MCP122, or a DMSO control for 96 h. Incubation with MCP1 significantly reduced cell number, especially in low serum conditions (Fig. 10.8A). Apoptotic cells (measured as binding of annexin V to exposed membrane phosphatidylserine and the exclusion of propidium iodide) were apparent within 12–16 h in the MCP1-treated cells, while incubation with control DMSO or MCP122 showed little effect in any cell line (Fig. 10.8B). As further confirmation of induction of apoptosis, the activity of caspase-3 (Fig. 10.8C) and the cleavage of poly ADP-ribose polymerase (PARP), a substrate of caspase-3 (Fig. 10.8D), were also analyzed in MCP-treated cells. MCP1 but not MCP122 significantly induced activation of caspase-3 and cleavage of PARP under low serum conditions in all cell lines, with a time course similar to induction of apoptosis. Similar results were obtained with the multiple myeloma cell line RPMI8226 (KRASA12) (data not shown). We also examined which caspases were responsible for induction of apoptosis by MCP1 in MM1.S cells, using various caspase inhibitors. Activation of caspase-3 and the upstream caspase, caspase-9, were required for MCP1 induction of apoptosis (Fig. 10.8E), while caspase-8 was dispensable. These results implied that MCP induces apoptosis through the mitochondrial pathway.

9.4. Efficacy of MCP110 and MCP1 in broad spectrum of human-transformed cell lines from NCI 60 panel

To more generally assess the activity of MCP compounds, we evaluated their potency in limiting the proliferation of the panel of 60 cell lines defined as a reference set by the NCI [87]. MCP1 and MCP110 significantly inhibited the growth of many of these cell lines. Eight out of 12 cell lines in which the MCP compounds were most potent (IC50 values <10 μM) (Table 10.1) either express a mutant form of RAS or the BRAFV600 mutation of BRAF; have constitutive activation of RAS based on upstream signaling changes; have overexpressed wild-type RAF1; and/or to possess a requirement for RAS activity for full oncogenic transformation by a

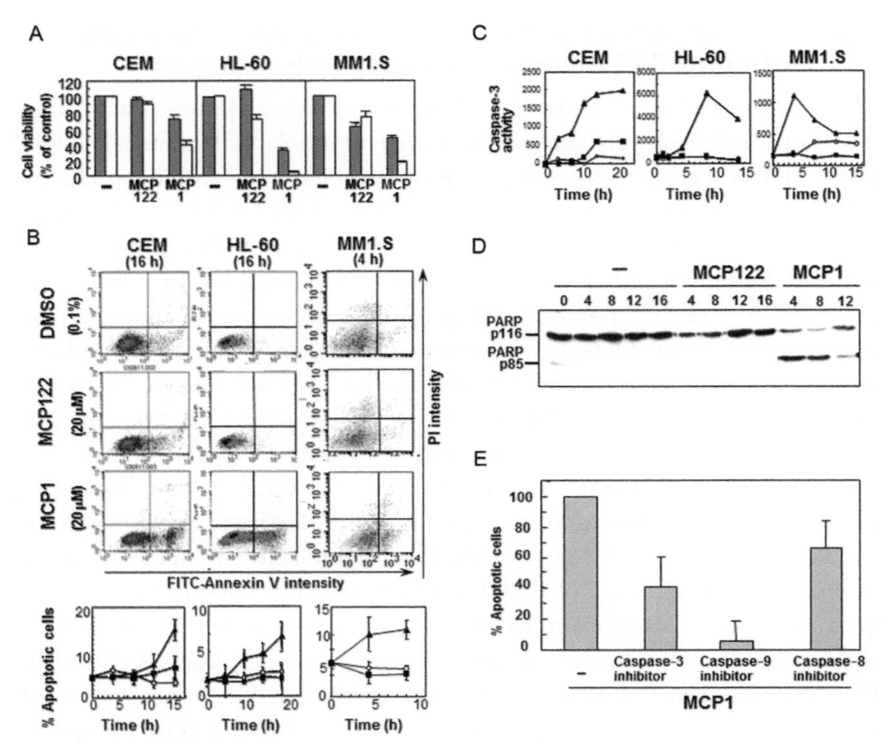

Figure 10.8 MCP1 induces apoptosis in hematopoietic cancer cell lines. (A) CEM, HL-60, and MM1.S cells were treated with DMSO (0.1%), MCP122 (20 μM), or MCP1 (20 μM) for 96 h in RPMI1640 containing 10% (solid bar) or 0.1% FBS (open bar). Cell proliferation was measured using a WST-1 assay. (B) The CEM, HL-60, and MM1.S cells were treated with DMSO (0.1%), MCP122 (20 μM), or MCP1 (20 μM) for indicated times in RPMI/0.1% FBS. Cells were double-stained with PI and FITC-annexin V and analyzed by flow cytometer. (Top) The lower left quadrant, the lower right quadrant, and the upper right quadrant indicate live cells, early apoptotic cells, and late apoptotic and/or necrotic cells, respectively. (Bottom) Time course of induction of early apoptotic population among cells treated as in top panels. (C) Cells were treated as in (B) and caspase-3 activity was measured using a fluorogenic substrate. Representative results from three independent experiments are shown. (D) MM1.S cells were treated with DMSO, MCP122, or MCP1 as described above and lysates analyzed by Western blotting to detect PARP. (E) MM1.S cells were pretreated with indicated caspase for 1 h, then incubated additional 4 h with MCP1 (20 μM) and subjected to flow cytometric analysis. Percentage of early apoptotic cells is shown. (For color version of this figure, the reader is referred to the online version of this chapter.)

synergizing oncogene (e.g., BCR-ABL in HL-60) [39]. MCP1 also effectively inhibited proliferation in some cell lines in which the RAS signaling pathway is thought not to be activated, such as T47D [88] and PC-3 [76]. These results suggest that either these cell lines are particularly susceptible to inhibition of proliferation or MCP compounds have at least some additional targets beyond the RAS–RAF PPI.

Table 10.1 The 12 NCI 60 cell lines most sensitive to treatment with MCP1, MCP110, and BAY43-9006 (sorafenib/nexavar) compounds, based on a proliferation assay

	Cell line	MCP1, GI50 (μM)	MCP110, GI50 (μM)	BAY43-9006, GI50 (μM)	Cancer type	Ras-pathway activation
1	T-47D	3.29	2.71	2.71	Breast cancer	PI3K (A3140G)
2	RPMI-8226	4.44	5.66	2.08	Myeloma	K-Ras(G12V)
3	MOLT-4	4.96	3.9	1.37	Leukemia	N-Ras(G12D)
4	A498	5.32	6.56	1.23	Renal cancer	EGFR
5	HL-60(TB)	6.52	6.71	1.14	Leukemia	N-Ras(Q61L)
6	UACC-62	6.93	7.42	0.96	Melanoma	B-Raf(V600E)
7	HCT-116	8.07	5.76	1.05	Colon cancer	K-Ras(G13V)
8	SR	8.33	6.75	1.04	Leukemia	N-Ras(G12D)
9	M14	8.35	10.1	1.31	Melanoma	B-Raf(V600E)
10	PC-3	8.77	14.9	1.71	Prostate cancer	PTEN (deletion)
11	MDA-MB-231/ATCC	8.82	9.8	1.03	Breast cancer	K-Ras(G13V)
12	NCI/ADR-RES	9.4	8.58	1.61	Breast cancer	Raf-1 overexpression

The cancer type and available information about the genetic activation of RAS/RAF, or other factors contributing to the activation of the RAS pathway are shown.

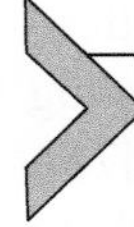

10. MCP COMPOUNDS SYNERGIZE WITH MAPK PATHWAY INHIBITORS AND MICROTUBULE-TARGETING CHEMOTHERAPEUTICS

If MCP compounds inhibit, but do not completely eliminate, RAS activation of RAF, combining MCP compounds with other compounds that also target the RAF–MEK–ERK cascade by different mechanisms would logically result in significant additive or synergistic effects. Similar principles underlie ongoing preclinical and clinical studies that attempt to more fully inhibit cancer-associated RTKs by combining antibody and small-molecule inhibitors to the same target (e.g., erlotinib with cetuximab for inhibition of EGFR [89]). We therefore evaluated the degree to which MCP compounds inhibited proliferation and soft-agar growth of NIH3T3–KRASV12 cells, either alone or in combination with multikinase inhibitor BAY43-9006 (sorafenib/nexavar), which inhibits the catalytic activity of RAF1 and other kinases (VEGFR1-3, PDGFRβ, FGFR1, c-kit, and others [90,91]) or U0126, a MEK kinase inhibitor. We observed a potent synergistic effect when cells were treated with 10 μM of MCP110 in combination with each of these inhibitors in anchorage-dependent proliferation assays and an even more pronounced effect in soft-agar colony formation assays (Fig. 10.9A and B) [39]. In contrast, combination of MCP110 with staurosporine, a very broad spectrum pan-kinase inhibitor [92], or the DNA synthesis inhibitor gemcitabine, did not exhibit synergy. A weak additive effect was observed from the combination of MCP110 and AACOCF3, which inhibits phospholipase A2, a mediator of feedback regulation mechanism of MAPK pathway (Fig. 10.9C). These results are compatible with the idea that the primary signaling activity of MCP110 is in blocking the RAS–RAF–MEK–ERK cascade in RAS-transformed cells.

Paclitaxel is a cytotoxic chemotherapeutic agent that is prescribed for the treatment of many cancers, including breast, lung, ovarian, and pancreatic ductal carcinomas, and Kaposi sarcoma. Interestingly, despite paclitaxel's primary mode of action in stabilizing microtubule polymerization and disrupting cell cycle progression, the efficacy of paclitaxel and other microtubule-targeting agents has also been associated with induction of RAF1 kinase activity [93–98]. This fact implies a possible foundation for synergistic activity with a combination of paclitaxel and MCP compounds. Indeed, the combination of paclitaxel with MCP110 potently reduced the anchorage-dependent and -independent growth of NIH3T3–KRASV12 cells (Fig. 10.9D and E). MCP110 also synergized strongly with two other

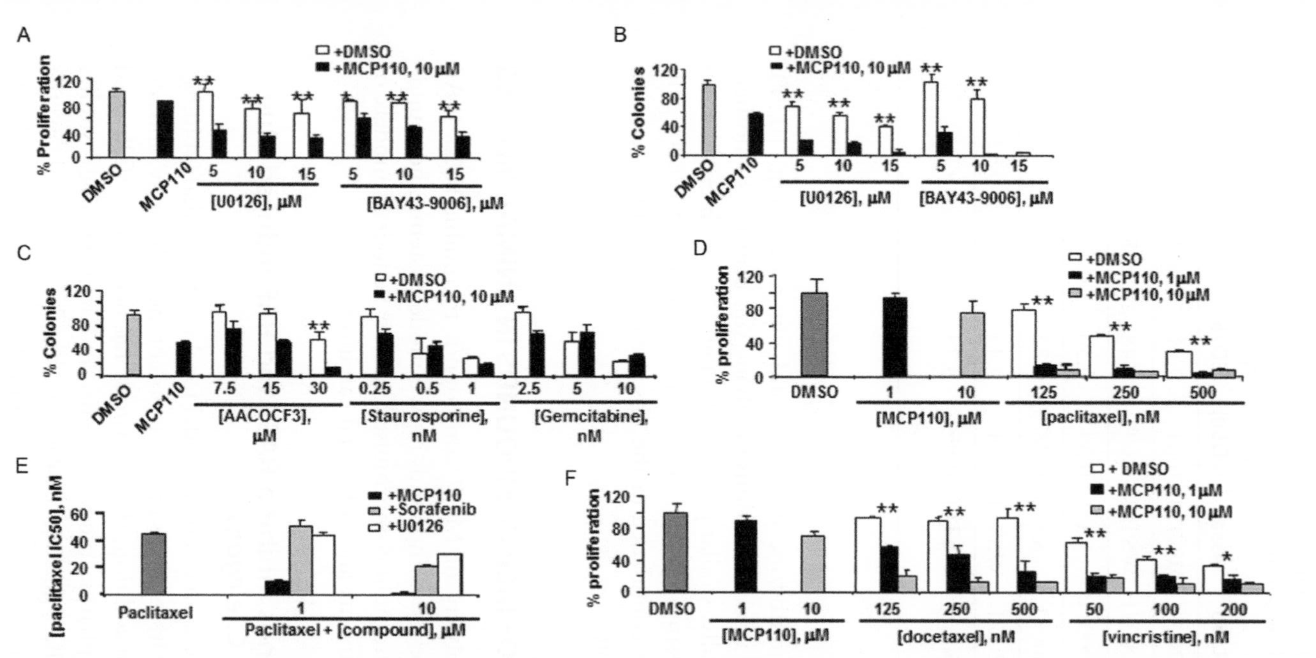

Figure 10.9 Synergy of MCP110 with RAF/MEK/ERK pathway targeted agents and paclitaxel in KRASV12-transformed NIH3T3 fibroblasts. (A) MCP110 synergized with a MEK1/2 inhibitor (U0126) and RAF1-kinase inhibitor BAY43-9006 (sorafenib/nexavar) in reducing proliferation of NIH3T3–KRASV12 cells in WST-1 assay. The number of DMSO-treated cells was taken as 100%. The proliferation of compounds treated cells was calculated as the percentage of the DMSO-treated cells. *, $P<0.05$; **, $P<0.001$. (B) MCP110 synergized with U0126 and BAY43-9006 in decreasing anchorage-independent colony formation growth of NIH3T3–KRASV12 cells. (C) MCP110 weakly interacted with AACOCF3 and did not interact with staurosporine or gemcitabine in reducing colony formation of NIH3T3–KRASV12 cells. (D) MCP110 strongly synergized with paclitaxel in a WST-1 proliferation assay. (E) MCP110 but not BAY43-9006 or U0126 sensitized NIH3T3–KRASV12 cells to paclitaxel treatment in a WST-1 proliferation assay. IC50 of paclitaxel is indicated for the treatment as a monoagent or in combinations with 1 or 10 μM of the inhibitors. (F) MCP110 synergized with docetaxel and vincristine to limit proliferation of NIH3T3–KRASV12 cells. *, $P<0.05$; **, $P<0.001$.

Table 10.2 MCP110 synergizes with Taxol, BAY43-9006 (sorafenib/nexavar), and U0126 in SW620 human colorectal carcinoma cells

Combination pair	Ratio	Assay	CI (ED50)	CI (ED75)	CI (ED90)	Averaged CI
MCP110–taxol	2200/1	WST-1	0.64	0.41	0.31	0.45
MCP110–taxol	1000/1	WST-1	0.69	0.48	0.35	0.51
MCP110–taxol	1000/1	SA	0.39	0.30	0.24	0.31
MCP110–taxol	300/1	SA	0.28	0.22	0.18	0.23
MCP110–U0126	2.1/1	WST-1	0.74	0.47	0.35	0.52
MCP110–BAY43-9006	5.5/1	WST-1	0.85	0.78	0.72	0.78

microtubule-targeting inhibitors, docetaxel and vincristine (Fig. 10.9F), implying general utility in combinations involving this class of chemotherapeutic agents [39]. Finally, similar patterns of synergy were observed using the KRASV12 expressing human colorectal carcinoma SW620 cells (Table 10.2), bolstering confidence in the observed relationships between MCP110 and other drugs.

11. ACTIVITY OF MCP COMPOUNDS IN ANIMAL MODELS

To explore the ability of MCP compounds to inhibit RAS–RAF interaction *in vivo*, we tested compounds in two model systems: the roundworm *C. elegans* and immunocompromised mouse xenograft models of human lung adenocarcinoma and colorectal carcinomas.

11.1. MCP compounds inhibit a RAS- and RAF-induced Muv phenotype in *C. elegans*

The nematode *C. elegans* is an essential model for the study of many biological processes, especially in the areas of development, neurobiology, and signal transduction [99]. In addition, this nematode has been successfully used to evaluate targets of pharmacological agents and in toxicology studies [100,101]. The MAPK signaling cascade is highly conserved in the *C. elegans* [102,103] and regulates development of vulval tissue [104]. Activating mutations of the nematode orthologs of RAS (Let-60), RAF

(Lin-45), and MEK/MAPK (MEK-2/MPK-1), as well as loss-of-function mutations of Ets-like transcription factor (Lin-1), cause hyperproliferation of vulval tissue that leads to a Muv phenotype [105]. Typically, worms with Muv phenotypes have up to four vulval protrusions (Fig. 10.10A).

The binding pocket on RAS we hypothesize to represent the site of MCP110 docking is completely conserved in Let-60, suggesting that MCP compounds may have activity in worms. Indeed, we found that treatment with MCP110 and MCP116 led to dose-dependent inhibition of Muv phenotypes associated with mutations in Let-60/RASV13, inducing disappearance of protrusions (Fig. 10.10A and B) [38]. The less active compounds MCP146 and MCP122 had limited potency in this Muv reversal assay. No MCP compounds reverted Muv phenotype induced by activating mek-2/mpk-1 mutations or inactivating lin-1 mutations (data not shown), independently confirming that the MCP compounds targeted the MAPK pathway in *C. elegans* on the level of RAS and RAF proteins. Interestingly, Muv protrusions triggered by activated RAF ortholog were also sensitive to the treatment by MCP110 and MCP116 (Fig. 10.10C), which implies that MCPs rather influence RAF protein [38]. However, despite its ability to independently induce a Muv phenotype, Lin-45 contains both RAS-binding RBD and CRD domains and it is presently unclear if Lin-45 also needs RAS to achieve complete activation.

11.2. MCP compounds are bioavailable and exhibit low toxicity profile *in vivo*

To be efficacious in mammals, MCP compounds must be bioavailable and have reasonable maximum tolerated and multiple toxicity doses to reach therapeutic concentrations in the blood. Preliminary pharmacokinetics studies using NMRI nu/nu mice indicated that MCP1 and MCP110 were well tolerated after p.o., i.v., and i.p. administration. Although there was low oral absorption, there was almost complete bioavailability for MCP110 via i.p. and i.v. routes (Fig. 10.11A and data not shown) [39]. Based on the absence of weight loss and kidney/liver toxicity profile the acute maximum tolerated dose for MCP110 via i.p. route was established as 1000 mg/kg and the maximum multiple tolerated dose (MMTD) as 900 mg/kg. Three hours after i.p. administration the level of MCP110 in the mouse plasma was determined as 16.6% of initially detected levels. In contrast, MCP1 levels in the plasma were strongly decreased 1 h after i.p. injection and reduced below 0% after 3 h (Fig. 10.11A). Based on these results, MCP110 was selected for efficacy studies in mouse xenograft models.

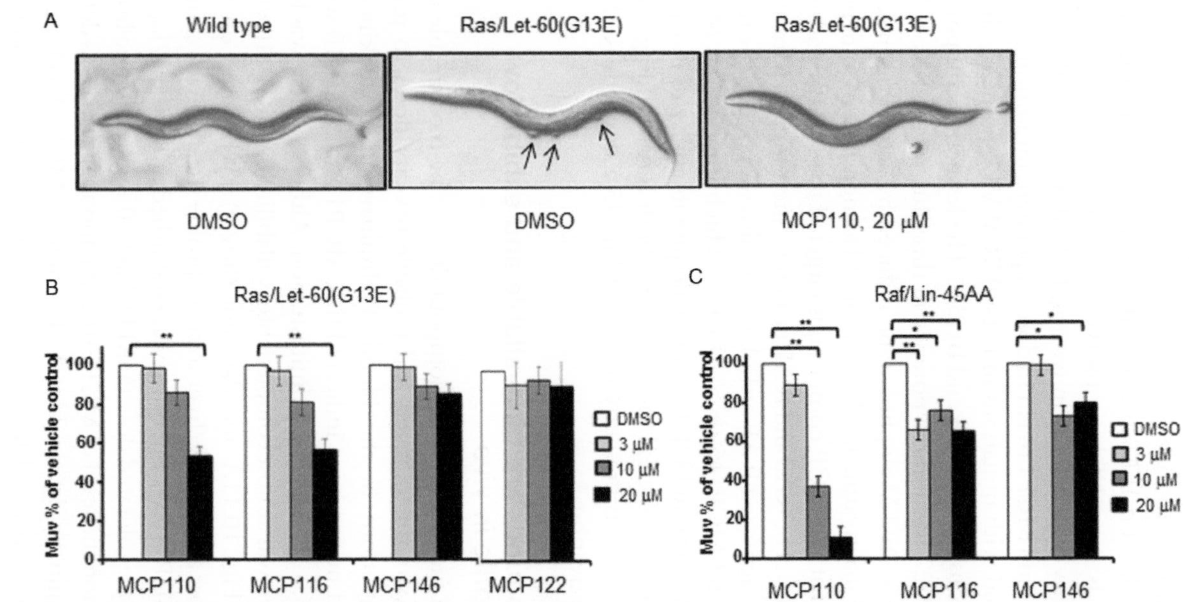

Figure 10.10 MCP compounds inhibit multivulva phenotype induced by activated RAS and RAF orthologs in *C. elegans*. (A) Wild-type worm with no ventral protrusions associated with RAS-induced Muv phenotype (left image), versus worm with Muv-associated protrusions (arrow heads) treated with vehicle control (central image). In worms treated with 20 μM of MCP110, Muv-associated protrusions were completely eradicated (right image). (B) MCP110 and MCP116 compounds produced dose-dependent reduction of number of worm with Muv phenotypes induced by activated RAS ortholog Let-60E13. Percent of Muv worms normalized to vehicle control group is shown. (C) MCP110-induced dose-dependent inhibition of Muv phenotype in worms with Lin45AA/RAFAA-activated mutation. *, $P<0.01$; **, $P<0.0001$.

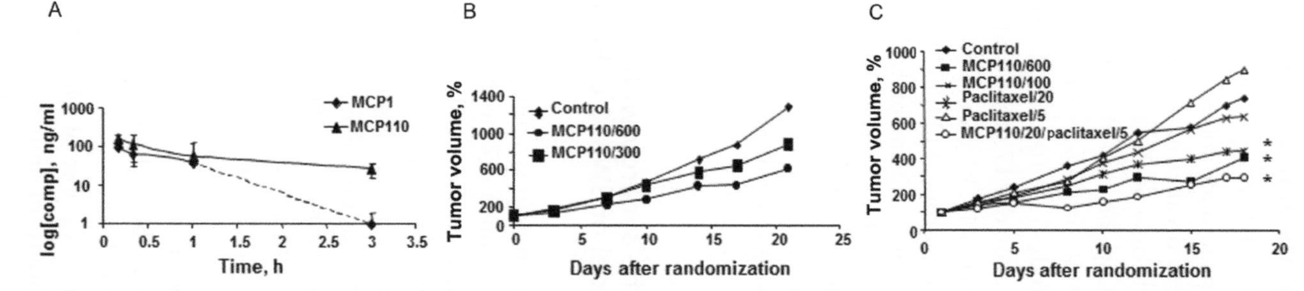

Figure 10.11 MCP bioavailability and activity in mouse xenograft models as a monoagent or in combination with paclitaxel. (A) Persistence of MCP1 and MCP110 in blood of nude mice 1 and 3 h after intraperitoneal injections. (B) MCP110 reduced tumor growth in LXFA629 nude mouse xenograft model. Animals were randomized to the three groups and treated with the vehicle (Control), a 600 mg/kg dose of MCP110 on days 0–5 and 14–18 (MCP110/600) or a 300 mg/kg dose on days 0–18. Tumor volumes were calculated using the formula: $(L \times W \times W)/2$. The tumor volume on day 0 was taken as 100%. (C) MCP110 inhibited growth of SW620 colorectal carcinoma tumors as a monoagent and synergized with paclitaxel in SCID mouse xenograft model. Following randomization, animals were treated with vehicle (Control), 100 and 600 mg/kg of MCP110 (MCP110/100 and MCP110/600), 20 and 5 mg/kg of paclitaxel (paclitaxel/20 and paclitaxel/5), or combination of MCP110 and paclitaxel (MCP110/20/paclitaxel/5). The tumor volumes were calculated as in (B). *, $P < 0.05$.

11.3. MCP110 slows the growth of LXFA 629 lung adenocarcinoma xenografts in nude mice

Nonsmall cell lung adenocarcinoma LXFA 629 cells were implanted subcutaneously in the right flank of the male nude mice. After tumors reached ~100 mm^3 in volume, animals were divided into three groups. Control group 1 was injected i.p. with vehicle only on days 0–20. Group 2 received 600 mg/kg MCP110, administered i.p. on days 0–5 and 14–18. Group 3 received 300 mg/kg doses of MCP110 on days 0–18. Tumors were measured every third day in a 21-day study (Fig. 10.11B). Based on calculated tumor growth inhibition (TGI) values on day 17 of the study, in comparison with control group, MCP110 induced a 49.3% reduction of tumor growth rate in group 2 and 73.8% reduction in group 3 [39]. The better efficacy in the group 3 probably reflects continuous treatment with MCP110, even though it was administered at a lower dose than in group 2: more work is required. In accordance with Whitney–Wilcoxon *U* test, the difference in average tumor sizes between experimental and control groups as well as the difference between MCP110-treated groups 2 and 3 were statistically significant.

11.4. MCP110 recapitulated the synergy with paclitaxel in mouse xenograft model of SW620 colorectal carcinoma

As MCP110 and paclitaxel showed a potent synergistic effect in inhibiting proliferation and colony formation of human colorectal carcinoma SW620 cells *in vitro*, this combination was tested *in vivo*. SW620 cells were subcutaneously injected in the right flanks of male SCID mice. After tumors reached ~100 mm^3 the animals were randomized into six groups. The control group mice were daily injected with the vehicle for 17 days. Mice in groups 2 and 3 received daily i.p. injections with 600 and 100 mg/kg doses of MCP110, respectively (MCP110/600 and MCP110/100). Similarly, animals in groups 4 and 5 were treated with 20 and 5 mg/kg doses of paclitaxel on days 1, 3, 5, 8, 10, 12, 15, and 17 (paclitaxel/20 and paclitaxel/5). Finally, animals in group 6 (MCP110/20/paclitaxel/5) were treated with 5 mg/kg of paclitaxel using the same schedule as the paclitaxel alone groups and daily i.p. injections of 20 mg/kg of the MCP100.

Treatment with either the 100 mg/kg dose of MCP110 or 5 mg/kg of paclitaxel did not induce a statistically meaningful reduction of tumor growth in comparison with animals in control group 1 (Fig. 10.11C).

However, tumor volumes in the animals from the group 2 dosed with 600 mg/kg of MCP110 showed strong and statistically significant TGI. 20 mg/kg treatment with paclitaxel (group 4) achieved some reduction in tumor progression but less pronounced than 600 mg/kg of MCP110. However, the animals from the group 4 exhibited signs of severe distress and weight loss. In comparison with animals in control and single treatment arms, animals in the combination group 6 achieved a dramatic reduction in the tumor growth rates, which was already discernible after day 9 of treatment. This synergistic effect far exceeded TGI determined in groups treated with single agents in much higher doses [39].

12. CONCLUSION

This work provides a proof-of-concept for the creation of hyperpermeable Y2H strains and describes their HTS application for selection of novel PPI inhibitors for the human RAS and RAF oncoproteins. Using an SAR approach, the initial hit molecules from HTS were optimized, yielding MCP110 as the best lead molecule in the MCP series. These compounds, predominantly represented by the best lead molecule MCP110, were extensively characterized by *in vitro* experiments in yeast and in mammalian cells, and *in vivo* evaluation of compound activity in *C. elegans* and mouse xenograft models. However, due to the lack of structural information, the success of optimization was limited and the compound class did not attain the potency needed for completion of formal preclinical development. Recent molecular docking efforts that placed MCP110 at a newly discovered druggable pocket on the RAS protein at RAS–SOS interface warrant creating new pharmacophores for modulating RAS–RAF interaction. Nevertheless, the success of our efforts to identify new PPI inhibitors of the RAS and RAF interaction proves that Y2H HTS is a viable and productive approach and worth to be revisiting in light of almost unlimited number of PPIs that could be explored for selecting new targeted therapeutics in drug discovery [106].

ACKNOWLEDGMENTS

We would like to thank the individuals in the Tamanoi's laboratory who helped to elucidate the effect of MCP compounds in hematopoietic cell lines. Dr. Tamanoi work was supported by NIH grant CA41996.

REFERENCES

[1] S. Fields, O. Song, A novel genetic system to detect protein–protein interactions, Nature 340 (6230) (1989) 245–246.
[2] I. Serebriiskii, V. Khazak, E.A. Golemis, A two-hybrid dual bait system to discriminate specificity of protein interactions, J. Biol. Chem. 274 (24) (1999) 17080–17087.
[3] P. Colas, et al., Genetic selection of peptide aptamers that recognize and inhibit cyclin-dependent kinase 2, Nature 380 (6574) (1996) 548–550.
[4] D.J. SenGupta, et al., A three-hybrid system to detect RNA–protein interactions in vivo, Proc. Natl. Acad. Sci. U.S.A. 93 (16) (1996) 8496–8501.
[5] F. Tirode, et al., A conditionally expressed third partner stabilizes or prevents the formation of a transcriptional activator in a three-hybrid system, J. Biol. Chem. 272 (37) (1997) 22995–22999.
[6] K. Young, et al., Identification of a calcium channel modulator using a high throughput yeast two-hybrid screen, Nat. Biotechnol. 16 (10) (1998) 946–950.
[7] M. Vidal, et al., Reverse two-hybrid and one-hybrid systems to detect dissociation of protein–protein and DNA–protein interactions, Proc. Natl. Acad. Sci. U.S.A. 93 (19) (1996) 10315–10320.
[8] G. Liu, et al., Cytoskeletal protein ABP-280 directs the intracellular trafficking of furin and modulates proprotein processing in the endocytic pathway, J. Cell Biol. 139 (7) (1997) 1719–1733.
[9] A.I. Archakov, et al., Protein–protein interactions as a target for drugs in proteomics, Proteomics 3 (4) (2003) 380–391.
[10] N. Lentze, D. Auerbach, The yeast two-hybrid system and its role in drug discovery, Expert Opin. Ther. Targets 12 (4) (2008) 505–515.
[11] G. Boguslawski, Effects of polymyxin B sulfate and polymyxin B nonapeptide on growth and permeability of the yeast *Saccharomyces cerevisiae*, Mol. Gen. Genet. 199 (3) (1985) 401–405.
[12] T. Bonner, et al., The human homologs of the raf (mil) oncogene are located on human chromosomes 3 and 4, Science 223 (4631) (1984) 71–74.
[13] J.L. Bos, Ras oncogenes in human cancer: a review, Cancer Res. 49 (17) (1989) 4682–4689.
[14] I.A. Prior, P.D. Lewis, C. Mattos, A comprehensive survey of Ras mutations in cancer, Cancer Res. 72 (10) (2012) 2457–2467.
[15] A. Demunter, et al., A novel N-ras mutation in malignant melanoma is associated with excellent prognosis, Cancer Res. 61 (12) (2001) 4916–4922.
[16] D.T. Bowen, et al., RAS mutation in acute myeloid leukemia is associated with distinct cytogenetic subgroups but does not influence outcome in patients younger than 60 years, Blood 106 (6) (2005) 2113–2119.
[17] M. Spoerner, et al., Conformational states of Ras complexed with the GTP analogue GppNHp or GppCH2p: implications for the interaction with effector proteins, Biochemistry 44 (6) (2005) 2225–2236.
[18] M. Geyer, et al., Structure of the Ras-binding domain of RalGEF and implications for Ras binding and signalling, Nat. Struct. Biol. 4 (9) (1997) 694–699.
[19] J. Downward, Targeting RAS signalling pathways in cancer therapy, Nat. Rev. Cancer 3 (1) (2003) 11–22.
[20] P.M. Campbell, et al., Genetic and pharmacologic dissection of Ras effector utilization in oncogenesis, Methods Enzymol. 407 (2006) 195–217.
[21] M. Malumbres, M. Barbacid, RAS oncogenes: the first 30 years, Nat. Rev. Cancer 3 (6) (2003) 459–465.
[22] R. Blum, A.D. Cox, Y. Kloog, Inhibitors of chronically active ras: potential for treatment of human malignancies, Recent Pat. Anticancer Drug Discov. 3 (1) (2008) 31–47.

[23] L. Santarpia, S.M. Lippman, A.K. El-Naggar, Targeting the MAPK–RAS–RAF signaling pathway in cancer therapy, Expert Opin. Ther. Targets 16 (1) (2012) 103–119.
[24] S.M. Maira, PI3K inhibitors for cancer treatment: five years of preclinical and clinical research after BEZ235, Mol. Cancer Ther. 10 (11) (2011) 2016.
[25] J.F. Lyons, et al., Discovery of a novel Raf kinase inhibitor, Endocr. Relat. Cancer 8 (3) (2001) 219–225.
[26] J. Llovet, Emerging agents for the medical therapy of hepatocellular carcinoma, Gastroenterol. Hepatol. (N. Y.) 3 (8) (2007) 600–602.
[27] K.T. Flaherty, et al., Inhibition of mutated, activated BRAF in metastatic melanoma, N. Engl. J. Med. 363 (9) (2010) 809–819.
[28] F.M. Kaplan, et al., Hyperactivation of MEK–ERK1/2 signaling and resistance to apoptosis induced by the oncogenic B-RAF inhibitor, PLX4720, in mutant N-RAS melanoma cells, Oncogene 30 (3) (2011) 366–371.
[29] A.M. Dingemans, et al., A phase II study of sorafenib in patients with platinum-pretreated, advanced (stage IIIb or IV) non-small cell lung cancer with a KRAS mutation, Clin. Cancer Res. 19 (3) (2013) 743–751.
[30] J.A. Sosman, et al., Survival in BRAF V600-mutant advanced melanoma treated with vemurafenib, N. Engl. J. Med. 366 (8) (2012) 707–714.
[31] V. Khazak, et al., Analysis of the interaction of the novel RNA polymerase II (pol II) subunit hsRPB4 with its partner hsRPB7 and with pol II, Mol. Cell. Biol. 18 (4) (1998) 1935–1945.
[32] V. Khazak, et al., Human RNA polymerase II subunit hsRPB7 functions in yeast and influences stress survival and cell morphology, Mol. Biol. Cell 6 (7) (1995) 759–775.
[33] V. Khazak, et al., Yeast screens for inhibitors of Ras–Raf interaction and characterization of MCP inhibitors of Ras–Raf interaction, Methods Enzymol. 407 (2006) 612–629.
[34] J.M. Shields, et al., Understanding Ras: 'it ain't over 'til it's over', Trends Cell Biol. 10 (4) (2000) 147–154.
[35] J. Kato-Stankiewicz, et al., Inhibitors of Ras/Raf-1 interaction identified by two-hybrid screening revert Ras-dependent transformation phenotypes in human cancer cells, Proc. Natl. Acad. Sci. U.S.A. 99 (22) (2002) 14398–14403.
[36] C.A. Lipinski, Drug-like properties and the causes of poor solubility and poor permeability, J. Pharmacol. Toxicol. Methods 44 (1) (2000) 235–249.
[37] K. Inouye, et al., Formation of the Ras dimer is essential for Raf-1 activation, J. Biol. Chem. 275 (6) (2000) 3737–3740.
[38] V. Gonzalez-Perez, et al., Genetic and functional characterization of putative Ras/Raf interaction inhibitors in *C. elegans* and mammalian cells, J. Mol. Signal. 5 (2010) 2.
[39] N. Skobeleva, et al., In vitro and in vivo synergy of MCP compounds with mitogen-activated protein kinase pathway- and microtubule-targeting inhibitors, Mol. Cancer Ther. 6 (3) (2007) 898–906.
[40] H. Hao, et al., Context-dependent roles of mutant B-Raf signaling in melanoma and colorectal carcinoma cell growth, Mol. Cancer Ther. 6 (8) (2007) 2220–2229.
[41] P.M. Campbell, et al., K-Ras promotes growth transformation and invasion of immortalized human pancreatic cells by Raf and phosphatidylinositol 3-kinase signaling, Cancer Res. 67 (5) (2007) 2098–2106.
[42] E. Farrelly, et al., A high-throughput assay for mitochondrial membrane potential in permeabilized yeast cells, Anal. Biochem. 293 (2) (2001) 269–276.
[43] A. Nourani, et al., Multiple-drug-resistance phenomenon in the yeast *Saccharomyces cerevisiae*: involvement of two hexose transporters, Mol. Cell. Biol. 17 (9) (1997) 5453–5460.
[44] A.L. Kruckeberg, The hexose transporter family of *Saccharomyces cerevisiae*, Arch. Microbiol. 166 (5) (1996) 283–292.

[45] M.D. Marger, M.H. Saier Jr., A major superfamily of transmembrane facilitators that catalyse uniport, symport and antiport, Trends Biochem. Sci. 18 (1) (1993) 13–20.
[46] M. Brendel, A simple method for the isolation and characterization of thymidylate uptaking mutants in *Saccharomyces cerevisiae*, Mol. Gen. Genet. 147 (2) (1976) 209–215.
[47] Y. Mahe, et al., The ATP-binding cassette multidrug transporter Snq2 of *Saccharomyces cerevisiae*: a novel target for the transcription factors Pdr1 and Pdr3, Mol. Microbiol. 20 (1) (1996) 109–117.
[48] D.J. Katzmann, et al., Transcriptional control of the yeast PDR5 gene by the PDR3 gene product, Mol. Cell. Biol. 14 (7) (1994) 4653–4661.
[49] G.W. Saunders, G.H. Rank, Allelism of pleiotropic drug resistance in *Saccharomyces cerevisiae*, Can. J. Genet. Cytol. 24 (5) (1982) 493–503.
[50] Y.M. Mamnun, et al., The yeast zinc finger regulators Pdr1p and Pdr3p control pleiotropic drug resistance (PDR) as homo- and heterodimers in vivo, Mol. Microbiol. 46 (5) (2002) 1429–1440.
[51] V. Khazak, E.A. Golemis, L. Weber, Development of a yeast two-hybrid screen for selection of human Ras–Raf protein interaction inhibitors, Methods Mol. Biol. 310 (2005) 253–271.
[52] S.G. Clark, J.P. McGrath, A.D. Levinson, Expression of normal and activated human Ha-ras cDNAs in *Saccharomyces cerevisiae*, Mol. Cell. Biol. 5 (10) (1985) 2746–2752.
[53] F.M. Ausubel, et al., Current Protocols in Molecular Biology, John Wiley & Sons, New York, 1994–present.
[54] C.J. Der, A.D. Cox, Isoprenoid modification and plasma membrane association: critical factors for ras oncogenicity, Cancer Cells 3 (9) (1991) 331–340.
[55] J. Estojak, R. Brent, E.A. Golemis, Correlation of two-hybrid affinity data with in vitro measurements, Mol. Cell. Biol. 15 (10) (1995) 5820–5829.
[56] C.S. Hill, et al., Functional analysis of a growth factor-responsive transcription factor complex, Cell 73 (2) (1993) 395–406.
[57] T.W. Schulte, W.G. An, L.M. Neckers, Geldanamycin-induced destabilization of Raf-1 involves the proteasome, Biochem. Biophys. Res. Commun. 239 (3) (1997) 655–659.
[58] Y. Lu, et al., Solution phase parallel synthesis and evaluation of MAPK inhibitory activities of close structural analogues of a Ras pathway modulator, Bioorg. Med. Chem. Lett. 14 (15) (2004) 3957–3962.
[59] M. Haeder, et al., Epidermal growth factor receptor expression in human lung cancer cell lines, Cancer Res. 48 (5) (1988) 1132–1136.
[60] R. Roskoski Jr., RAF protein-serine/threonine kinases: structure and regulation, Biochem. Biophys. Res. Commun. 399 (3) (2010) 313–317.
[61] E. Chuang, et al., Critical binding and regulatory interactions between Ras and Raf occur through a small, stable N-terminal domain of Raf and specific Ras effector residues, Mol. Cell. Biol. 14 (8) (1994) 5318–5325.
[62] A.B. Vojtek, S.M. Hollenberg, J.A. Cooper, Mammalian Ras interacts directly with the serine/threonine kinase Raf, Cell 74 (1) (1993) 205–214.
[63] S.J. Leevers, H.F. Paterson, C.J. Marshall, Requirement for Ras in Raf activation is overcome by targeting Raf to the plasma membrane, Nature 369 (6479) (1994) 411–414.
[64] L.K. Rushworth, et al., Regulation and role of Raf-1/B-Raf heterodimerization, Mol. Cell. Biol. 26 (6) (2006) 2262–2272.
[65] J. de Rooij, J.L. Bos, Minimal Ras-binding domain of Raf1 can be used as an activation-specific probe for Ras, Oncogene 14 (5) (1997) 623–625.
[66] T.G. Bivona, S. Quatela, M.R. Philips, Analysis of Ras activation in living cells with GFP-RBD, Methods Enzymol. 407 (2006) 128–143.
[67] Q. Sun, et al., Discovery of small molecules that bind to K-Ras and inhibit Sos-mediated activation, Angew. Chem. Int. Ed. Engl. 51 (25) (2012) 6140–6143.

[68] M. Roring, T. Brummer, Aberrant B-Raf signaling in human cancer – 10 years from bench to bedside, Crit. Rev. Oncog. 17 (1) (2012) 97–121.
[69] A. Fernandez-Medarde, E. Santos, Ras in cancer and developmental diseases, Genes Cancer 2 (3) (2011) 344–358.
[70] J.L. Bos, et al., Three different mutations in codon 61 of the human N-ras gene detected by synthetic oligonucleotide hybridization, Nucleic Acids Res. 12 (23) (1984) 9155–9163.
[71] M. Davis, et al., Localisation of the human N-ras oncogene to chromosome 1cen - p21 by in situ hybridisation, EMBO J. 2 (12) (1983) 2281–2283.
[72] S.H. Hansen, et al., Induced expression of Rnd3 is associated with transformation of polarized epithelial cells by the Raf–MEK-extracellular signal-regulated kinase pathway, Mol. Cell. Biol. 20 (24) (2000) 9364–9375.
[73] H. Paterson, et al., Activated N-ras controls the transformed phenotype of HT1080 human fibrosarcoma cells, Cell 51 (5) (1987) 803–812.
[74] K.G. Billingsley, et al., Pulmonary metastases from soft tissue sarcoma: analysis of patterns of diseases and postmetastasis survival, Ann. Surg. 229 (5) (1999) 602–610 discussion 610–2.
[75] N. Shimizu, M. Ohtsubo, S. Minoshima, MutationView/KMcancerDB: a database for cancer gene mutations, Cancer Sci. 98 (3) (2007) 259–267.
[76] S.A. Forbes, et al., COSMIC: mining complete cancer genomes in the Catalogue of Somatic Mutations in Cancer, Nucleic Acids Res. 39 (Database issue) (2011) D945–D950.
[77] C. Marshall, How do small GTPase signal transduction pathways regulate cell cycle entry? Curr. Opin. Cell Biol. 11 (6) (1999) 732–736.
[78] H. Davies, et al., Mutations of the BRAF gene in human cancer, Nature 417 (6892) (2002) 949–954.
[79] V. Emuss, et al., Mutations of C-RAF are rare in human cancer because C-RAF has a low basal kinase activity compared with B-RAF, Cancer Res. 65 (21) (2005) 9719–9726.
[80] M.S. Brose, et al., BRAF and RAS mutations in human lung cancer and melanoma, Cancer Res. 62 (23) (2002) 6997–7000.
[81] H. Rajagopalan, et al., Tumorigenesis: RAF/RAS oncogenes and mismatch-repair status, Nature 418 (6901) (2002) 934.
[82] G. Singer, et al., Mutations in BRAF and KRAS characterize the development of low-grade ovarian serous carcinoma, J. Natl. Cancer Inst. 95 (6) (2003) 484–486.
[83] Y. Cohen, et al., BRAF mutation in papillary thyroid carcinoma, J. Natl. Cancer Inst. 95 (8) (2003) 625–627.
[84] P.T. Wan, et al., Mechanism of activation of the RAF–ERK signaling pathway by oncogenic mutations of B-RAF, Cell 116 (6) (2004) 855–867.
[85] T. Abbas, A. Dutta, p21 in cancer: intricate networks and multiple activities, Nat. Rev. Cancer 9 (6) (2009) 400–414.
[86] M.S. Song, L. Salmena, P.P. Pandolfi, The functions and regulation of the PTEN tumour suppressor, Nat. Rev. Mol. Cell Biol. 13 (5) (2012) 283–296.
[87] J.N. Weinstein, et al., An information-intensive approach to the molecular pharmacology of cancer, Science 275 (5298) (1997) 343–349.
[88] L.B. Eckert, et al., Involvement of Ras activation in human breast cancer cell signaling, invasion, and anoikis, Cancer Res. 64 (13) (2004) 4585–4592.
[89] A.J. Weickhardt, et al., Dual targeting of the epidermal growth factor receptor using the combination of cetuximab and erlotinib: preclinical evaluation and results of the phase II DUX study in chemotherapy-refractory, advanced colorectal cancer, J. Clin. Oncol. 30 (13) (2012) 1505–1512.
[90] S.M. Wilhelm, et al., BAY 43–9006 exhibits broad spectrum oral antitumor activity and targets the RAF/MEK/ERK pathway and receptor tyrosine kinases involved in tumor progression and angiogenesis, Cancer Res. 64 (19) (2004) 7099–7109.

[91] F. Carlomagno, et al., BAY 43-9006 inhibition of oncogenic RET mutants, J. Natl. Cancer Inst. 98 (5) (2006) 326–334.
[92] M.A. Fabian, et al., A small molecule-kinase interaction map for clinical kinase inhibitors, Nat. Biotechnol. 23 (3) (2005) 329–336.
[93] M.V. Blagosklonny, et al., Taxol-induced apoptosis and phosphorylation of Bcl-2 protein involves c-Raf-1 and represents a novel c-Raf-1 signal transduction pathway, Cancer Res. 56 (8) (1996) 1851–1854.
[94] M.V. Blagosklonny, et al., Raf-1/bcl-2 phosphorylation: a step from microtubule damage to cell death, Cancer Res. 57 (1) (1997) 130–135.
[95] M.V. Blagosklonny, Sequential activation and inactivation of G2 checkpoints for selective killing of p53-deficient cells by microtubule-active drugs, Oncogene 21 (41) (2002) 6249–6254.
[96] R.A. Britten, et al., Paclitaxel is preferentially cytotoxic to human cervical tumor cells with low Raf-1 kinase activity: implications for paclitaxel-based chemoradiation regimens, Radiother. Oncol. 48 (3) (1998) 329–334.
[97] A. Rasouli-Nia, et al., High Raf-1 kinase activity protects human tumor cells against paclitaxel-induced cytotoxicity, Clin. Cancer Res. 4 (5) (1998) 1111–1116.
[98] R.A. Britten, et al., Raf-1 kinase activity predicts for paclitaxel resistance in TP53mut, but not TP53wt human ovarian cancer cells, Oncol. Rep. 7 (4) (2000) 821–825.
[99] G. Lettre, M.O. Hengartner, Developmental apoptosis in *C. elegans*: a complex CEDnario, Nat. Rev. Mol. Cell Biol. 7 (2) (2006) 97–108.
[100] M.C. Leung, et al., *Caenorhabditis elegans*: an emerging model in biomedical and environmental toxicology, Toxicol. Sci. 106 (1) (2008) 5–28.
[101] S.E. Hulme, G.M. Whitesides, Chemistry and the worm: *Caenorhabditis elegans* as a platform for integrating chemical and biological research, Angew. Chem. Int. Ed. Engl. 50 (21) (2011) 4774–4807.
[102] D.J. Reiner, et al., Use of *Caenorhabditis elegans* to evaluate inhibitors of Ras function in vivo, Methods Enzymol. 439 (2008) 425–449.
[103] N. Moghal, P.W. Sternberg, The epidermal growth factor system in *Caenorhabditis elegans*, Exp. Cell Res. 284 (1) (2003) 150–159.
[104] P.W. Sternberg, H.R. Horvitz, Pattern formation during vulval development in *C. elegans*, Cell 44 (5) (1986) 761–772.
[105] G.J. Beitel, S.G. Clark, H.R. Horvitz, *Caenorhabditis elegans* ras gene let-60 acts as a switch in the pathway of vulval induction, Nature 348 (6301) (1990) 503–509.
[106] U. Stelzl, et al., A human protein–protein interaction network: a resource for annotating the proteome, Cell 122 (6) (2005) 957–968.
[107] G.M. Morris, et al., AutoDock4 and AutoDockTools4: automated docking with selective receptor flexibility, J. Comput. Chem. 30 (16) (2009) 2785–2791.
[108] D. Filchtinski, et al., What makes Ras an efficient molecular switch: a computational, biophysical, and structural study of Ras–GDP interactions with mutants of Raf, J. Mol. Biol. 399 (3) (2010) 422–435.
[109] J.S. Fraser, et al., Accessing protein conformational ensembles using room-temperature X-ray crystallography, Proc. Natl. Acad. Sci. U.S.A. 108 (39) (2011) 16247–16252.
[110] G. Holzapfel, G. Buhrman, C. Mattos, Shift in the equilibrium between on and off states of the allosteric switch in Ras-GppNHp affected by small molecules and bulk solvent composition, Biochemistry 51 (31) (2012) 6114–6126.
[111] A. Moll, et al., BALLView: an object-oriented molecular visualization and modeling framework, J. Comput. Aided Mol. Des. 19 (11) (2005) 791–800.

Inhibitors of K-Ras Plasma Membrane Localization

Kwang-jin Cho*[,1], Dharini van der Hoeven†[,1], John F. Hancock*[,2]
*Department of Integrative Biology and Pharmacology, The University of Texas Medical School at Houston, Houston, Texas, USA
†Department of Diagnostic and Biomedical Sciences, The University of Texas School of Dentistry at Houston, Houston, Texas, USA
[1]Contributed equally.
[2]Corresponding author: e-mail address: john.f.hancock@uth.tmc.edu

Contents

Abstract

Oncogenic mutant K-Ras is highly prevalent in multiple human tumors. Despite significant efforts to directly target Ras activity, no K-Ras-specific inhibitors have been developed and taken into the clinic. Since Ras proteins must be anchored to the inner leaflet of the plasma membrane (PM) for full biological activity, we devised a high-content screen to identify molecules with ability to displace K-Ras from the PM. Here we summarize the biochemistry and biology of three classes of compound identified by this screening method that inhibit K-Ras PM targeting: staurosporine and analogs, fendiline, and metformin. All three classes of compound significantly abrogate cell proliferation and Ras signaling in K-Ras-transformed cancer cells. Taken together, these studies provide an important proof of concept that blocking PM localization of K-Ras is a tractable therapeutic target.

The Enzymes, Volume 33
ISSN 1874-6047
http://dx.doi.org/10.1016/B978-0-12-416749-0.00011-7

1. INTRODUCTION

Ras proteins operate as membrane-bound molecular switches in signal transduction pathways regulating proliferation, differentiation, and apoptosis. There are three main Ras isoforms, H-Ras, N-Ras, and K-Ras, that are ubiquitously expressed in mammalian cells. Although all three Ras isoforms interact with common sets of regulators and effectors, evidence from cell biology, transgenic mice studies, and oncology indicates that the isoforms are biologically distinct [1]. The Ras G-domains (amino acids 1–165) that bind guanine nucleotides and interact with effectors and exchange factors are nearly identical. In contrast, the C-terminal 23 or 24 amino acids of K- and H-/N-Ras, respectively, known as the hypervariable region (HVR), exhibit significant divergence in sequence and account for the observed biological differences. The HVR contains two signal sequences that allow Ras proteins to translocate to and interact with the inner leaflet of the plasma membrane (PM) [2]. The first of these sequences, the CAAX motif, is common to all Ras isoforms (where C=Cys, A=aliphatic amino acid, and X=Ser or Met) [2]. The CAAX motif undergoes sequential processing by protein farnesyltransferase, Rce1 (Ras and a-factor converting enzyme), and Icmt (isoprenylcysteine carboxyl-methyltransferase) [1,2] to leave a farnesyl carboxy-methyl ester attached to the now C-terminal cysteine. The correctly modified CAAX motif is sufficient to target Ras to the endoplasmic reticulum (ER) and Golgi but acts in concert with a second C-terminal signal motif for maximal membrane affinity and PM localization [1]. The second signal consists of palmitoylation of Cys181 and Cys184 in H-Ras, Cys181 in N-Ras, and a polybasic sequence comprising six contiguous lysine residues (Lys175–180) in K-Ras [2,3]. Palmitoylation by Ras palmitoyltransferase of H- and N-Ras occurs in the ER and Golgi, allowing H- and N-Ras to traffic through the classical secretory pathway to the PM [4]. Palmitoylation is an unstable modification with a half-life of <20 min [1,5,6]. Interestingly, recent studies have shown that a depalmitoylation/repalmitoylation cycle is critical to maintain the fidelity of H-Ras PM localization. Depalmitoylation by widely distributed, albeit poorly characterized thiosterases effectively counteracts entropic redistribution of H-Ras through endomembrane compartments by returning depalmitoylated H-Ras to the ER and Golgi for repalmitoylation and vectorial trafficking back to the PM. Strikingly, when depalmitoylation of H-Ras is inhibited, H-Ras protein is redistributed from the PM to all cellular endomembranes [6,7].

In contrast, the mechanism used by K-Ras to access the PM from the ER is unknown, although several scenarios have been proposed. A key feature of K-Ras membrane binding is interaction of the C-terminal polybasic domain (PBD) with anionic lipid head groups on the strongly electronegative inner leaflet of the PM; thus K-Ras may simply diffuse down an electrostatic gradient [8–10]. Proteins such as PDEδ and PRA1 bind the prenylated C-terminus of multiple small GTPases, including K-Ras, and enhance the dissociation rate of K-Ras from the PM [11]. PDEδ may therefore help maintain the fidelity of PM targeting by increasing the diffusion rate of soluble, prenylated K-Ras [12–14]. Other works suggest K-Ras transport via a microtubule-dependent process [15,16] and genetic screens in *Saccharomyces cerevisiae* have implicated a role for mitochondria [17]. K-Ras PM binding is also reversible: phosphorylation of S181 in the PBD drives K-Ras redistribution from the PM to the ER and mitochondria [18,19], whereas acute neutralization of PM inner leaflet surface charge by calcium influx drives K-Ras into the cytosol [8].

Constitutively active mutant K-Ras is involved in 90% of pancreatic, 45% of colorectal, and 35% of lung carcinomas [20,21]. To date, no isoform-specific anti-Ras drugs have been successfully developed. Point mutations in the CAAX motif, which prevent all posttranslational processing, block PM localization and completely inhibit all biological activities of oncogenic mutant Ras [22]. Potent farnesyltransferase inhibitors (FTIs) were therefore developed to phenocopy this mode of Ras inhibition. However, in cells treated with FTIs, an alternative prenyltransferase, geranylgeranyltransferase 1, efficiently geranylgeranylates K-Ras and N-Ras. Geranylgeranylated K- and N-Ras localize normally to the PM and are equipotent with farnesylated K- and N-Ras in transforming assays [23,24], effectively subverting the therapeutic potential of FTIs. Despite the clinical failure of FTIs, since K-Ras must be on the PM for its biological activity inhibition of PM localization remains a valid therapeutic approach to abrogate K-Ras oncogenic activity. Finding compounds that mislocalize K-Ras from the PM will give new insights into the uncharacterized K-Ras trafficking pathway, and such compounds may offer a starting point to develop novel anti-K-Ras cancer therapies. In recent work, we have identified multiple classes of compound that inhibit K-Ras PM localization and K-Ras signaling. In this chapter, we describe the discovery process for inhibitors of Ras PM localization and discuss the molecular mechanisms of action of three of these compound classes.

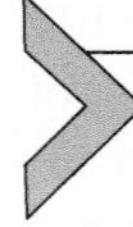

2. DISCOVERY PROCESS FOR INHIBITORS OF Ras PM LOCALIZATION

To search for inhibitors of K-Ras PM localization we established a simple high-content screen using Madin–Darby canine kidney (MDCK) cell lines stably expressing monomeric green fluorescence protein (mGFP) linked to the HVR of K-Ras (=CTK, K-Ras residues 166–188) or H-Ras (=CTH, H-Ras residues 166–189) (Fig. 11.1). MDCK cells were chosen as they show clear Ras PM localization and the columnar morphology is well suited for automated imaging. The isolated HVR of Ras undergoes full posttranslational processing and targets mGFP-fusion proteins efficiently to the PM [4,25]. mGFP–CTK and mGFP–CTH are biologically inert and best preserve MDCK morphology and were therefore preferred for primary screening. In general, cell lines were treated with compounds for 48 h to allow Ras turnover and visualization of effects on the trafficking of newly synthesized Ras. Images were taken by automated confocal microscopy and scored for the extent of Ras mislocalization. Mislocalization of mGFP–CTK or mGFP–CTH from the PM indicates disruption of posttranslational modification or trafficking. In secondary screens to accurately quantify Ras mislocalization from the PM, MDCK cells stably expressing mGFP–K-RasG12V or mGFP–H-RasG12V together with mCherry-CAAX, a red fluorescence fusion protein that decorates endomembranes [26], were used. Manders coefficients were then calculated by measuring the fraction of mCherry-CAAX colocalizing with mGFP–RasG12V [26]. We identified multiple compounds that selectively and potently inhibited K-Ras PM localization while having no or minimal activity on H-Ras PM localization. The characterization of three of these compounds will be described here.

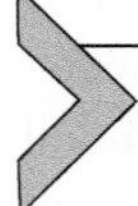

3. STAUROSPORINES INHIBIT Ras PM LOCALIZATION BY BLOCKING PHOSPHATIDYLSERINE TRAFFICKING

Staurosporine (STS) and three analogs, 7-oxostaurosporine (OSS), UCN-01, and UCN-02, significantly mislocalize oncogenic mutant K-Ras (K-RasG12V) from the PM [26] (Fig. 11.2). The relative potency of these compounds in the K-Ras mislocalization assay is: STS($IC_{50}=0.42\pm0.03$ nM) > OSS($IC_{50}=8.21\pm2.11$ nM) > UCN-02($IC_{50}=52.6\pm14.13$ nM) > UCN-01($IC_{50}=840\pm51$ nM). STSs also weakly mislocalize H-RasG12V from the PM but are ~4–10-fold less active than against K-RasG12V. STSs are well

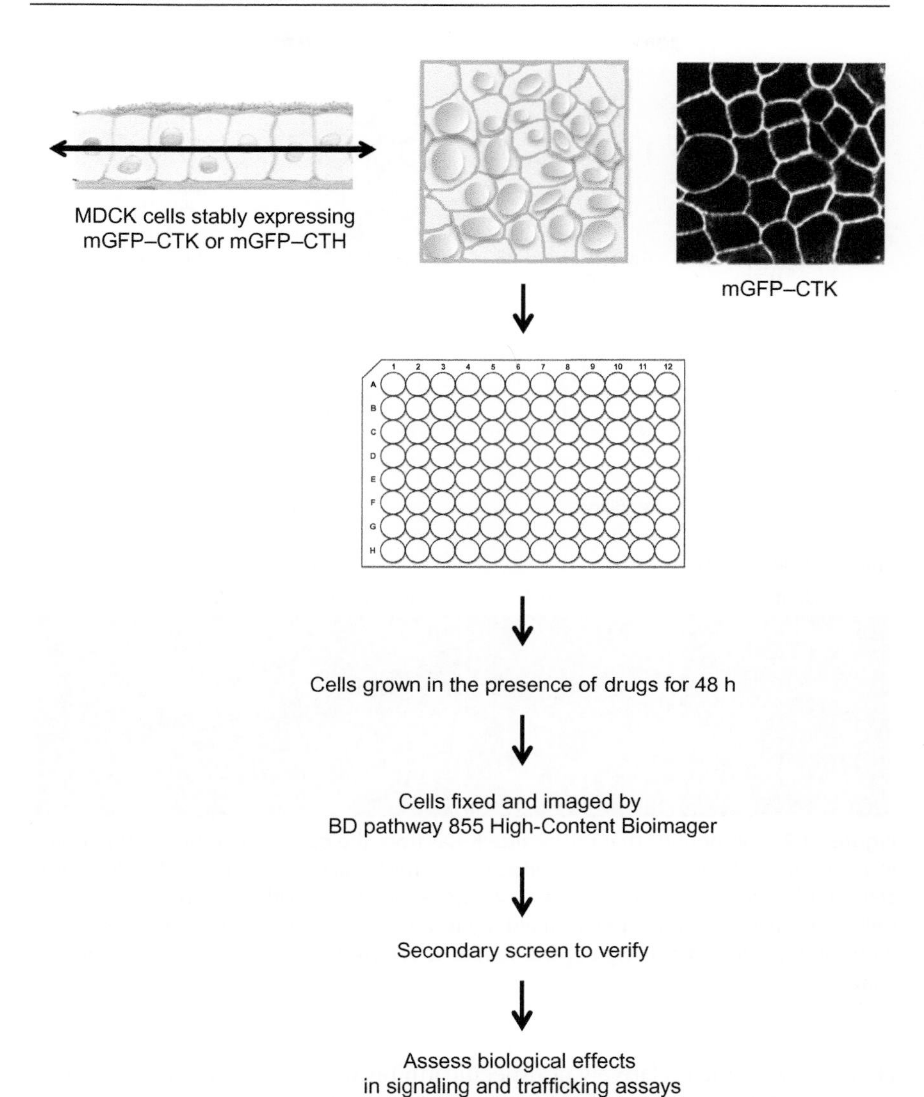

Figure 11.1 Flowchart of a high-content screening assay for inhibitors of Ras plasma membrane localization. MDCK cells stably expressing mGFP–CTK or –CTH are plated at a density of 30,000 cells/well in 96-well plates. Cells are incubated with compounds for 48 h and fixed with 4% paraformaldehyde. Images are taken in BD pathway 855 High-Content Bioimager and compounds that show mislocalization of CTK and/or CTH are taken into a secondary screening for verification. Compounds showing Ras mislocalization are further studied in signaling and trafficking assays. (For color version of this figure, the reader is referred to the online version of this chapter.)

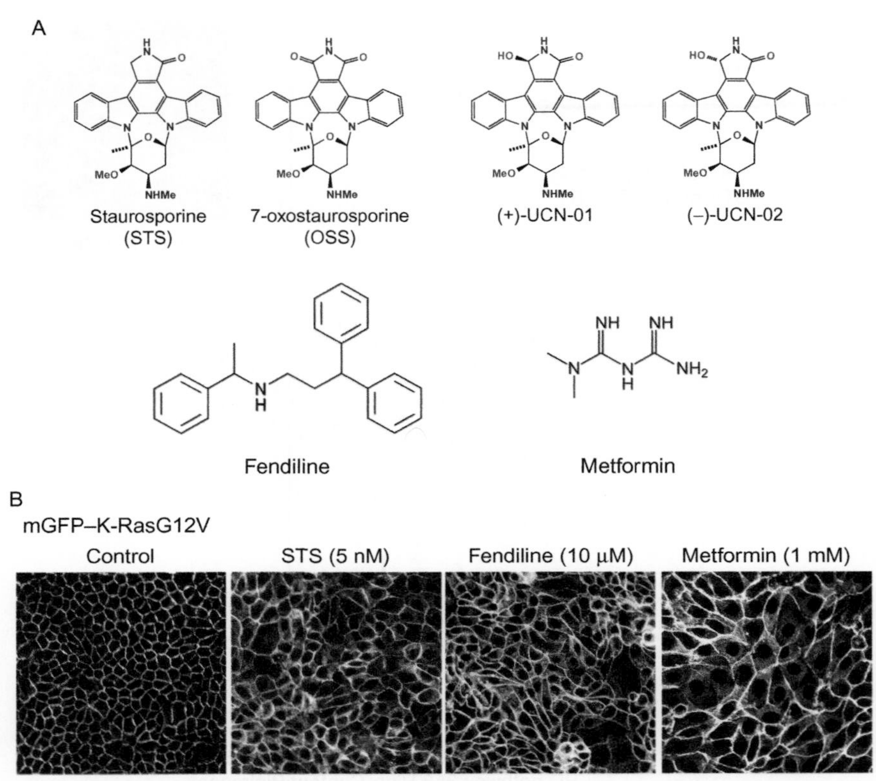

Figure 11.2 Compounds that mislocalize K-Ras from the plasma membrane. (A) Chemical structures of compounds that induce K-Ras mislocalization from the PM. (B) MDCK cells stably expressing mGFP–K-RasG12V were incubated with compounds for 48 h. Cells were fixed with 4% paraformaldehyde and imaged in a confocal microscope. (For color version of this figure, the reader is referred to the online version of this chapter.)

characterized albeit relatively nonspecific inhibitors of protein kinase C (PKC), but several lines of evidence strongly suggest that Ras mislocalization is unrelated to PKC inhibition. First, the rank order potency of STSs for PKC inhibition is UCN-01 > STS ≈ OSS > UCN-02 [27–29], which is inconsistent with that for Ras mislocalization, indicating a different structure–activity relationship. Secondly, STSs do not inhibit phosphorylation of MARCKS, a PKC substrate, at concentrations that mislocalize K-Ras. And finally, whereas PKC has been implicated in the regulation of K-Ras PM association by phosphorylating S181 in the PBD, it is PKC activation not inhibition that mediates the effect on K-Ras. Taken together, these data show that the mechanism of action for STSs on K-Ras mislocalization is independent of PKC inhibition [26].

Phosphatidylserine (PS) is an anionic lipid that is asymmetrically concentrated on the inner leaflet of the PM, which confers significant electrostatic potential to the PM [30]. Acute removal of PS from the inner PM or acute neutralization of PM electrostatic potential by calcium influx results in rapid K-Ras mislocalization [8,31]. The mechanism of action of the STSs involves perturbation of PS subcellular distribution. OSS treatment reduces the level of PS on the inner leaflet of the PM but without externalizing PS to the outer leaflet. A working hypothesis supported by cell imaging and PS supplementation experiments is that STSs block endosomal recycling of PS. Thus, PS internalized from the PM accumulates in the recycling endosome, resulting in a reduced electrostatic potential of the PM and concomitant redistribution of K-Ras to early and late endosomes, lysosomes, mitochondria, the Golgi, and the ER [26] (Fig. 11.3). In support of this mechanism, OSS treatment also mislocalizes mGFP–CTK, mGFP–tK (mGFP targeted by the minimal membrane anchor of K-Ras, residues 175–188), wild-type K-Ras, a CAAX box mutant K-Ras CCIL that is geranylgeranylated, not farnesylated. In addition, other small GTPases that are targeted to the PM by C-terminal PBDs, including Rac1 and Rap1, are sensitive to OSS treatment. Although H-Ras does not have a classical PBD, H-Ras membrane anchoring is modulated by multiple, mutually exclusive electrostatic interactions between basic residues in the HVR or helix $\alpha4$ of the G-domain and anionic lipids in the PM [32–34]. Thus as expected, mGFP-anchored by the full length HVR (mGFP–CTH), wild-type H-Ras and H-RasG12V, but not mGFP–tH (mGFP targeted by the minimal membrane anchor of H-Ras, residues 179–189) are also weakly sensitive to OSS. Taken together, these results strongly implicate electrostatic interactions between C-terminal basic residues in both the K-Ras and H-Ras HVR and PM anionic lipids as the basis of a common mechanism of action for STSs [26].

UCN-01 (7-hydroxystaurosporine) inhibits several protein kinases including PKC and cyclin-dependent kinases, resulting in dysregulation of cell cycle and induction of apoptosis [35,36]. Clinical trials with UCN-01 as a single anticancer agent or in combination with other anticancer agents have been completed or are currently active for a wide range of cancers including T-cell lymphomas, leukemia, advanced kidney cancers, melanoma, and advanced solid tumors such as pancreatic, colorectal, and non-small cells lung cancers [37–41]. However, the precise mechanism of action for its anticancer activity is still not fully understood [37]. The plasma

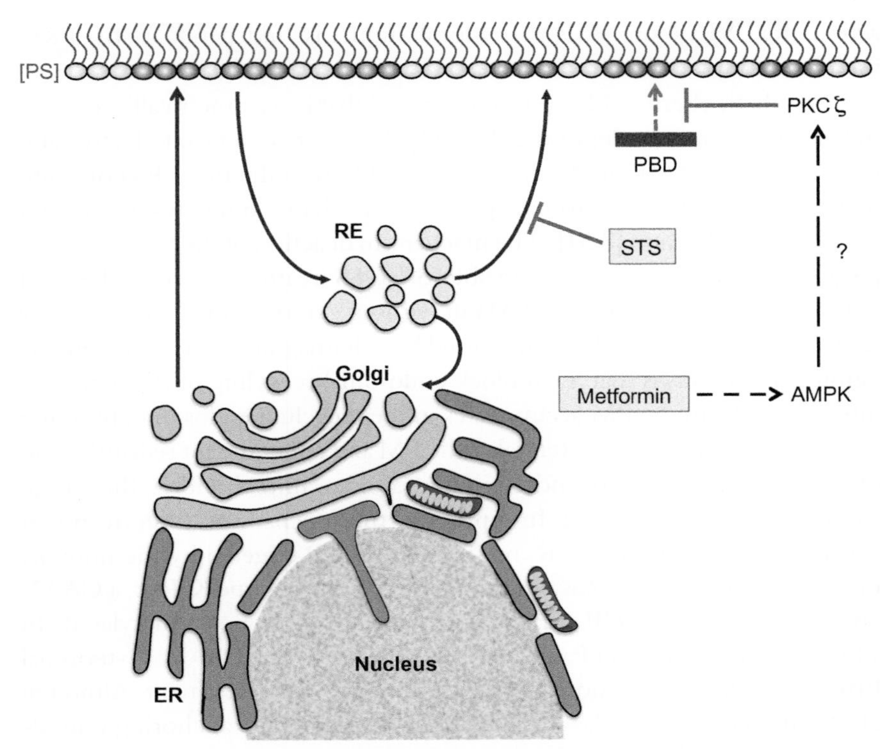

Figure 11.3 Mechanisms of action for compounds that induce K-Ras mislocalization. The diagram illustrates pathways of PS trafficking. PS is synthesized on the ER and mitochondrial-associated membranes and is transported to the PM via the Golgi. Although PS is a relatively minor cellular phospholipid, it is actively concentrated on the inner leaflet of the PM by phospholipid flippases and contributes significantly to the electronegative zeta potential of the PM. PS undergoes endocytic recycling, and internalized PS is sorted and recycled back to the PM through the recycling endosomal compartment. Staurosporines block the sorting/recycling of PS back to the PM such that internalized PS is redistributed to endosomal and endomembrane compartments. The PM electrostatic potential is consequently reduced, resulting in dissociation of the K-Ras of polybasic domain (PBD) from the PM. Metformin induces the activation of AMPK which via a poorly characterized signaling cascade activates PKCζ. PKCζ in turn triggers mislocalization of K-Ras by phosphorylating S181 in the PBD of K-Ras, activating the farnesyl-electrostatic switch. Thus, both drugs inhibit PBD/PM electrostatic inetractions but do so by targeting different components of the biophysics. (See color plate.)

concentrations of UCN-01 in patients are between 0.2 and 0.6 μM [38], similar to the IC_{50} of UCN-01 for K-RasG12V PM mislocalization (0.8 μM). Therefore, it is plausible that the anticancer effects of UCN-01 are at least partially mediated by blocking PS endosomal recycling and Ras mislocalization. Several lines of evidence support this proposal. First,

OSS treatment significantly disrupts not only the total amount of PS and K-Ras on the PM, but also the nanoscale spatial organization of K-RasG12V; as PS levels on the PM fall the fraction of K-Ras that assembles into nanoclusters also decreases [26]. These are important observations, since perturbation of Ras nanocluster organization profoundly dysregulates Ras-mediated cellular signaling [42–45]. Consistent with these results, STS and OSS treatment significantly decrease cell proliferation and MAPK signaling in K-Ras-transformed cells. A third contribution to impaired K-Ras signaling is a significant reduction in total K-Ras levels (~50%) in OSS-treated cells. The mechanism does not involve proteasomal degradation but may be a consequence of perturbed endosomal function that increases K-Ras delivery to lysosomes. Taken together, these results suggest that the anticancer activities of UCN-01 may be accounted for by disruption of the spatial organization of oncogenic mutant K-Ras and a significant decrease in K-Ras protein levels [26]. In this context it will be informative to reevaluate the largely unremarkable clinical trial results with UCN-01 for any evidence of efficacy against K-Ras-driven tumors. The molecular target of STSs remains to be elucidated, but since STSs are relatively promiscuous kinase inhibitors, attractive candidates include the various kinases involved in endosomal function and/or lipid sorting [46].

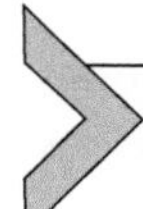

4. FENDILINE INHIBITS K-Ras PM TARGETING BY AN OFF-TARGET MECHANISM THAT IS UNRELATED TO CALCIUM CHANNEL BLOCKADE

Fendiline hydrochloride is a clinically approved, but now obsolete inhibitor of voltage-gated L-type calcium channel, which was used as an antianginal agent for the treatment of coronary heart disease. Together with other phenylalkylamines, it is subclassified in the group of lipophilic calcium antagonists. However, fendiline is rather unique in that unlike other phenylalkylamines, it blocks calcium channels in their open state, a property that has been attributed to the relatively lower potency of fendiline as a calcium channel blocker [47]. Somewhat paradoxically, in a few cell types fendiline induces a rise in intracellular calcium level, which is believed to be by causing extracellular calcium influx and intracellular calcium release from the ER [48–51].

Recently, high-content screening of the Prestwick Chemical library, which consists of 1120 small molecule drugs, led to the discovery of fendiline as a specific inhibitor of PM localization of K-Ras [52] (Fig. 11.2). Fendiline does not alter the localization of H- and N-Ras at concentrations that significantly mislocalize K-Ras. The IC_{50} of fendiline on K-Ras mislocalization is 9.64 ± 0.42 μM, which is close to its therapeutic concentration as a calcium channel blocker [52]. However, other calcium channel blockers, including verapamil, which belongs to the same phenylalkylamine class as fendiline, and members of other classes such as diltizatem and nifedipine, at high concentrations do not alter K-Ras localization, suggesting that the K-Ras inhibitory function of fendiline is independent of its role as a calcium channel blocker. Moreover, in multiple cell lines where K-Ras mislocalization was observed, no concomitant changes in intracellular calcium levels were detected with either acute or chronic exposure to fendiline, indicating that fendiline does not cause K-Ras mislocalization by inducing a rise in intracellular calcium [52]. Taken together, these observations strongly suggest that the mechanism of action of fendiline on K-Ras targeting is calcium independent, is unrelated to the known pharmacology of the drug, and is therefore a novel off-target effect of the compound.

Upon fendiline treatment, the majority of K-Ras displaced from the PM associates with intracellular membranes, including the Golgi, ER, and endosomes, while some remains cytosolic. The loss of K-Ras from the PM with fendiline treatment has been confirmed using electron microscopy. Using this technique, it is also clear that in addition to causing disassociation of K-Ras from the PM, fendiline significantly reduces the nanoclustering of K-Ras that remain PM associated. Interestingly, although fendiline does not inhibit PM association of oncogenic H-RasG12V (in fact, it increases H-Ras membrane association), it does modestly decrease the extent of H-RasG12V nanoclustering [52]. These observations suggest that fendiline may alter PM lipid composition that regulates Ras protein clustering.

The effects of fendiline on K-Ras PM association and nanoclustering correlate with reduced signal output. Fendiline inhibits oncogenic K-RasG12V activation of both the MAPK and PI3K/Akt signaling axes, with IC_{50}s that are similar to the IC_{50} for K-Ras mislocalization. Fendiline does not affect signaling downstream of constitutively active CRaf, providing strong evidence that the effect of fendiline on K-Ras signaling occurs at the level of Ras. Fendiline partially inhibits signaling downstream of constitutively active BRaf [52], consistent with the finding that constitutively active BRaf must be recruited to the PM by Ras for full activation of MAPK [53]. Fendiline also inhibits the signaling downstream of oncogenic H-Ras (but not N-Ras) even though it does not mislocalize H-Ras [52]. This unexpected effect of fendiline likely reflects a significant contribution of endogenous K-Ras to the signaling of oncogenic H-Ras via an autocrine feed forward loop.

Fendiline more potently inhibits the proliferation of pancreatic, colon, lung, and endometrial tumor cells that express oncogenic mutant K-Ras than tumor cells that express wild-type K-Ras, suggesting that fendiline is a selective inhibitor of oncogenic K-Ras function, regardless of the origin of the tumor. Fendiline is also more potent than a MEK inhibitor at inhibiting proliferation of tumor cells expressing oncogenic K-Ras, even though the MEK inhibitor is more potent at inhibiting MEK and ERK activation [52]. This important proof of concept experiment shows that the inhibition of Ras function achieves concordant inhibition of parallel Ras effector pathways and delivers important synergistic effects in abrogating tumor cell growth, rendering fendiline a potentially effective anticancer therapeutic.

The mechanism of fendiline action on K-Ras trafficking and/or PM association is yet to be characterized, although a few likely mechanisms have been ruled out. Fendiline does not alter prenylation or methylation of K-Ras and therefore must modify K-Ras transport or PM interactions distal to CAAX processing. Mislocalization of K-Ras by fendiline does not involve the farnesyl-electrostatic switch. K-Ras does not localize to the mitochondria upon treatment with fendiline and a K-Ras mutant, K-RasS181A with the major PKC phosphorylation site removed, is equally responsive to fendiline [52].

C-terminal mapping experiments and sensitivity of other Ras-related proteins to fendiline provide clues to the biophysical mechanism of fendiline action. Fendiline mislocalizes not only mGFP targeted by the wild-type and mutant K-Ras, but also mGFP–tK and mGFP–CTK implicating the C-terminal polybasic region of K-Ras as being required for sensitivity to fendiline. Farnesylated and geranylgeranylated forms of K-Ras are mislocalized by fendiline suggesting that the nature of the prenoid group attached to the CAAX box does not influence the effect of fendiline. Fendiline also causes mislocalization of other proteins such as Rap1A and Rac1, which are anchored to the PM by a prenylated PBD [52]. Taken together, these observations identify a PM anchor comprising a prenylated PBD as the minimal determinant for fendiline sensitivity. In the case of K-Ras, the weak membrane binding provided by the farnesylated cysteine is strengthened by electrostatic interaction between the polybasic sequence and anionic lipid head groups enriched in the inner leaflet of the PM. Given that PS is a major contributor to PM electrostatic potential, and recent experiments showing that STS redistributes PS from the PM to endomembranes [26], we are currently exploring whether fendiline has a similar mechanism of action.

Our preliminary studies indicate that fendiline has potential as an anticancer agent, especially in the treatment of oncogenic K-Ras expressing tumors. It also has potential in combination therapy, specifically to abrogate the undesirable Ras-dependent signaling that occurs as a side effect with agents that are in current clinical use, with a view to improving treatment efficacy. For example, fendiline prevents paradoxical activation of MEK/ERK signaling in Ras-transformed cells treated with BRaf inhibitors, because fendiline reduces K-Ras levels on the PM and abrogates K-Ras nanoclustering [43]. mTOR inhibitors such as rapamycin and derivatives CCI-779, AP23573, and RAD001, which are currently in clinical trial, also activate the MAPK cascade through a loop that depends on an S6K-PI3K-Ras pathway [54,55]. Our unpublished experiments show that fendiline abrogates such signaling and potentiates the inhibitory effect of RAD001 on K-Ras-transformed endometrial cancer cells.

5. METFORMIN: A NEW USE AS AN ANTICANCER THERAPEUTIC TARGETING K-Ras

Metformin (*N'*,*N'*-dimethylbiguanide) is the most widely prescribed hypoglycemic agent for the management of type II diabetes mellitus. Metformin reduces hyperglycemia primarily by suppressing hepatic gluconeogenesis through activation of AMP-activated protein kinase (AMPK). In addition to treatment of diabetes, metformin has been also approved for the treatment of polycystic ovarian syndrome and metabolic syndrome.

We recently identified metformin as a small molecule inhibitor of K-Ras PM localization with an IC_{50} of 261.03 ± 57.91 μM (Fig. 11.2). K-Ras mislocalizes to endomembranes and cytosol in metformin-treated cells. Metformin selectively inhibits MAPK activation in endometrial cancer cells expressing oncogenic K-Ras but has no effect on MAPK kinase signaling in endometrial cancer cells expressing wild-type K-Ras. Similarly, metformin preferentially inhibits cell proliferation and induces apoptosis of K-Ras-transformed endometrial cancer cells. Metformin sensitivity of K-Ras-transformed cancer cells is abrogated when K-Ras expression is silenced, further validating K-Ras as the key metformin target, suggesting that the induction of apoptosis in K-Ras-transformed cells may be associated with metformin-induced K-Ras redistribution [56].

The mechanism of metformin-mediated mislocalization of K-Ras is likely mediated through activation of AMPK. Aminoimidazole-4-carboxamide riboside (AICAR), an analog of AMP that directly activates

AMPK, also causes mislocalization of K-Ras (D.V. Hoeven, K-j. Cho, J.F. Hancock, unpublished data). The subsequent signaling pathway is unclear but both metformin and AICAR activate atypical PKCζ in an AMPK-dependent manner in muscle cells [57]. PKCζ in turn could trigger mislocalization of K-Ras through phosphorylation of S181 in the PBD of K-Ras thus activating the farnesyl-electrostatic switch [18] (Fig. 11.3). Concurrently, in support of this mechanism, the K-RasS181A mutant is largely resistant to metformin-induced mislocalization [56].

In contrast to STSs and fendiline, there is a growing body of epidemiological studies linking metformin with reduced incidence, recurrence, and mortality of a variety of cancers in type II diabetes mellitus patients [58–61]. While the majority of evidence supporting a role for metformin as an anticancer agent has been derived from studies involving individuals with diabetes, more recent clinical trials show that metformin is associated with a decrease in cancer risk in nondiabetics as well. It is important to note that metformin reduces the risk of pancreatic and colorectal cancers, two cancers with high incidence of K-Ras mutation [62,63]. In addition to the epidemiological studies, there is compelling evidence from *in vitro* and *in vivo* studies suggesting that metformin acts as an anticancer agent through inhibition of K-Ras and could be used for the treatment of oncogenic K-Ras expressing tumors. Metformin inhibits cell proliferation and induces apoptosis of pancreatic, lung, and colon cancer cell lines and it inhibits the growth of K-Ras-transformed PANC-1 pancreatic cancer cell xenografts in nude mice in a dose-dependent manner [64,65]. Metformin also decreases tumor burden by 72% in a mouse model of lung cancer, which correlates with decreased cellular proliferation [66].

Mechanistic pathways suggested in the past by which metformin might inhibit tumor growth or proliferation include: (1) eradication of cancer stem cells targets, (2) activation of LKB1/AMPK pathway leading to inhibition of mTOR, (3) induction of cell cycle arrest and/or apoptosis, (4) inhibition of protein synthesis, (5) inhibition of the unfolded protein response, and (6) inhibition of inflammatory response required for cancer cell transformation [67–69]. However, in addition to the effect of metformin on K-Ras-transformed endometrial cancer cells discussed previously, a recent study has shown that metformin also inhibits ERK activation in K-Ras-transformed pancreatic cancer cells [70]. In light of the new role of metformin as a K-Ras inhibitor, the epidemiological studies need to be revisited to determine the correlation between K-Ras mutation status in cancers sensitive to metformin treatment.

6. CONCLUSION AND FUTURE DIRECTIONS

In this chapter, we describe a discovery process that has to date identified three classes of compound as new K-Ras PM inhibitors: STS and analogs, fendiline, and metformin. Although the precise mechanisms of action for K-Ras mislocalization are yet to be fully unraveled, all three classes of compound selectively inhibit K-Ras signaling and cellular proliferation in K-Ras-driven cancer cells. These results provide an important proof of concept that blocking K-Ras PM localization is a valid and chemically tractable approach to block K-Ras biological activity. Our ongoing screening of different compound libraries is enriching our collection of K-Ras PM inhibitors. The high-content screening assay described here could also be modified to identify proteins that regulate K-Ras PM localization using RNAi knock down. In combination with compound screening such assays will generate new insights into the uncharacterized K-Ras trafficking pathway and interaction with the PM, and identify specific molecular targets for further drug development assays.

ACKNOWLEDGMENT

This work was supported by a grant from the Cancer Prevention and Research Institute of Texas (RP100483).

REFERENCES

[1] J.F. Hancock, Ras proteins: different signals from different locations, Nat. Rev. Mol. Cell Biol. 4 (5) (2003) 373–384.
[2] J.F. Hancock, et al., All ras proteins are polyisoprenylated but only some are palmitoylated, Cell 57 (7) (1989) 1167–1177.
[3] J.F. Hancock, H. Paterson, C.J. Marshall, A polybasic domain or palmitoylation is required in addition to the CAAX motif to localize p21ras to the plasma membrane, Cell 63 (1) (1990) 133–139.
[4] A. Apolloni, et al., H-ras but not K-ras traffics to the plasma membrane through the exocytic pathway, Mol. Cell Biol. 20 (7) (2000) 2475–2487.
[5] T.L. Baker, M.A. Booden, J.E. Buss, *S*-Nitrosocysteine increases palmitate turnover on Ha-Ras in NIH 3T3 cells, J. Biol. Chem. 275 (29) (2000) 22037–22047.
[6] O. Rocks, et al., An acylation cycle regulates localization and activity of palmitoylated Ras isoforms, Science 307 (5716) (2005) 1746–1752.
[7] O. Rocks, et al., The palmitoylation machinery is a spatially organizing system for peripheral membrane proteins, Cell 141 (3) (2010) 458–471.
[8] T. Yeung, et al., Membrane phosphatidylserine regulates surface charge and protein localization, Science 319 (5860) (2008) 210–213.
[9] N.M. Okeley, M.H. Gelb, A designed probe for acidic phospholipids reveals the unique enriched anionic character of the cytosolic face of the mammalian plasma membrane, J. Biol. Chem. 279 (21) (2004) 21833–21840.

[10] M.O. Roy, R. Leventis, J.R. Silvius, Mutational and biochemical analysis of plasma membrane targeting mediated by the farnesylated, polybasic carboxy terminus of K-ras4B, Biochemistry 39 (28) (2000) 8298–8307.
[11] P. Bhagatji, et al., Multiple cellular proteins modulate the dynamics of K-ras association with the plasma membrane, Biophys. J. 99 (10) (2010) 3327–3335.
[12] S.A. Ismail, et al., Arl2-GTP and Arl3-GTP regulate a GDI-like transport system for farnesylated cargo, Nat. Chem. Biol. 7 (12) (2011) 942–949.
[13] A. Chandra, et al., The GDI-like solubilizing factor PDEdelta sustains the spatial organization and signalling of Ras family proteins, Nat. Cell Biol. 14 (2) (2012) 148–158.
[14] M.R. Philips, Ras hitchhikes on PDE6delta, Nat. Cell Biol. 14 (2) (2012) 128–129.
[15] Z. Chen, et al., The C-terminal polylysine region and methylation of K-Ras are critical for the interaction between K-Ras and microtubules, J. Biol. Chem. 275 (52) (2000) 41251–41257.
[16] J.A. Thissen, et al., Prenylation-dependent association of Ki-Ras with microtubules. Evidence for a role in subcellular trafficking, J. Biol. Chem. 272 (48) (1997) 30362–30370.
[17] G. Wang, R.J. Deschenes, Plasma membrane localization of Ras requires class C Vps proteins and functional mitochondria in *Saccharomyces cerevisiae*, Mol. Cell Biol. 26 (8) (2006) 3243–3255.
[18] T.G. Bivona, et al., PKC regulates a farnesyl-electrostatic switch on K-Ras that promotes its association with Bcl-XL on mitochondria and induces apoptosis, Mol. Cell 21 (4) (2006) 481–493.
[19] M. Fivaz, T. Meyer, Reversible intracellular translocation of KRas but not HRas in hippocampal neurons regulated by Ca^{2+}/calmodulin, J. Cell Biol. 170 (3) (2005) 429–441.
[20] A.T. Baines, D. Xu, C.J. Der, Inhibition of Ras for cancer treatment: the search continues, Future Med. Chem. 3 (14) (2011) 1787–1808.
[21] I.A. Prior, P.D. Lewis, C. Mattos, A comprehensive survey of Ras mutations in cancer, Cancer Res. 72 (10) (2012) 2457–2467.
[22] B.M. Willumsen, et al., The p21 ras C-terminus is required for transformation and membrane association, Nature 310 (5978) (1984) 583–586.
[23] J.F. Hancock, et al., A CAAX or a CAAL motif and a second signal are sufficient for plasma membrane targeting of ras proteins, EMBO J. 10 (13) (1991) 4033–4039.
[24] A.D. Cox, et al., Specific isoprenoid modification is required for function of normal, but not oncogenic, Ras protein, Mol. Cell Biol. 12 (6) (1992) 2606–2615.
[25] E. Choy, et al., Endomembrane trafficking of ras: the CAAX motif targets proteins to the ER and Golgi, Cell 98 (1) (1999) 69–80.
[26] K.J. Cho, et al., Staurosporines disrupt phosphatidylserine trafficking and mislocalize ras proteins, J. Biol. Chem. 287 (52) (2012) 43573–43584.
[27] H. Osada, et al., A new inhibitor of protein kinase C, RK-1409 (7-oxostaurosporine). I. Taxonomy and biological activity, J. Antibiot. 45 (2) (1992) 189–194.
[28] C. Courage, J. Budworth, A. Gescher, Comparison of ability of protein kinase C inhibitors to arrest cell growth and to alter cellular protein kinase C localisation, Br. J. Cancer 71 (4) (1995) 697–704.
[29] C.M. Seynaeve, et al., Differential inhibition of protein kinase C isozymes by UCN-01, a staurosporine analogue, Mol. Pharmacol. 45 (6) (1994) 1207–1214.
[30] P.A. Leventis, S. Grinstein, The distribution and function of phosphatidylserine in cellular membranes, Annu. Rev. Biophys. 39 (2010) 407–427.
[31] T. Yeung, et al., Receptor activation alters inner surface potential during phagocytosis, Science 313 (5785) (2006) 347–351.
[32] D. Abankwa, et al., Ras membrane orientation and nanodomain localization generate isoform diversity, Proc. Natl. Acad. Sci. U.S.A. 107 (3) (2010) 1130–1135.

[33] D. Abankwa, et al., A novel switch region regulates H-ras membrane orientation and signal output, EMBO J. 27 (5) (2008) 727–735.

[34] A.A. Gorfe, A. Babakhani, J.A. McCammon, H-ras protein in a bilayer: interaction and structure perturbation, J. Am. Chem. Soc. 129 (40) (2007) 12280–12286.

[35] M.M. Facchinetti, et al., UCN-01-induced cell cycle arrest requires the transcriptional induction of p21(waf1/cip1) by activation of mitogen-activated protein/extracellular signal-regulated kinase kinase/extracellular signal-regulated kinase pathway, Cancer Res. 64 (10) (2004) 3629–3637.

[36] I. Takahashi, et al., UCN-01 and UCN-02, new selective inhibitors of protein kinase C. II. Purification, physico-chemical properties, structural determination and biological activities, J. Antibiot. (Tokyo) 42 (4) (1989) 571–576.

[37] S.J. Hotte, et al., Phase I trial of UCN-01 in combination with topotecan in patients with advanced solid cancers: a Princess Margaret Hospital Phase II Consortium study, Ann. Oncol. 17 (2) (2006) 334–340.

[38] E.A. Sausville, et al., Phase I trial of 72-hour continuous infusion UCN-01 in patients with refractory neoplasms, J. Clin. Oncol. 19 (8) (2001) 2319–2333.

[39] T. Li, et al., A phase II study of cell cycle inhibitor UCN-01 in patients with metastatic melanoma: a California Cancer Consortium trial, Invest. New Drugs 30 (2) (2012) 741–748.

[40] M.J. Edelman, et al., Phase I and pharmacokinetic study of 7-hydroxystaurosporine and carboplatin in advanced solid tumors, Clin. Cancer Res. 13 (9) (2007) 2667–2674.

[41] G.E. Marti, et al., Phase I trial of 7-hydroxystaurosporine and fludararbine phosphate: in vivo evidence of 7-hydroxystaurosporine induced apoptosis in chronic lymphocytic leukemia, Leuk. Lymphoma 52 (12) (2011) 2284–2292.

[42] T. Tian, et al., Plasma membrane nanoswitches generate high-fidelity Ras signal transduction, Nat. Cell Biol. 9 (8) (2007) 905–914.

[43] K.J. Cho, et al., Raf inhibitors target ras spatiotemporal dynamics, Curr. Biol. 22 (11) (2012) 945–955.

[44] B.N. Kholodenko, J.F. Hancock, W. Kolch, Signalling ballet in space and time, Nat. Rev. Mol. Cell Biol. 11 (6) (2010) 414–426.

[45] S.J. Plowman, et al., H-ras, K-ras, and inner plasma membrane raft proteins operate in nanoclusters with differential dependence on the actin cytoskeleton, Proc. Natl. Acad. Sci. U.S.A. 102 (43) (2005) 15500–15505.

[46] L. Pelkmans, et al., Genome-wide analysis of human kinases in clathrin- and caveolae/raft-mediated endocytosis, Nature 436 (7047) (2005) 78–86.

[47] H. Nawrath, et al., Open state block by fendiline of L-type Ca++ channels in ventricular myocytes from rat heart, J. Pharmacol. Exp. Ther. 285 (2) (1998) 546–552.

[48] C. Huang, et al., Fendiline-evoked $[Ca^{2+}]_i$ rises and non-Ca^{2+}-triggered cell death in human oral cancer cells, Hum. Exp. Toxicol. 28 (1) (2009) 41–48.

[49] C.R. Jan, et al., Fendiline, an anti-anginal drug, increases intracellular Ca^{2+} in PC3 human prostate cancer cells, Cancer Chemother. Pharmacol. 48 (1) (2001) 37–41.

[50] M.C. Lin, C.R. Jan, The anti-anginal drug fendiline elevates cytosolic Ca(2+) in rabbit corneal epithelial cells, Life Sci. 71 (9) (2002) 1071–1079.

[51] J. Wang, et al., The anti-anginal drug fendiline increases intracellular Ca(2+) levels in MG63 human osteosarcoma cells, Toxicol. Lett. 119 (3) (2001) 227–233.

[52] D. van der Hoeven, et al., Fendiline inhibits K-Ras plasma membrane localization and blocks K-Ras signal transmission, Mol. Cell Biol. 33 (2) (2013) 237–251.

[53] A. Harding, et al., Identification of residues and domains of Raf important for function in vivo and in vitro, J. Biol. Chem. 278 (46) (2003) 45519–45527.

[54] S.A. Wander, B.T. Hennessy, J.M. Slingerland, Next-generation mTOR inhibitors in clinical oncology: how pathway complexity informs therapeutic strategy, J. Clin. Invest. 121 (4) (2011) 1231–1241.

[55] A. Carracedo, et al., Inhibition of mTORC1 leads to MAPK pathway activation through a PI3K-dependent feedback loop in human cancer, J. Clin. Invest. 118 (9) (2008) 3065–3074.
[56] D.A. Iglesias, et al., Another surprise from metformin: novel mechanism of action via K-Ras influences endometrial cancer response to therapy, (2013), submitted.
[57] M.P. Sajan, et al., AICAR and metformin, but not exercise, increase muscle glucose transport through AMPK-, ERK-, and PDK1-dependent activation of atypical PKC, Am. J. Physiol. Endocrinol. Metab. 298 (2) (2010) E179–E192.
[58] I. Ben Sahra, et al., Metformin in cancer therapy: a new perspective for an old anti-diabetic drug? Mol. Cancer Ther. 9 (5) (2010) 1092–1099.
[59] J.M. Evans, et al., Metformin and reduced risk of cancer in diabetic patients, BMJ 330 (7503) (2005) 1304–1305.
[60] A. Emami Riedmaier, et al., Metformin and cancer: from the old medicine cabinet to pharmacological pitfalls and prospects, Trends Pharmacol. Sci. 34 (2) (2013) 126–135.
[61] A. Aljada, S.A. Mousa, Metformin and neoplasia: implications and indications, Pharmacol. Ther. 133 (1) (2012) 108–115.
[62] P. Zhang, et al., Association of metformin use with cancer incidence and mortality: a meta-analysis, Cancer Epidemiol. 37 (3) (2013) 207–218.
[63] C.J. Currie, C.D. Poole, E.A. Gale, The influence of glucose-lowering therapies on cancer risk in type 2 diabetes, Diabetologia 52 (9) (2009) 1766–1777.
[64] K. Kisfalvi, et al., Metformin disrupts crosstalk between G protein-coupled receptor and insulin receptor signaling systems and inhibits pancreatic cancer growth, Cancer Res. 69 (16) (2009) 6539–6545.
[65] K. Kisfalvi, et al., Metformin inhibits the growth of human pancreatic cancer xenografts, Pancreas. 42 (5) (2013) 781–785.
[66] C.A. Granville, et al., Identification of a highly effective rapamycin schedule that markedly reduces the size, multiplicity, and phenotypic progression of tobacco carcinogen-induced murine lung tumors, Clin. Cancer Res. 13 (7) (2007) 2281–2289.
[67] M.N. Pollak, Investigating metformin for cancer prevention and treatment: the end of the beginning, Cancer Discov. 2 (9) (2012) 778–790.
[68] R.J. Dowling, et al., Metformin in cancer: translational challenges, J. Mol. Endocrinol. 48 (3) (2012) R31–R43.
[69] T.V. Kourelis, R.D. Siegel, Metformin and cancer: new applications for an old drug, Med. Oncol. 29 (2) (2012) 1314–1327.
[70] H.P. Soares, et al., Different patterns of akt and ERK feedback activation in response to rapamycin, active-site mTOR inhibitors and metformin in pancreatic cancer cells, PLoS One 8 (2) (2013) e57289.

CHAPTER TWELVE

Ras Chaperones: New Targets for Cancer and Immunotherapy

Yoel Kloog[*,1], **Galit Elad-Sfadia**[*], **Roni Haklai**[*], **Adam Mor**[†,‡]

[*]Department of Neurobiology, Faculty of Life Sciences, Tel-Aviv University, Ramat-Aviv, Israel
[†]Department of Medicine, New York University School of Medicine, New York, New York, USA
[‡]Department of Pathology, New York University School of Medicine, New York, New York, USA
[1]Corresponding author: e-mail address: kloog@post.tau.ac.il

Contents

Abstract

The Ras inhibitor *S-trans,trans*-farnesylthiosalicylic acid (FTS, Salirasib®) interferes with Ras membrane interactions that are crucial for Ras-dependent signaling and cellular transformation. FTS had been successfully evaluated in clinical trials of cancer patients. Interestingly, its effect is mediated by targeting Ras chaperones that serve as key coordinators for Ras proper folding and delivery, thus offering a novel target for cancer therapy. The development of new FTS analogs has revealed that the specific modifications to the FTS carboxyl group by esterification and amidation yielded compounds with improved growth inhibitory activity. When FTS was combined with additional therapeutic agents its activity toward Ras was significantly augmented. FTS should be tested not only in cancer but also for genetic diseases associated with abnormal Ras signaling, as well as for various inflammatory and autoimmune disturbances, where Ras plays a major role. We conclude that FTS has a great potential both as a safe anticancer drug and as a promising immune modulator agent.

The Enzymes, Volume 33
ISSN 1874-6047
http://dx.doi.org/10.1016/B978-0-12-416749-0.00012-9

1. INTRODUCTION

The small GTPases H-, K-, and N-Ras control many cellular functions including growth, differentiation, motility, and survival [1–4]. They function as molecular switches and as such alternate between a GDP-bound (inactive) and a GTP-bound (active) state through the action of guanine nucleotide exchange factors (GEFs) and GTPase activating proteins (GAPs). In as many as 33% of human cancers Ras genes are mutated at positions 12, 13, or 61, rendering the mutant Ras proteins insensitive to the GAPs activity [1–3,5–7]. This leads to expression of constitutively active Ras and to carcinogenesis. For example, more than 90% of pancreatic tumors and about 50% of non-small-cell lung and colon cancers exhibit mutated K-Ras. Wild-type Ras proteins could also be activated by overexpressed or mutated upstream tyrosine kinase receptors such as epidermal growth factor receptor, fibroblast growth factor receptor, platelet-derived growth factor receptor, and vascular endothelial growth factor receptor [1–3,8,9], or by mutation or deletions of the Ras GAP neurofibromin 1 (NF1). Active Ras proteins transmit signals to several downstream signaling pathways such as the Raf–MEK–ERK, PI3-kinase, Rheb–mTORC1, and the Ral-GDS cascades that operate in well-coordinated networks that regulate cellular behavior [1–3]. In cancer cells this coordination is lost, a process that may lead to a state of oncogenes addiction, a Ras-dependent upregulation of additional genes [10].

Ras proteins play important roles not only in cancer, but also in other biological arenas including the immune system [10]. In mast cells Ras is critical for IgE-mediated degranulation [11]. In lymphocytes Ras proteins regulate signaling downstream the antigen receptor, leading to cytokine secretion, increased adhesion and clonal proliferation [12]. Animals transgenic for activated Ras present with autoimmune phenotypes due to unregulated immune responses, supporting the notion that Ras is an important target for treating inflammatory conditions [12].

Although Ras was among the first oncogenes discovered (almost three decades ago) and its role in tumor formation and maintenance is well established, we still do not benefit from anti-Ras drugs in clinical use. Numerous attempts to design Ras inhibitors have been undertaken, including the farnesyl transferase inhibitors (FTIs) that block Ras farnesylation, an absolute requirement for Ras functions [13,14]. These drugs failed to block oncogenic Ras because of the ability of K- and N-Ras oncoproteins to resort an alternative prenylation pathway, allowing them to remain active in the presence of FTIs. The failure of these Ras inhibitors to yield anticancer drugs [15,16] led to the unfortunate assumption that Ras was not targetable and kept the important field of Ras inhibitors a long way behind other targeted proteins, such as receptor tyrosine kinases and soluble kinases [17]. In this work we will describe a different approach to inhibit Ras by interfering with its chaperones. We will describe our efforts to design farnesylthiosalicylic acid (FTS) and its new derivatives, and their applications both in cancer and in emerging novel immunotherapies.

2. Ras CHAPERONES ARE TARGETS FOR Ras INHIBITION

We have proposed an entirely different approach to inhibit Ras based on the notion that Ras proteins are stabilized at cell membranes through their interaction with chaperon proteins. Chaperones are proteins that assist the noncovalent folding, or unfolding, of other proteins, but do not occur in these structures when the proteins are performing their normal biological functions. Chaperones also prevent both newly synthesized polypeptide chains and assembled subunits from aggregating into nonfunctional structures. The interaction between Ras and its chaperones is dependent on the farnesylcysteine moiety of Ras, which is required for its membrane association and functions. We have suggested that the chaperones that stabilize Ras actually facilitate its signaling from the cell membrane and therefore designed a series of farnesylcysteine analogs that would inhibit active Ras proteins signaling and Ras-dependent tumor growth [18–21]. One of these compounds, FTS, turned out to have a potent anticancer activity, targeting primarily the active isoforms of Ras [22,23]. FTS has been developed as an oral agent and has been used successfully in phase II clinical trials in patients with pancreatic and non-small-cell lung carcinomas [24].

The concept of targeting the Ras chaperones is novel. Previously, merely the Ras GEFs, GAPs, and downstream effectors were considered as functionally important Ras-interacting proteins [1–3,5–7]. More recently, three distinct Ras chaperones were identified, each with high specificity toward a different isoform in its GTP-bound state. Galectin-1 (Gal-1) binds and stabilizes primarily activated H-Ras, while galectin-3 (Gal-3) binds to activated K-Ras [25–27]. It has been shown that Gal-1 regulates H-Ras nanocluster formation and signaling [28–32] (Fig. 12.1), and likewise Gal-3 controls K-Ras nanoclustering and downstream signaling [33,34]. Consistent with these results, biophysical experiments revealed that Gal-3 reduced the rate

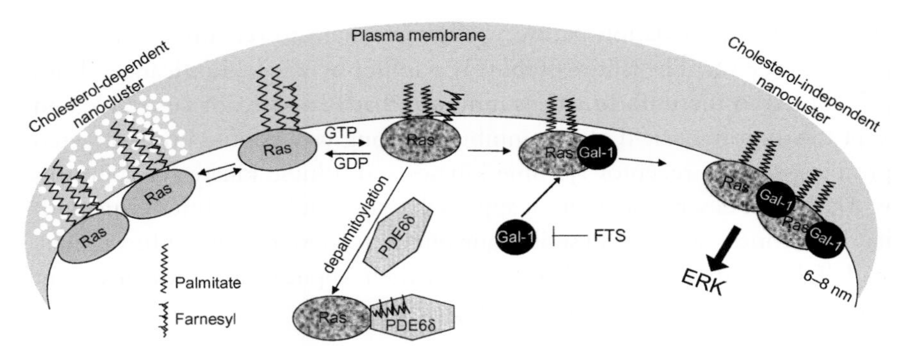

Figure 12.1 Gal-1 is a novel structural component and a major regulator of H-Ras nanoclusters. The double palmitoylated and farnesylated H-Ras in its GDP-bound state forms nanoclusters in cholesterol-dependent microdomains. Exchange of GDP for GTP induces conformational change in Ras, rendering its lipid moieties to become closer to the inner leaflet of the plasma membrane. GTP-bound H-Ras can undergo depalmitoylation that promotes its dissociation from the plasma membrane and association with PDE6δ that acts as a Ras-solubilizing factor that allows proper Ras trafficking. Alternatively, the GTP-bound H-Ras binds to Gal-1 in the cell membrane and together diffuses to cholesterol-independent microdomains. These nanoclusters promote robust Ras signaling to the Raf–MEK–ERK cascade, acting in a positive cooperatively manner. The formation of these complexes is prevented by FTS that interacts with Gal-I and thus prevents H-Ras–GTP nanoclustering.

of plasma membrane dissociation of activated K-Ras, an effect that was blocked by FTS [35]. Recent studies have established phosphodiesterase 6 delta (PDE6δ) as additional Ras chaperone that can solubilize all Ras isoforms and contribute to their subcellular trafficking [36–38]. In this sense PDE6δ and galectins operate in opposing directions; the former solubilizes Ras, while the later stabilize its interactions with the plasma membrane [33,34].

To further understand how the interaction between K-Ras and Gal-3 is regulated, mouse embryonic fibroblasts (MEFs) isolated from Gal-1$^{-/-}$, Gal-3$^{-/-}$, and Gal-3$^{-/-}$/Gal-1$^{-/-}$ knockout mice were studied [39]. We found that the activation of K-Ras, ERK, and Akt was strongly reduced in the absence of Gal-3 [39]. Reexpression of Gal-3 reversed the phenotype of Gal-3$^{-/-}$ MEFs [40]. The importance of Gal-3/K-Ras interactions has been recently documented in thyroid carcinoma cell lines [39]. These studies showed that the most aggressive thyroid cell line expressed the highest levels of Gal-3 in correlation with the levels of activated K-Ras [39]. Recently another Ras chaperon, nucleolin, which binds and stabilizes activated N-Ras, has been identified [41]. How does FTS modulate nucleolin activity and how does it control Ras signaling under active investigation in our laboratory.

Our leading compound, FTS, interferes with GTP-bound Ras chaperones interactions, preventing effectors signaling [18,19]. This seems to account for the high selectivity of FTS, which does not affect the inactive (GDP-bound) forms of Ras proteins and might explain the lack of clinical toxicity in mice and in humans [22,23]. The realization of Ras chaperones has facilitated the discovery of new mechanisms of Ras signaling and, as described here, the development of a new class of anticancer drugs.

3. FTS ANALOGS FOR CANCER THERAPY

Recently, efforts have made to design FTS derivatives that would enhance the anticancer effects of the drug. These derivatives were based on the FTS backbone and were modified on different positions such as the benzene ring or different lengths of the isoprenoid chain. These studies established that the minimal length of the isoprenoid chain required for Ras inhibition is C15 (farnesyl), since the C10 (geranyl) was inactive [20,42]. Some compounds such as 5-fluoro-FTS and 5-chloro-FTS were superior inhibitors of Ras-transformed cells growth over the original FTS [20]. More recently we investigated and developed additional FTS derivatives with modifications on their carboxyl group in an attempt to enhance its anti-Ras activity. During this process an important consideration was the degree of applicability to brain tumor pathophysiology. Unmodified FTS is able to penetrate the blood–brain barrier (BBB) despite its free carboxyl end [43]. We wanted the new compounds to have similar properties, having the ability to penetrate the BBB by increasing its lipophilicity.

The potency of the new derivatives was examined in terms of inhibition of Ras-dependent human cell lines growth. The most potent compounds were FTS-methoxymethyl ester (FTS-MOME) and FTS-amide (FTS-A) (Fig. 12.2). However, selectivity toward the active GTP-loaded Ras was apparent only with the amide derivatives (Table 12.1) with FTS-A exhibiting overall the highest ability. Interestingly, the FTS-carboxymethyl ester (FTS-ME) and FTS-MOME inhibited growth of both Panc-1 and U87 cells but not in a Ras-dependent manner [44]. Thus only the amide derivatives provided potent cell growth inhibitors without loss of selectivity toward the active Ras proteins [45,46]. Intriguingly, FTS-isopropyl ester (FTS-IE) and FTS-benzyl ester (FTS-BE) were far less active than native

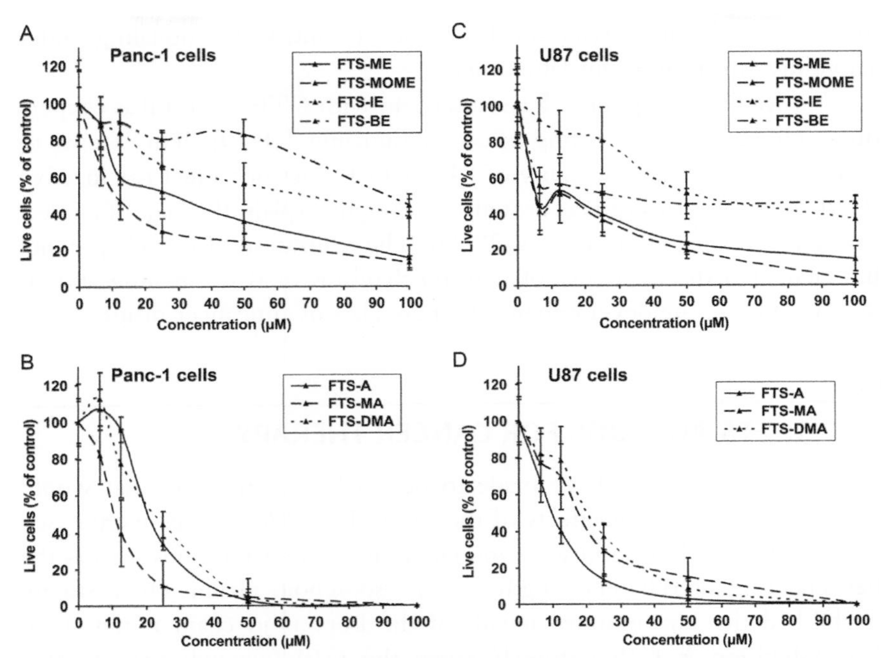

Figure 12.2 Inhibition of cell growth by FTS derivatives. Panc-1 and U87 cells were treated with different FTS derivatives at different concentrations (6.25, 12.5, 25, 50, 100 μM). Five days later, the cells were collected and counted. Growth inhibition curves are presented for: (A) Panc-1 cells treated with FTS-carboxymethyl ester (FTS-ME), FTS-methoxymethyl ester (FTS-MOME), FTS-isopropyl ester (FTS-IE), and FTS-benzyl ester (FTS-BE); (B) Panc-1 cells treated with FTS-amide (FTS-A), FTS-methyl amide (FTS-MA), and FTS-dimmethyl amide (FTS-DMA); (C) U87 cells treated with FTS-ME, FTS-MOME, FTS-IE, and FTS-BE; (D) U87 cells treated with FTS-A, FTS-MA, and FTS-DMA. *Data are from Ref. [44] added by permission from ACS publisher.*

FTS. This indicates that the bulky group (isopropyl and benzyl) of these esters interferes with FTS-growth inhibition activity.

FTS-As reduced both the levels of active Ras and the cells growth in cultures while FTS esters, even those that exhibited strong growth inhibitory activity, had a much lower effect on Ras activation (Table 12.1). Based on that we suggest that the growth inhibitory action of the FTS-ester derivatives (FTS-ME and FTS-MOME) is the result of interference with Ras-independent cell growth pathways. Thus, not all of farnesyl derivatives of FTS block Ras signaling [44]. The mild effect of FTS-MOME on Ras inhibition is more surprising in light of the notion that all Ras proteins possess a carboxy-terminal farnesylcysteine carboxy methyl ester [47,48].

Table 12.1 Inhibition of active Ras by FTS derivatives

	Inhibition of active Ras level (%)							
	Panc-1				U87			
FTS derivatives	Total Ras–GTP	K-Ras–GTP	N-Ras–GTP	H-Ras–GTP	Total Ras–GTP	K-Ras–GTP	N-Ras–GTP	H-Ras–GTP
FTS-ME	5.3 ± 4.1	11.9 ± 13.8	–	–	–	23.8 ± 18.6	18.7 ± 2.8	–
FTS-MOME	2.6 ± 3.8	19.9 ± 41.2	–	–	–	4.8 ± 14.8	5.1 ± 23.6	–
FTS-A	$^{*}37.4 \pm 8.6$	$^{**}56 \pm 11.2$	8.6 ± 33.1	–	$^{*}33 \pm 12.7$	$^{**}50.8 \pm 9.2$	$^{*}37.3 \pm 8.9$	–
FTS-MA	$^{**}39.7 \pm 6.1$	$^{**}52 \pm 12.6$	18.8 ± 19.3	–	$^{*}34.7 \pm 7.8$	$^{*}32.6 \pm 9.4$	$^{*}22.5 \pm 6.5$	–
FTS-DMA	$^{**}38.6 \pm 6.3$	$^{**}48 \pm 12.4$	13.6 ± 7.1	–	$^{**}45.4 \pm 8.8$	$^{*}28 \pm 16.5$	$^{**}41 \pm 4.6$	14 ± 12.7

$^{*}P < 0.05$.
$^{**}P < 0.001$.
Panc-1 and U87 cells were treated with different FTS derivatives at a concentration of 50 μM for 24 h. The cells were then lysed and aliquots of the lysates were used to determine total Ras–GTP, K-Ras–GTP, H-Ras–GTP, and N-Ras–GTP levels by pull-down assays. The percentage of active Ras inhibition was calculated as the subtraction of the percentage of active Ras levels after drug treatment from the percentage of active Ras levels in a control. Values are presented for Panc-1 cells (left panel) and U87 cells (right panel) as means ± SD.
Data are from Ref. [44] added by permission from ACS publisher.

Remarkably, and similar to FTS, *N*-acetyl farnesylcystine (AFC) that mimics the carboxy terminal of Ras proteins also inhibits ERK activation [49]. However, AFC and FTS appear to differ in their mode of action on Ras proteins and on isoprenylcysteine methyltransferase (Icmt), the enzyme that methylates Ras proteins. On the one hand, AFC inhibits Icmt, by serving as a competitive substrate for the enzyme [50]. FTS, on the other hand, is a direct inhibitor of Icmt membrane association and does not serve as a substrate for the enzyme [18]. At its growth inhibitory concentrations (10–50 μM), FTS dislodges active Ras from the cell membrane without an effect on Ras methylation [18,19]. In contrast, AFC inhibits Ras methylation and by doing that also blocks its trafficking to the cell membrane [49]. Nevertheless, when AFC (but not FTS) enters the cell it inevitably becomes methylated and accumulates in that compartment. When FTS enters the cell it retains its free carboxyl group. Since FTS-methyl ester has a diminished effect on Ras, we reason that AFC-methyl ester does not dislodge Ras from the cell membrane because it does not bind to Ras chaperones such as Gal-1 and Gal-3 [44]. Accordingly, the hydrophobic nature of FTS-MOME and of AFC-methyl ester, compared to their free carboxylic acid derivatives, might inhibit the interactions with Ras scaffolds other than Gal-1, Gal-3, or nucleolin. However, this is not the case of the FTS-As that accommodate better in the putative prenyl-binding pockets of Gal-1 and Gal-3 [51,52].

These observations strongly suggest that carboxyl methylated prenyl analogs, including FTS-MOME, inhibit cell growth by interfering with prenyl-binding proteins other than Ras escorts. It is tempting to consider that FTS-MOME would interfere with the action of the well-described prenyl-binding proteins RhoGDIs that bind the geranylgeranyl isoprenoid moieties of Rac/Rho GTPases, known to be involved in cell growth and cell transformation [44,53,54]. Consistent with this model were the results of the experiments that characterized the binding of isoprenylated acids to RhoGDI [55]. These experiments showed that isoprenylated carboxy methyl esters exhibited strong interactions with RhoGDI, unlike their corresponding free carboxylic acid derivatives that exhibited weak interactions with this protein [55]. For example, AFC-methyl ester exhibited high affinity to RhoGDI, while AFC and FTS barely interacted with it [55]. Interestingly, the isoprenoid moiety of Rac/Rho GTPases accommodates precisely within a hydrophobic pocket in RhoGDI, and indeed *N*-acetyl geranylgeranyl cysteine binds strongly to RhoGDI [53–55]. Thus it seems that the strong association between prenylated compounds and RhoGDI depends on both the length of the isoprenoid group and the presence of a carboxymethyl ester. Important questions that remain to be answered include whether or not such compounds block the hydrophobic pocket of RhoGDI *in vivo*, and if so, how does this affect cell growth. In light of the important functions of RhoGDI as a specific regulator of Rac/Rho GTPases trafficking to and from the cell membrane, it is reasonable to consider that such blockage will result in mislocalization and signaling of the appropriate GTPases [56].

While methylation abrogated anti-Ras activity, amidation strengthened it. The relatively strong anti-Ras activity of FTS-As, particularly FTS-A, suggests that amidation, unlike methylation, of the carboxyl group is favorable for the interaction with Ras escort proteins. Importantly, FTS-As retained the selectivity toward active GTP-bound Ras, similar to unmodified FTS. This was apparent in experiments exploiting panc-1 and U87 cells. In panc-1 cells, the most prevalent active Ras isoform is K-Ras, as the cells typically express K-RasG12V [45]. Similarly, we showed that N-Ras–GTP, and to a lesser extent K-Ras–GTP or H-Ras–GTP, was inhibited by FTS in these cells [57]. Among all the prenylated derivatives of FTS, FTS-A appears to be the most potent inhibitor of active Ras having IC50 of 10 μM.

The results described in this section, along with previous reports that considered prenylated small molecules, show that the structure of these molecules is a critical functional determinant [58,59]. While FTS and its amide derivatives act as Ras inhibitors, FTS esters show lesser activity as Ras inhibitors and probably interfere with the function of other prenyl-binding proteins such as RhoGDI or PDE6δ and thereby inhibit cell growth. Finally, it is interesting to note that mice deficient of Icmt develop tumors [personal communication], suggesting that Ras carboxy methylation may not be required for tumorigenesis.

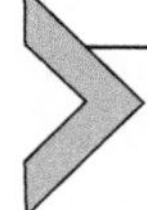

4. FTS IN COMBINATION WITH OTHER AGENTS FOR CANCER THERAPY

Cancer is a disease with multigenetic and epigenetic aberrations and it is not likely that a single drug treatment will be able to cure the disease [60]. Cases such as chronic myeloid leukemia where a single aberrant chromosomal rearrangement can be overcome with a single agent (e.g., imatinib, dasatinib, or nilotinib) are the exception [61]. It seems that only a rationally designed drug combinations would be an applied approach in our attempts to treat cancer more effectively. In our studies, we attempted to design such combination of FTS with other anticancer drugs that would act by mechanisms that are independent of FTS activities on Ras.

The first demonstration of this approach was the combination of FTS with gemcitabine in pancreatic cancer [62]. The combination was not more toxic in animals and in humans, compared to each drug given separately [62]. The results of these studies confirmed strong synergistic growth inhibitory effects [62]. We have also studied FTS combined with 2-deoxy-glucose (hampers cell growth) in glioblastoma and panc-1 cells, and showed that the combined treatment resulted in HIF1α downregulation [57]. The combination of FTS with the histone deacetylase 1 inhibitor valproic acid in A549 non-small lung carcinoma and DLD1 colon carcinoma cells resulted in downregulation of survivin and aurora [63]. When FTS was combined with the proteasomal inhibitor bortezomib and tested in multiple myeloma cells, growth was significantly inhibited [64]. When FTS was combined with pemetrexed and given to colon and lung cancer cells, significant decline in cells growth was noticed. Another interesting combination was FTS with the epidermal growth factor receptor inhibitor Erbitux. When tested in DLD1 and SW480 cells, growth was further inhibited (Fig. 12.3) (unpublished data). We believe that this drug combination could be translated into clinical trials for colon cancer patients with mutated K-Ras that failed to respond to epidermal growth factor receptor inhibitor mono therapy [65]. As shown in Table 12.2, most of the drug combinations resulted in synergistic effect on cell growth, likely due to different mechanisms of action between FTS and the additional drug.

5. INHIBITION OF Ras-RELATED PROTEINS BY FTS

FTS capacity to inhibit cell growth is not limited to its interaction with chaperones. The ability of FTS to reverse the transformed phenotype of NF1 associated tumor cell lines of malignant peripheral nerve sheath tumor (MPNST) was recently studied. The Ras GAP-related domain (GRD) of NF1 acts as Ras GAP and thus we have demonstrated that NF1 deficiency, or loss of a functional GRD, resulted in higher levels of activated Ras in MPNST and in similar cell lines such as ST88-14, S265P2, and 90-8 [66]. This was not the case of MPNST expressing wild-type NF1 [66]. It was shown that FTS treatment led to lower steady state levels of GTP-bound Ras and its activated targets and shortened the relatively long duration of Ras activation and signaling to ERK, Akt, and RalA in the NF1-deficient cells [66]. Additional supporting evidence is the fact that oral FTS

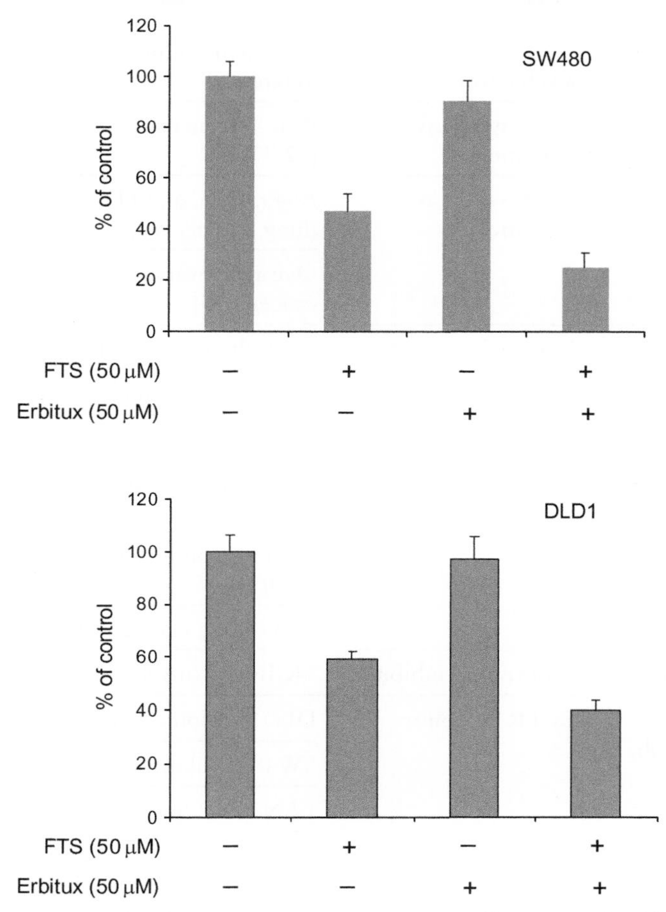

Figure 12.3 Bar graphs demonstrating the synergistic effects of FTS and Erbitux in SW480 and DLD1 cells.

also attenuated ST88-14 tumor growth in nude mice [66]. NF1 cells were found to possess strong actin stress fibers, and this phenotype was also reverted by FTS. NF1 tumor growth in a nude mouse model was inhibited by oral FTS as FTS treatment has normalized Ras–GTP levels, resulting in reversal of the transformed phenotype and inhibition of tumor growth. Thus, FTS should be considered as a potential drug for the treatment of NF1.

It has been shown that neurofibromin regulates cell motility via several GTPase pathways acting through two different domains, the GRD and the pre-GRD domain [67]. First, the GRD domain inhibits Ras-dependent

Table 12.2 FTS-based combination treatments

#	Drug	Mechanism	Cell line used (model) [reference]	Observed effect
1	Gemcitabine	DNA synthesis inhibition	Panc-1 (pancreatic cancer) [62,97]	Synergism
2	Cisplatinum	DNA synthesis inhibition	A549, H23, and HTB (lung cancer) [98]	Synergism
			Ovcar-3 (ovarian cancer) [99]	Additive
3	Valporic acid	Inhibits histone acetylase	A549 (lung cancer) [63]	Synergism
			DLD1 (colon cancer) [63]	Synergism
			Aro (thyroid cancer)	Synergism
4	Doxorubicin	DNA synthesis inhibition	Panc-1 (pancreatic cancer) [100]	Synergism
			SW480 (colon cancer) [100]	Additive
			A549 (lung cancer) [101]	Synergism
5	Velcade	Proteasome inhibitor	NCIH929 (myeloma) [64]	Synergism
6	Traceva (elrotinib)	EGFR inhibitor	DLD1 (colon cancer)	Synergism
			SW480 (colon cancer)	Synergism
			LAN-CAP (prostate adenocarcinoma)	Additive
7	Erbitux (cetuximab)	EGFR inhibitor	DLD1 (colon cancer) [41]	Synergism
8	Colchicine	Inhibits micritubuli organization	NF1 (neurofibromatosis type 1) [102]	Synergism
9	Abraxane (taxol)	Micritubuli inhibitor	NF1 (neurofibromatosis type 1) [102]	Synergism
			A549 (lung cancer) [98]	Synergism
10	Alimta (pemetrexed)	Inhibits DNA metabolic enzymes	A549 (lung cancer) [63]	Additive
11	Glivec (imatinib)	Bcr-Abl, PDGFR (c-Kit) inhibitor	K562 (chronic myelogenous leukemia)	Synergism

changes in cell motility through the mitogen-activated protein kinase cascade. Second, it also regulates Rho-dependent changes by activating the LIM kinase 2 (LIMK2), an enzyme that phosphorylates and inactivates cofilin (an actin-depolymerizing factor). Third, the pre-GRD domain acts through Rac1, which activates the P21-activated kinase 1-LIMK1/cofilin cascade [67]. We were able to identify a novel compound, T56-LIMKi, which inhibits LIMK1/2 activities. Importantly, T56-LIMKi in combination with FTS synergistically inhibited cell growth in neurofibromin-deficient cells. We therefore suggest that these drug combinations should be tested for treatment of neurofibromatosis and other types of cancers associated with defective NF1 signaling [68].

As pointed out above, not all FTS analogs inhibit active Ras. To our surprise, the various FTS esters did not inhibit Ras, and some of them actually inhibited cell growth via alternative mechanisms. For example, these compounds inhibited the GTPase Rap1 [69]. Both FTS-MOME and FTS-A inhibited Rap1 activation more than native FTS. In a pull-down assay FTS-MOME was superior to FTS-A in terms of GTP-bound Rap1 levels [69].

In related studies we found that the Ras homologue enriched in brain (Rheb), which is highly homologous to K-Ras, was inhibited by FTS in lymphangioleiomyomatosis (LAM) cells [70]. Rheb, like Ras, serves as a molecular switch regulating cell proliferation, differentiation, and apoptosis. Ras also regulates Rheb by inactivating the tuberous sclerosis complexes (TSC), which includes products of the TSC1 and TSC2 genes, and encoding hamartin and tuberin, respectively [71]. These protein complexes act as a Rheb-specific GAP. Loss of function of TSC1 or TSC2 results in an increase in Rheb-GTP levels and excessive cell proliferation characteristic of the genetic disorders TSC and LAM. To determine whether inactivation of Rheb, Ras, or both might be a potential treatment for LAM we used TSC2-null ELT3 cells as a LAM model and treated them with FTS and with FTS-A. Untreated, these cells expressed significant amounts of activated Rheb but not GTP-bound Ras. This phenotype was reversed by TSC2 reexpression [70]. Treatment with FTS or FTS-A slightly decreased the levels of GTP-bound Ras but has a significant reduction in the levels of activated Rheb and cell proliferation, migration, and growth [70]. The effect of FTS-A was significantly stronger ($\sim$2-fold) than that of FTS. Notably, TSC2 expression in these cells rescued them from FTS or FTS-A induced growth inhibition [70]. Evidently FTS and FTS-A blocked active Rheb in TSC2-null ELT3 cells and may have therapeutic potential for LAM and TSC.

Table 12.3 Rasopathies and associated Ras isoform [1]

Syndrome	Ras isoform
Costello syndrome	H-Ras
Cardio–facio–cutaneous	K-Ras
Noonan syndrome	K-Ras

6. Ras IN ORPHAN DISEASES

More recent studies have shown that germ line mutations, both in Ras and in other upstream and downstream signaling elements, are associated with a class of developmental syndromes referred to as Rasopathies (Table 12.3). Germ line H-Ras mutations were identified in Costello syndrome and K-Ras mutations were identified in cardio–facio–cutaneous and in Noonan syndromes [72–75]. These syndromes exhibit unique features, but because of a common genetic basis in Ras signaling, they share overlapping characteristics. These include craniofacial dysmorphology, cardiac malformations, and ocular abnormalities. Finally, whereas patients with Costello syndrome are at increased risk for malignant tumors, patients suffering from Noonan or cardio–facio–cutaneous syndrome have very slight increased risk of cancer [76]. It should be further investigated whether these patients would benefit from treatment with FTS.

7. FTS IN THE IMMUNE SYSTEM

The ability of FTS to modulate different aspects of the immune system has been described recently. Biochemically it was confirmed that FTS inhibits the Ras/MAPK signaling cascade as well as other downstream effectors in immune cells. At the cellular level, the ability of FTS to inhibit immune cells activation was examined [77]. Proliferation of T cells, secretion of cytokines, and the adhesion process were greatly inhibited by FTS [77,78]. Release of inflammatory mediators by mast cells and the ability of neutrophils to produce reactive oxygen species were all altered by FTS. Most importantly, the ability of FTS to improve clinical outcome in several animal models of inflammatory diseases was examined (Table 12.4). Some of the models represent autoimmune conditions while others are associated with allergy and end organs fibrosis. Both humoral and cellular aspects of the immune response, as well as acquired and innate aspects, were involved in these models.

Table 12.4 FTS and inflammatory animal models [10]

Syndrome	**Animal model**
Systemic lupus erythematosis	MRL/lpr
Antiphospholipid syndrome	Balb/c mice immunized with beta-2-glycoprotein I
Guillain–Barré syndrome	Experimental autoimmune neuritis
Multiple sclerosis	Experimental autoimmune encephalitis
Myocarditis	Experimental autoimmune myocarditis (immunizing with cardiac myosin)
Inflammatory bowel disease	Dextran sodium sulfate
Type I diabetes	NOD mice
Type II diabetes	High-fat-induced diabetic mice
Contact dermatitis	Delayed cutaneous hypersensitivity
Allergy	Cutaneous anaphylaxis
Atherosclerosis	High-fat diet
Ischemia/reperfusion	Induced ischemia and reperfusion
Liver cirrhosis	Intraperitoneal administration of thioacetamide
Muscular dystrophy	Congenital muscular dystrophy
Proliferative nephritis	Thy-1 nephritis

Data are from Ref. [44] added by permission from ACS publisher.

8. FTS AND AUTOIMMUNITY

Systemic lupus erythematous (SLE) is a prototype autoimmune disease in which the immune system attacks the body's cells and tissues, resulting in inflammation and tissue damage. It is caused mainly by antibody-immune complex formation. Activation and proliferation of lymphocytes, the key event in the pathogenesis of SLE, require the activation of Ras [79]. Inhibiting Ras modifies the activation of lymphocytes and could serve as immunosuppressive therapy for patients with SLE. To test this hypothesis we studied the effect of FTS on MRL/lpr mice [80], which is a genetic model of a generalized autoimmune disease sharing many features and organ pathology with SLE. FTS treatment resulted in a 50% decrease in splenocyte proliferation and in a significant decrease in the levels of serum antibodies directed against dsDNA. Also, proteinuria, lymphadenopathy, and spleen weights were reduced in FTS-treated MRL/lpr mice. These findings suggest that inhibition of Ras activation with FTS has a significant impact on the MRL/lpr model and that it could be useful for treating autoimmune diseases such as SLE.

Similar results were obtained with FTS and antiphospholipid syndrome (APS). In this case hypercoagulable state is caused by autoantibodies against cell membrane phospholipids that provoke blood clots in arteries and veins [81,82]. Antibody production by lymphocytes is stimulated by activation of Ras; therefore, inhibition of Ras by FTS might decrease autoantibody levels in APS [80]. Moreover, antiphospholipid antibodies activate endothelial cells, which is also Ras-dependent. The impact of Ras inhibition by FTS was studied in an animal model of APS [80], in which female Balb/c mice were immunized with beta2-glycoprotein in complete Freund's adjuvant. FTS treatment resulted in decreases in antiphospholipids and anti-beta2-glycoprotein I antibodies levels as in the MRL/lpr model. APS can affect multiple organs, including the brain vasculature, and in a follow-up study brain vascular endothelial cells were treated with IgG purified from APS patients. The expected response in these cells, phosphorylation of ERK, was significantly blocked by FTS, suggesting that Ras inhibition is amenable to FTS therapy.

Guillain–Barré syndrome (GBS) is one of several peripheral autoimmune neuropathies in which progressive muscle weakness and paralysis are seen [83]. Animals with experimental autoimmune neuritis serve as models for GBS, and we used them to test the hypothesis that FTS may function as a therapeutic agent for this disease [84]. Rats were treated by immunization with peripheral bovine myelin and then given FTS. This intervention significantly attenuated the clinical severity of the disease and enhanced recovery. The beneficial effects of FTS were confirmed by nerve conduction studies [84]. There is no specific treatment for the above-debilitating condition and the management is mainly supportive. FTS treatment may be used as a protecting drug in the three syndromes described above.

Multiple sclerosis (MS) is an inflammatory disease in which the fatty myelin sheaths around axons in the brain and spinal cord are damaged, leading to demyelization, scarring, and a wide range of neurological signs and symptoms [85]. Treatment of MS is based on immunosuppression aimed at downregulation of proliferating myelin-reactive lymphocytes. Ras is critical for the activation of these cells, and since FTS modulates their function we tested the compound in experimental autoimmune encephalitis (EAE), the animal model for MS [86]. We found a strong protective effect of FTS. More recently it was found that glatiramer acetate (GA) synergized with FTS in the EAE model [78]. Thus, combination of FTS and GA (that act through different mechanisms) may be used to ameliorate MS.

9. FTS IN OTHER DISEASE MODELS

To conclude this work on FTS and immunity we shall refer to additional manuscripts describing the beneficial effects of FTS in additional disease models. These include atherosclerosis [87], ischemia/reperfusion [88], liver fibrosis [89,90], congenital muscular dystrophy [91], glomerular nephritis [92], type 1 [93,94] and type 2 diabetes [95], hypersensitivity inflammation [69], and allergic inflammation [11,96].

10. CONCLUSIONS

In this work we have reviewed our remarkable journey exploring FTS and its biological outcomes, from the early stages of the structural designs to the most recent clinical trials. FTS was developed as a direct Ras inhibitor; however, it is now mostly appreciated as Ras chaperones' modulator. The development of this agent enabled us to discover new aspects of Ras biology that were underappreciated beforehand. We predict that the most promising clinical outcomes will arrive from combination protocols involving FTS and other drugs. Although we were able to clearly characterize the effects of FTS on the immune system, it is still unclear whether tumor immunotherapy mediates the augmented anticancer effect of FTS. Since Ras aberrant signaling is not limited to cancer pathogenesis, FTS shows potential for treating patients with congenital Rasopathies as well as autoimmune disorders.

REFERENCES

[1] A.D. Cox, C.J. Der, Ras history: the saga continues, Small Gtpases 1 (1) (2011) 2–27.
[2] C.J. Der, T. Van Dyke, Stopping ras in its tracks, Cell 129 (5) (2007) 855–857.
[3] Y. Kloog, A.D. Cox, RAS inhibitors: potential for cancer therapeutics, Mol. Med. Today 6 (10) (2000) 398–402.
[4] A.D. Cox, C.J. Der, The dark side of Ras: regulation of apoptosis, Oncogene 22 (56) (2003) 8999–9006.
[5] M. Barbacid, Ras oncogenes: their role in neoplasia, Eur. J. Clin. Invest. 20 (3) (1990) 225–235.

[6] J.L. Bos, Ras oncogenes in human cancer: a review, Cancer Res. 49 (17) (1989) 4682–4689.

[7] J.L. Bos, p21ras: an oncoprotein functioning in growth factor-induced signal transduction, Eur. J. Cancer 31 (7–8) (1995) 1051–1054.

[8] Y. Yarden, A. Ullrich, Growth factor receptor tyrosine kinases, Annu. Rev. Biochem. 57 (1988) 443–478.

[9] A. Levitzki, EGF receptor as a therapeutic target, Lung Cancer 41 (Suppl. 1) (2003) S9–S14.

[10] A. Mor, E. Aizman, J. Chapman, Y. Kloog, Immunomodulatory properties of farnesoids: the new steroids? Curr. Med. Chem. 20 (10) (2013) 1218–1224.

[11] A. Mor, O. Ben-Moshe, Y.A. Mekori, Y. Kloog, Inhibitory effect of farnesylthiosalicylic acid on mediators release by mast cells: preferential inhibition of prostaglandin D(2) and tumor necrosis factor-alpha release, Inflammation 34 (5) (2011) 314–318.

[12] A. Mor, M.R. Philips, M.H. Pillinger, The role of Ras signaling in lupus T lymphocytes: biology and pathogenesis, Clin. Immunol. 125 (3) (2007) 215–223.

[13] T.N. Seagroves, H.E. Ryan, H. Lu, B.G. Wouters, M. Knapp, P. Thibault, K. Laderoute, R.S. Johnson, Transcription factor HIF-1 is a necessary mediator of the Pasteur effect in mammalian cells, Mol. Cell Biol. 21 (10) (2001) 3436–3444.

[14] S.M. Sebti, A.D. Hamilton, Farnesyltransferase and geranylgeranyltransferase I inhibitors and cancer therapy: lessons from mechanism and bench-to-bedside translational studies, Oncogene 19 (56) (2000) 6584–6593.

[15] T. Whalley, M. Terasaki, M.S. Cho, S.S. Vogel, Direct membrane retrieval into large vesicles after exocytosis in sea urchin eggs, J. Cell Biol. 131 (5) (1995) 1183–1192.

[16] D.B. Whyte, P. Kirschmeier, T.N. Hockenberry, I. Nunez Oliva, L. James, J.J. Catino, W.R. Bishop, J.K. Pai, K- and N-Ras are geranylgeranylated in cells treated with farnesyl protein transferase inhibitors, J. Biol. Chem. 272 (22) (1997) 14459–14464.

[17] T.M. Brand, M. Iida, C. Li, D.L. Wheeler, The nuclear epidermal growth factor receptor signaling network and its role in cancer, Discov. Med. 12 (66) (2011) 419–432.

[18] M. Marom, R. Haklai, G. Ben Baruch, D. Marciano, Y. Egozi, Y. Kloog, Selective inhibition of Ras-dependent cell growth by farnesylthiosalicylic acid, J. Biol. Chem. 270 (38) (1995) 22263–22270.

[19] R. Haklai, G. Gana-Weisz, G. Elad, A. Paz, D. Marciano, Y. Egozi, G. Ben Baruch, Y. Kloog, Dislodgment and accelerated degradation of Ras, Biochemistry 37 (5) (1998) 1306–1314.

[20] Z. Aharonson, M. Gana-Weisz, T. Varsano, R. Haklai, D. Marciano, Y. Kloog, Stringent structural requirements for anti-Ras activity of S-prenyl analogues, Biochim. Biophys. Acta 1406 (1) (1998) 40–50.

[21] D. Marciano, Z. Aharonson, T. Varsano, R. Haklai, Y. Kloog, Novel inhibitors of the prenylated protein methyltransferase reveal distinctive structural requirements, Bioorg. Med. Chem. Lett. 7 (13) (1997) 1709–1714.

[22] R. Haklai, G. Elad-Sfadia, Y. Egozi, Y. Kloog, Orally administered FTS (salirasib) inhibits human pancreatic tumor growth in nude mice, Cancer Chemother. Pharmacol. 61 (1) (2008) 89–96.

[23] B. Rotblat, M. Ehrlich, R. Haklai, Y. Kloog, The Ras inhibitor farnesylthiosalicylic acid (salirasib) disrupts the spatiotemporal localization of active Ras: a potential treatment for cancer, Methods Enzymol. 439 (2008) 467–489.

[24] E. Bustinza-Linares, R. Kurzrock, A.M. Tsimberidou, Salirasib in the treatment of pancreatic cancer, Future Oncol. 6 (6) (2010) 885–891.

[25] G. Elad-Sfadia, R. Haklai, E. Ballan, H.J. Gabius, Y. Kloog, Galectin-1 augments Ras activation and diverts Ras signals to Raf-1 at the expense of phosphoinositide 3-kinase, J. Biol. Chem. 277 (40) (2002) 37169–37175.
[26] A. Paz, R. Haklai, G. Elad-Sfadia, E. Ballan, Y. Kloog, Galectin-1 binds oncogenic H-Ras to mediate Ras membrane anchorage and cell transformation, Oncogene 20 (51) (2001) 7486–7493.
[27] G. Elad-Sfadia, R. Haklai, E. Balan, Y. Kloog, Galectin-3 augments K-Ras activation and triggers a Ras signal that attenuates ERK but not phosphoinositide 3-kinase activity, J. Biol. Chem. 279 (33) (2004) 34922–34930.
[28] B. Rotblat, L. Belanis, H. Liang, R. Haklai, G. Elad-Zefadia, J.F. Hancock, Y. Kloog, S.J. Plowman, H-Ras nanocluster stability regulates the magnitude of MAPK signal output, PLoS One 5 (8) (2010) e11991.
[29] L. Belanis, S.J. Plowman, B. Rotblat, J.F. Hancock, Y. Kloog, Galectin-1 is a novel structural component and a major regulator of h-ras nanoclusters, Mol. Biol. Cell 19 (4) (2008) 1404–1414.
[30] B. Rotblat, I.A. Prior, C. Muncke, R.G. Parton, Y. Kloog, Y.I. Henis, J.F. Hancock, Three separable domains regulate GTP-dependent association of H-ras with the plasma membrane, Mol. Cell Biol. 24 (15) (2004) 6799–6810.
[31] I.A. Prior, J.F. Hancock, Compartmentalization of Ras proteins, J. Cell Sci. 114 (Pt 9) (2001) 1603–1608.
[32] A.A. Gorfe, M. Hanzal-Bayer, D. Abankwa, J.F. Hancock, J.A. McCammon, Structure and dynamics of the full-length lipid-modified H-Ras protein in a 1,2-dimyristoylglycero-3-phosphocholine bilayer, J. Med. Chem. 50 (4) (2007) 674–684.
[33] R. Shalom-Feuerstein, S.J. Plowman, B. Rotblat, N. Ariotti, T. Tian, J.F. Hancock, Y. Kloog, K-ras nanoclustering is subverted by overexpression of the scaffold protein galectin-3, Cancer Res. 68 (16) (2008) 6608–6616.
[34] T. Tian, S.J. Plowman, R.G. Parton, Y. Kloog, J.F. Hancock, Mathematical modeling of K-Ras nanocluster formation on the plasma membrane, Biophys. J. 99 (2) (2010) 534–543.
[35] P. Bhagatji, R. Leventis, R. Rich, C.J. Lin, J.R. Silvius, Multiple cellular proteins modulate the dynamics of K-ras association with the plasma membrane, Biophys. J. 99 (10) (2010) 3327–3335.
[36] O. Rocks, A. Peyker, M. Kahms, P.J. Verveer, C. Koerner, M. Lumbierres, J. Kuhlmann, H. Waldmann, A. Wittinghofer, P.I. Bastiaens, An acylation cycle regulates localization and activity of palmitoylated Ras isoforms, Science 307 (5716) (2005) 1746–1752.
[37] H. Zheng, J. McKay, J.E. Buss, H-Ras does not need COP I- or COP II-dependent vesicular transport to reach the plasma membrane, J. Biol. Chem. 282 (35) (2007) 25760–25768.
[38] A. Chandra, H.E. Grecco, V. Pisupati, D. Perera, L. Cassidy, F. Skoulidis, S.A. Ismail, C. Hedberg, M. Hanzal-Bayer, A.R. Venkitaraman, A. Wittinghofer, P.I. Bastiaens, The GDI-like solubilizing factor PDEdelta sustains the spatial organization and signalling of Ras family proteins, Nat. Cell Biol. 14 (2) (2012) 148–158.
[39] R. Levy, A. Biran, F. Poirier, A. Raz, Y. Kloog, Galectin-3 mediates cross-talk between K-Ras and Let-7c tumor suppressor microRNA, PLoS One 6 (11) (2011) e27490.
[40] R. Levy, M. Grafi-Cohen, Z. Kraiem, Y. Kloog, Galectin-3 promotes chronic activation of K-Ras and differentiation block in malignant thyroid carcinomas, Mol. Cancer Ther. 9 (8) (2011) 2208–2219.
[41] K. Farin, S. Schokoroy, R. Haklai, I. Cohen-Or, G. Elad-Sfadia, M.E. Reyes-Reyes, P.J. Bates, A.D. Cox, Y. Kloog, R. Pinkas-Kramarski, Oncogenic synergism between ErbB1, nucleolin, and mutant Ras, Cancer Res. 71 (6) (2011) 2140–2151.

[42] D. Marciano, G. Ben Baruch, M. Marom, Y. Egozi, R. Haklai, Y. Kloog, Farnesyl derivatives of rigid carboxylic acids-inhibitors of ras-dependent cell growth, J. Med. Chem. 38 (8) (1995) 1267–1272.

[43] E. Shohami, I. Yatsiv, A. Alexandrovich, R. Haklai, G. Elad-Sfadia, R. Grossman, A. Biegon, Y. Kloog, The Ras inhibitor *S-trans*, *trans*-farnesylthiosalicylic acid exerts long-lasting neuroprotection in a mouse closed head injury model, J. Cereb. Blood Flow Metab. 23 (6) (2003) 728–738.

[44] L. Goldberg, R. Haklai, V. Bauer, A. Heiss, Y. Kloog, New derivatives of farnesylthiosalicylic acid (salirasib) for cancer treatment: farnesylthiosalicylamide inhibits tumor growth in nude mice models, J. Med. Chem. 52 (1) (2009) 197–205.

[45] G. Berrozpe, J. Schaeffer, M.A. Peinado, F.X. Real, M. Perucho, Comparative analysis of mutations in the p53 and K-ras genes in pancreatic cancer, Int. J. Cancer 58 (2) (1994) 185–191.

[46] A. Guha, M.M. Feldkamp, N. Lau, G. Boss, A. Pawson, Proliferation of human malignant astrocytomas is dependent on Ras activation, Oncogene 15 (23) (1997) 2755–2765.

[47] A.D. Cox, C.J. Der, Protein prenylation: more than just glue? Curr. Opin. Cell Biol. 4 (6) (1992) 1008–1016.

[48] A.D. Cox, M.M. Hisaka, J.E. Buss, C.J. Der, Specific isoprenoid modification is required for function of normal, but not oncogenic Ras protein, Mol. Cell Biol. 12 (6) (1992) 2606–2615.

[49] V.K. Chiu, J. Silletti, V. Dinsell, H. Wiener, K. Loukeris, G. Ou, M.R. Philips, M.H. Pillinger, Carboxyl methylation of Ras regulates membrane targeting and effector engagement, J. Biol. Chem. 279 (8) (2004) 7346–7352.

[50] R.R. Rando, Chemical biology of protein isoprenylation/methylation, Biochim. Biophys. Acta 1300 (1) (1996) 5–16.

[51] B. Rotblat, H. Niv, S. Andre, H. Kaltner, H.J. Gabius, Y. Kloog, Galectin-1(L11A) predicted from a computed galectin-1 farnesyl-binding pocket selectively inhibits Ras–GTP, Cancer Res. 64 (9) (2004) 3112–3118.

[52] U. Ashery, O. Yizhar, B. Rotblat, G. Elad-Sfadia, B. Barkan, R. Haklai, Y. Kloog, Spatiotemporal organization of Ras signaling: rasosomes and the galectin switch, Cell. Mol. Neurobiol. 26 (2006) 469–493.

[53] G.R. Hoffman, N. Nassar, R.A. Cerione, Structure of the Rho family GTP-binding protein Cdc42 in complex with the multifunctional regulator RhoGDI, Cell 100 (3) (2000) 345–356.

[54] K. Scheffzek, I. Stephan, O.N. Jensen, D. Illenberger, P. Gierschik, The Rac–RhoGDI complex and the structural basis for the regulation of Rho proteins by RhoGDI, Nat. Struct. Biol. 7 (2) (2000) 122–126.

[55] M.S. Mondal, Z. Wang, A.M. Seeds, R.R. Rando, The specific binding of small molecule isoprenoids to rhoGDP dissociation inhibitor (rhoGDI), Biochemistry 39 (2) (2000) 406–412.

[56] D. Michaelson, J. Silletti, G. Murphy, P. D'Eustachio, M. Rush, M.R. Philips, Differential localization of Rho GTPases in live cells: regulation by hypervariable regions and RhoGDI binding, J. Cell Biol. 152 (1) (2001) 111–126.

[57] R. Blum, J. Jacob-Hirsch, N. Amariglio, G. Rechavi, Y. Kloog, Ras inhibition in glioblastoma down-regulates hypoxia-inducible factor-1alpha, causing glycolysis shutdown and cell death, Cancer Res. 65 (3) (2005) 999–1006.

[58] J. Ding, D.J. Lu, D. Perez Sala, Y.T. Ma, J.F. Maddox, B.A. Gilbert, J.A. Badwey, R.R. Rando, Farnesyl-L-cysteine analogs can inhibit or initiate superoxide release by human neutrophils, J. Biol. Chem. 269 (24) (1994) 16837–16844.

[59] E.W. Tan, D. Perez Sala, F.J. Canada, R.R. Rando, Identifying the recognition unit for G protein methylation, J. Biol. Chem. 266 (17) (1991) 10719–10722.

[60] M. Rebucci, C. Michiels, Molecular aspects of cancer cell resistance to chemotherapy, Biochem. Pharmacol. 85 (9) (2013) 1219–1226.
[61] J.R. Westin, R. Kurzrock, It's about time: lessons for solid tumors from chronic myelogenous leukemia therapy, Mol. Cancer Ther. 11 (12) (2012) 2549–2555.
[62] B. Weisz, K. Giehl, M. Gana-Weisz, Y. Egozi, G. Ben-Baruch, D. Marciano, P. Gierschik, Y. Kloog, A new functional Ras antagonist inhibits human pancreatic tumor growth in nude mice, Oncogene 18 (16) (1999) 2579–2588.
[63] A. Biran, M. Brownstein, R. Haklai, Y. Kloog, Downregulation of survivin and aurora a by histone deacetylase and RAS inhibitors: a new drug combination for cancer therapy, Int. J. Cancer 128 (2010) 691–701.
[64] S. Yaari-Stark, Y. Nevo-Caspi, J. Jacob-Hircsh, G. Rechavi, A. Nagler, Y. Kloog, Combining the Ras inhibitor salirasib and proteasome inhibitor: a potential treatment for multiple myeloma, J. Cancer Sci. Ther. 3 (8) (2011) 187–194.
[65] A. Lievre, J.B. Bachet, D. Le Corre, V. Boige, B. Landi, J.F. Emile, J.F. Cote, G. Tomasic, C. Penna, M. Ducreux, P. Rougier, F. Penault-Llorca, P. Laurent-Puig, KRAS mutation status is predictive of response to cetuximab therapy in colorectal cancer, Cancer Res. 66 (8) (2006) 3992–3995.
[66] B. Barkan, S. Starinsky, E. Friedman, R. Stein, Y. Kloog, The Ras inhibitor farnesylthiosalicylic acid as a potential therapy for neurofibromatosis type 1, Clin. Cancer Res. 12 (18) (2006) 5533–5542.
[67] S. Starinsky-Elbaz, L. Faigenbloom, E. Friedman, R. Stein, Y. Kloog, The pre-GAP-related domain of neurofibromin regulates cell migration through the LIM kinase/cofilin pathway, Mol. Cell. Neurosci. 42 (4) (2009) 278–287.
[68] E. Mashiach-Farkash, R. Rak, G. Elad-Sfadia, R. Haklai, S. Carmeli, Y. Kloog, H.J. Wolfson, Computer-based identification of a novel LIMK1/2 inhibitor that synergizes with salirasib to destabilize the actin cytoskeleton, Oncotarget 3 (6) (2012) 629–639.
[69] A. Mor, R. Haklai, O. Ben-Moshe, Y.A. Mekori, Y. Kloog, Inhibition of contact sensitivity by farnesylthiosalicylic acid-amide, a potential Rap1 inhibitor, J. Invest. Dermatol. 131 (10) (2011) 2040–2048.
[70] V. Makovski, R. Haklai, Y. Kloog, Farnesylthiosalicylic acid (salirasib) inhibits Rheb in TSC2-null ELT3 cells: a potential treatment for lymphangioleiomyomatosis, Int. J. Cancer 130 (6) (2012) 1420–1429.
[71] A.F. Castro, J.F. Rebhun, G.J. Clark, L.A. Quilliam, Rheb binds tuberous sclerosis complex 2 (TSC2) and promotes S6 kinase activation in a rapamycin- and farnesylation-dependent manner, J. Biol. Chem. 278 (35) (2003) 32493–32496.
[72] Y. Aoki, T. Niihori, H. Kawame, K. Kurosawa, H. Ohashi, Y. Tanaka, M. Filocamo, K. Kato, Y. Suzuki, S. Kure, Y. Matsubara, Germline mutations in HRAS proto-oncogene cause Costello syndrome, Nat. Genet. 37 (10) (2005) 1038–1040.
[73] T. Niihori, Y. Aoki, Y. Narumi, G. Neri, H. Cave, A. Verloes, N. Okamoto, R.C. Hennekam, G. Gillessen-Kaesbach, D. Wieczorek, M.I. Kavamura, K. Kurosawa, H. Ohashi, L. Wilson, D. Heron, D. Bonneau, et al., Germline KRAS and BRAF mutations in cardio–facio–cutaneous syndrome, Nat. Genet. 38 (3) (2006) 294–296.
[74] C. Carta, F. Pantaleoni, G. Bocchinfuso, L. Stella, I. Vasta, A. Sarkozy, C. Digilio, A. Palleschi, A. Pizzuti, P. Grammatico, G. Zampino, B. Dallapiccola, B.D. Gelb, M. Tartaglia, Germline missense mutations affecting KRAS Isoform B are associated with a severe Noonan syndrome phenotype, Am. J. Hum. Genet. 79 (1) (2006) 129–135.
[75] S. Schubbert, M. Zenker, S.L. Rowe, S. Boll, C. Klein, G. Bollag, I. van der Burgt, L. Musante, V. Kalscheuer, L.E. Wehner, H. Nguyen, B. West, K.Y. Zhang, E. Sistermans, A. Rauch, C.M. Niemeyer, et al., Germline KRAS mutations cause Noonan syndrome, Nat. Genet. 38 (3) (2006) 331–336.

[76] K.W. Gripp, Tumor predisposition in Costello syndrome, Am. J. Med. Genet. C Semin. Med. Genet. 137C (1) (2005) 72–77.
[77] A. Mor, G. Keren, Y. Kloog, J. George, N-Ras or K-Ras inhibition increases the number and enhances the function of Foxp3 regulatory T cells, Eur. J. Immunol. 38 (6) (2008) 1493–1502.
[78] E. Aizman, A. Mor, J. Chapman, Y. Assaf, Y. Kloog, The combined treatment of copaxone and salirasib attenuates experimental autoimmune encephalomyelitis (EAE) in mice, J. Neuroimmunol. 229 (1–2) (2010) 192–203.
[79] M.J. Rapoport, O. Bloch, M. Amit-Vasina, E. Yona, Y. Molad, Constitutive abnormal expression of RasGRP-1 isoforms and low expression of PARP-1 in patients with systemic lupus erythematosus, Lupus 20 (14) (2011) 1501–1509.
[80] A. Katzav, Y. Kloog, A.D. Korczyn, H. Niv, D.M. Karussis, N. Wang, R. Rabinowitz, M. Blank, Y. Shoenfeld, J. Chapman, Treatment of MRL/lpr mice, a genetic autoimmune model, with the Ras inhibitor, farnesylthiosalicylate (FTS), Clin. Exp. Immunol. 126 (3) (2001) 570–577.
[81] J. Chapman, J.H. Rand, R.L. Brey, S.R. Levine, I. Blatt, M.A. Khamashta, Y. Shoenfeld, Non-stroke neurological syndromes associated with antiphospholipid antibodies: evaluation of clinical and experimental studies, Lupus 12 (7) (2003) 514–517.
[82] R.L. Brey, J. Chapman, S.R. Levine, G. Ruiz-Irastorza, R.H. Derksen, M. Khamashta, Y. Shoenfeld, Stroke and the antiphospholipid syndrome: consensus meeting Taormina 2002, Lupus 12 (7) (2003) 508–513.
[83] N. Yuki, H.P. Hartung, Guillain–Barre syndrome, N. Engl. J. Med. 366 (24) (2012) 2294–2304.
[84] M. Kafri, Y. Kloog, A.D. Korczyn, R. Ferdman-Aronovich, V. Drory, A. Katzav, I. Wirguin, J. Chapman, Inhibition of Ras attenuates the course of experimental autoimmune neuritis, J. Neuroimmunol. 168 (1–2) (2005) 46–55.
[85] A. Compston, A. Coles, Multiple sclerosis, Lancet 359 (9313) (2002) 1221–1231.
[86] D. Karusis, O. Abramsky, N. Grigoriadis, J. Chapman, R. Mizrachi-Koll, H. Niv, Y. Kloog, The Ras-pathway inhibitor *S,trans-trans* farnesylthiosalicylic acid (FTS) suppresses experimental allergic encephalomyelitis, J. Neuroimmunol. 120 (1–2) (2001) 1–9.
[87] J. George, A. Afek, P. Keren, I. Goldberg, R. Shapiro, R. Haklai, Y. Kloog, G. Keren, Functional inhibition of Ras by farnesylthiosalicylic acid (FTS) attenuates atherosclerosis in apolipoproteinE knockout mice, Circulation 105 (2002) 2416–2422.
[88] R. Pando, Y. Cheporko, R. Haklai, S. Maysel-Auslender, G. Keren, J. George, E. Porat, A. Sagie, Y. Kloog, E. Hochhauser, Ras inhibition attenuates myocardial ischemia–reperfusion injury, Biochem. Pharmacol. 77 (10) (2009) 1593–1601.
[89] S. Reif, B. Weisz, H. Aeed, M. Gana-Weisz, L. Zaidl, Y. Kloog, R. Bruck, The Ras antagonist, farnesyl thiosalicylic acid (FTS) inhibits experimentally-induced liver cirrhosis in rats, J. Hepatol. 31 (6) (1999) 1053–1061.
[90] S. Reif, H. Aeed, Y. Shilo, R. Reich, Y. Kloog, Y.O. Kweon, R. Bruck, Treatment of thioacetamide-induced liver cirrhosis by the Ras antagonist, farnesylthiosalicylic acid, J. Hepatol. 41 (2) (2004) 235–241.
[91] Y. Nevo, S. Aga-Mizrachi, E. Elmakayes, N. Yanay, K. Ettinger, M. Elbaz, Z. Brunschwig, O. Dadush, G. Elad-Sfadia, R. Haklai, Y. Kloog, J. Chapman, S. Reif, The Ras antagonist, farnesylthiosalicylic acid (FTS), decreases fibrosis and improves muscle strength in dy/dy mouse model of muscular dystrophy, PLoS One 6 (3) (2011) e18049.
[92] H.C. Clarke, H.M. Kocher, A. Khwaja, Y. Kloog, H.T. Cook, B.M. Hendry, Ras antagonist farnesylthiosalicylic acid (FTS) reduces glomerular cellular proliferation and macrophage number in rat thy-1 nephritis, J. Am. Soc. Nephrol. 14 (4) (2003) 848–854.

[93] A. Mor, Y. Kloog, G. Keren, J. George, Ras inhibition increases the frequency and function of regulatory T cells and attenuates type-1 diabetes in non-obese diabetic mice, Eur. J. Pharmacol. 616 (1–3) (2009) 301–305.
[94] E. Aizman, A. Mor, J. George, Y. Kloog, Ras inhibition attenuates pancreatic cell death and experimental type 1 diabetes: possible role of regulatory T cells, Eur. J. Pharmacol. 643 (1) (2010) 139–144.
[95] A. Mor, E. Aizman, J. George, Y. Kloog, Ras inhibition induces insulin sensitivity and glucose uptake, PLoS One 6 (6) (2011) e21712.
[96] S. Rotshenker, F. Reichert, M. Gitik, R. Haklai, G. Elad-Sfadia, Y. Kloog, Galectin-3/MAC-2, Ras and PI3K activate complement receptor-3 and scavenger receptor-AI/II mediated myelin phagocytosis in microglia, Glia 56 (15) (2008) 1607–1613.
[97] R. Haklai, G. Elad-Sfadia, Y. Egozi, Y. Kloog, Orally administered FTS (salirasib) inhibits human pancreatic tumor growth in nude mice, Cancer Chemother. Pharmacol. 61 (1) (2007) 89–96.
[98] A. Zundelevich, G. Elad-Sfadia, R. Haklai, Y. Kloog, Suppression of lung cancer tumor growth in a nude mouse model by the Ras inhibitor salirasib (farnesylthiosalicylic acid), Mol. Cancer Ther. 31 (2007) 31.
[99] M.E. Beiner, H. Niv, R. Haklai, G. Elad-Sfadia, Y. Kloog, G. Ben-Baruch, Ras antagonist inhibits growth and chemosensitizes human epithelial ovarian cancer cells, Int. J. Gynecol. Cancer 16 (Suppl. 1) (2006) 200–206.
[100] M. Gana-Weisz, J. Halaschek-Wiener, B. Jansen, G. Elad, R. Haklai, Y. Kloog, The Ras inhibitor *S-trans*, *trans*-farnesylthiosalicylic acid chemosensitizes human tumor cells without causing resistance, Clin. Cancer Res. 8 (2) (2002) 555–565.
[101] S. Yaari-Stark, M. Shaked, Y. Nevo-Caspi, J. Jacob-Hircsh, R. Shamir, G. Rechavi, Y. Kloog, Ras inhibits endoplasmic reticulum stress in human cancer cells with amplified Myc, Int. J. Cancer 126 (10) (2010) 2268–2281.
[102] S. Shapira, B. Barkan, E. Fridman, Y. Kloog, R. Stein, The tumor suppressor neurofibromin confers sensitivity to apoptosis by Ras-dependent and Ras-independent pathways, Cell Death Differ. 5 (2007) 895–906.

AUTHOR INDEX

Numbers in regular font are page numbers and indicate that an author's work is referred to although the name is not cited in the text. Numbers in italics refer to the page numbers on which the complete reference appears.

A

C

D

G

H

I

J

K

L

M

S

T

U

X

Y

SUBJECT INDEX

Note: Page numbers followed by "*f*" indicate figures, and "*t*" indicate tables.

D

E

F

L

M

N

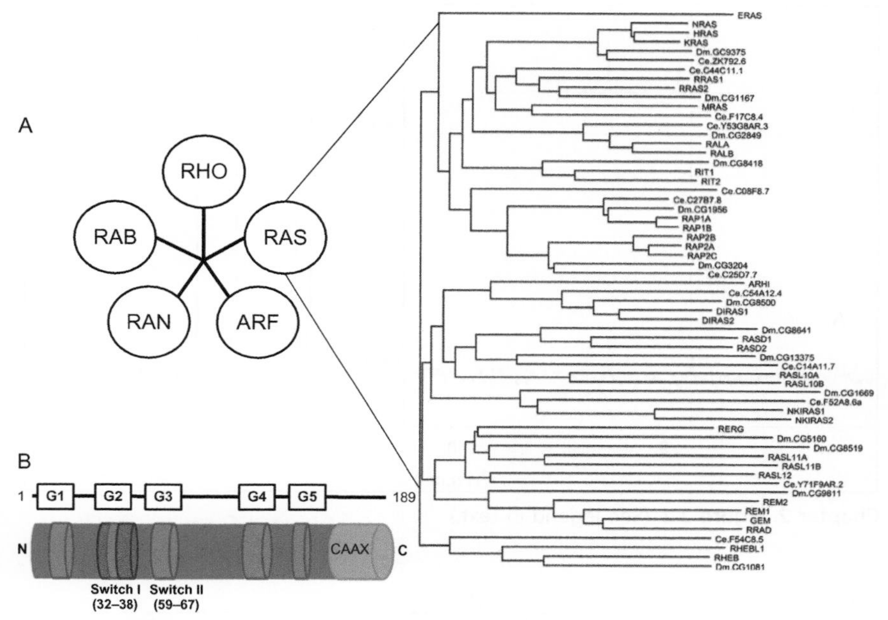

Chapter 1. Figure 1.1 (See legend in text.)

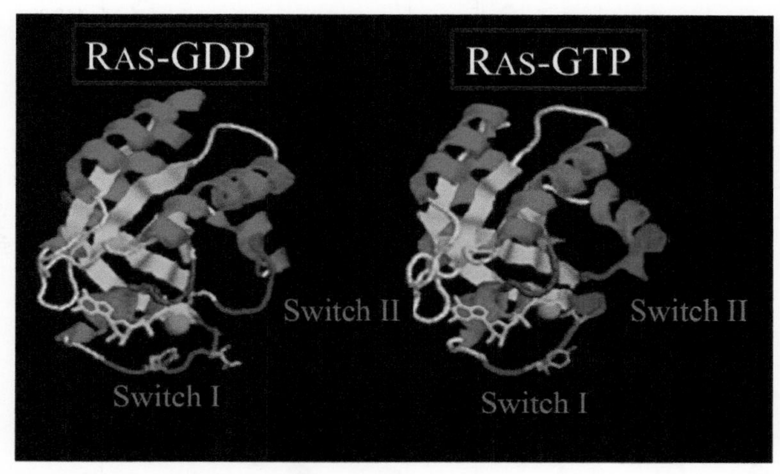

Chapter 1. Figure 1.2 (See legend in text.)

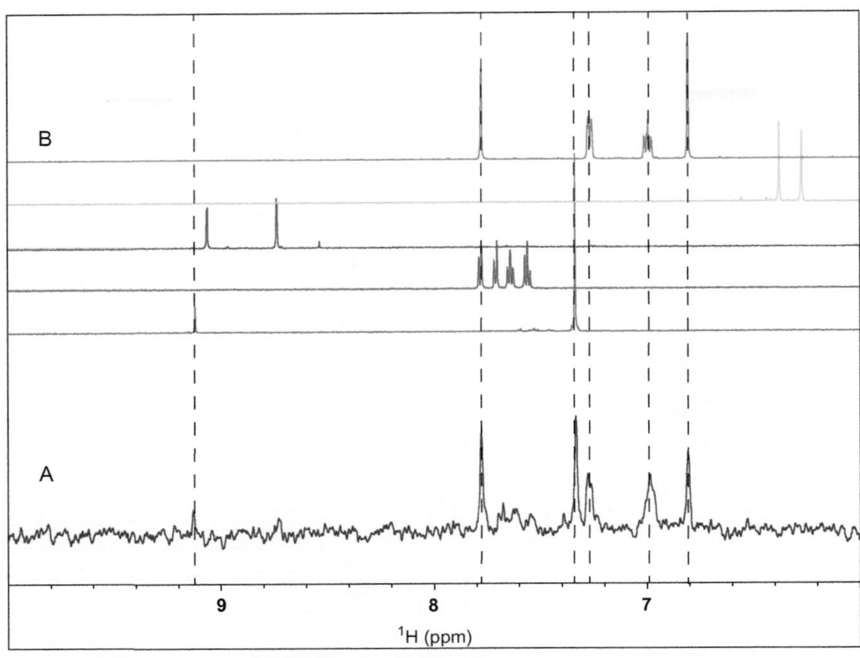

Chapter 2. Figure 2.2 (See legend in text.)

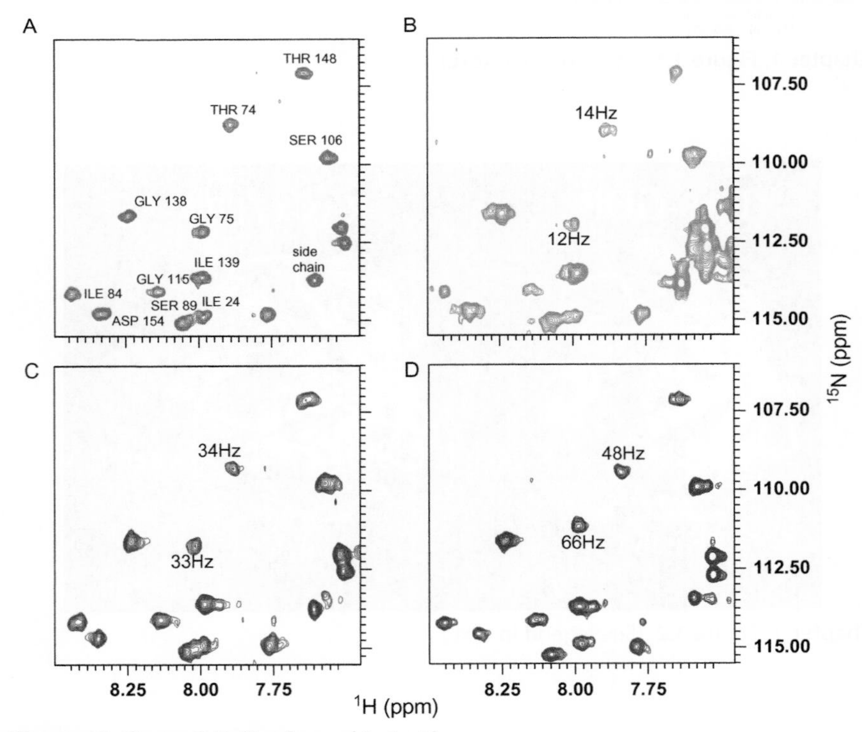

Chapter 2. Figure 2.3 (See legend in text.)

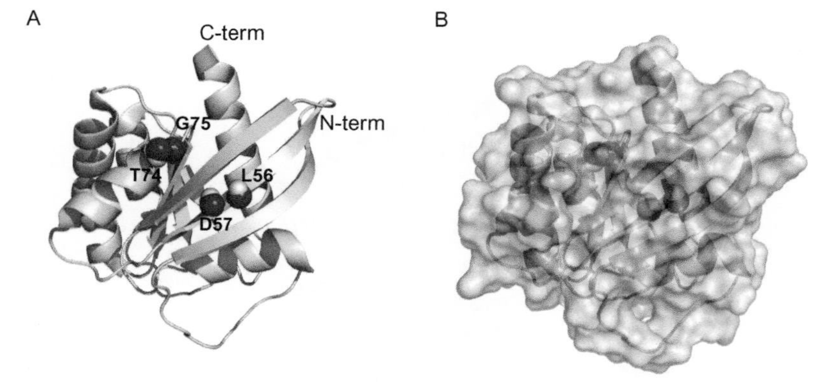

Chapter 2. Figure 2.4 (See legend in text.)

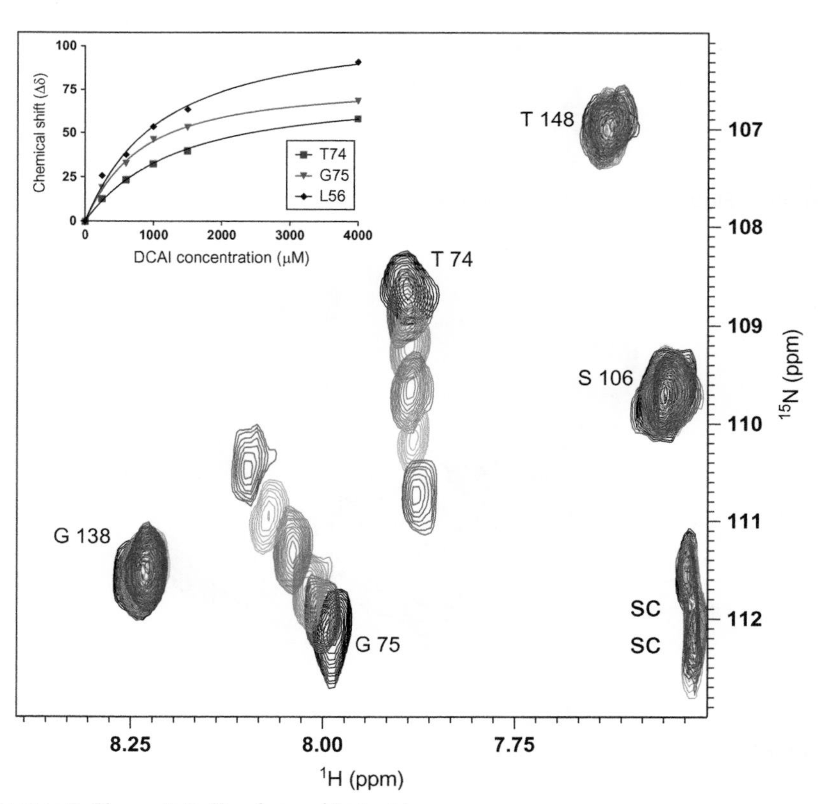

Chapter 2. Figure 2.5 (See legend in text.)

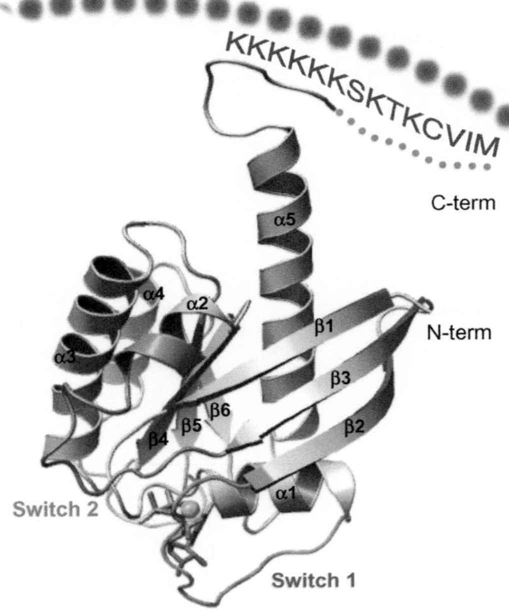

Chapter 2. Figure 2.6 (See legend in text.)

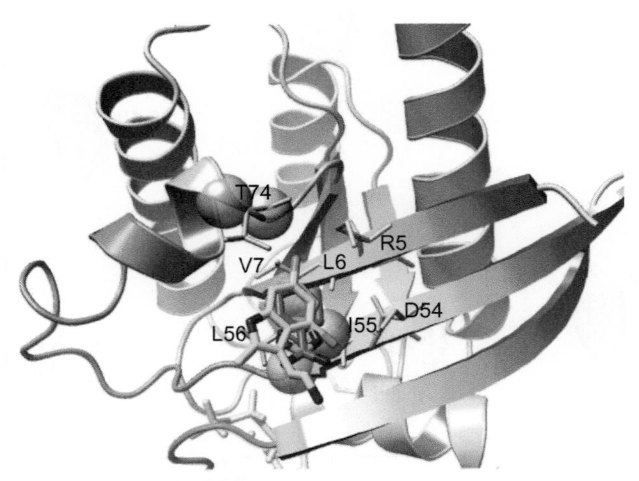

Chapter 2. Figure 2.7 (See legend in text.)

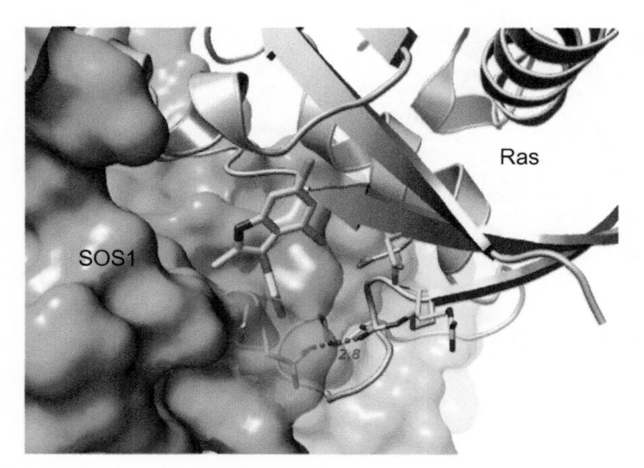

Chapter 2. Figure 2.9 (See legend in text.)

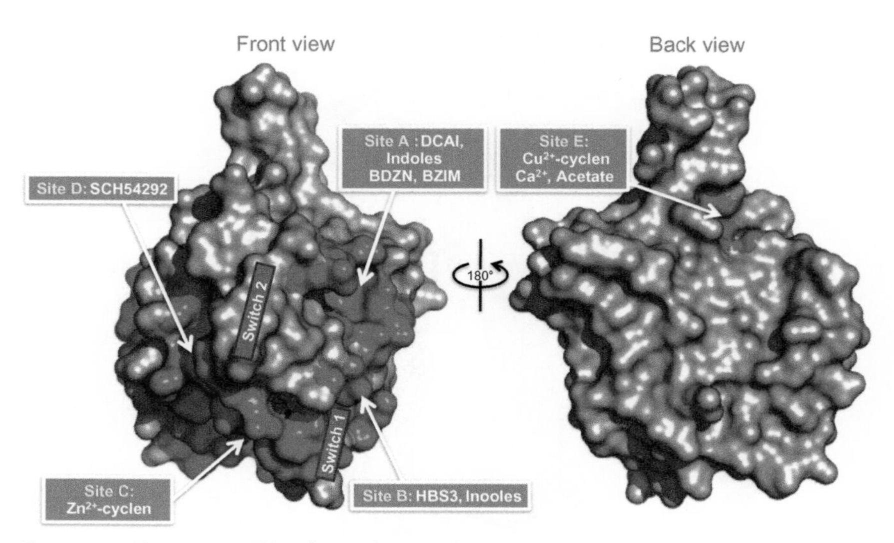

Chapter 2. Figure 2.10 (See legend in text.)

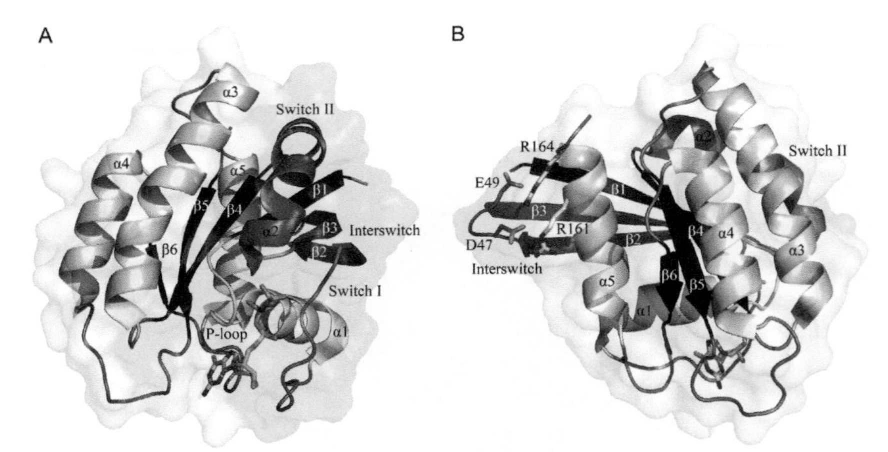

Chapter 3. Figure 3.1 (See legend in text.)

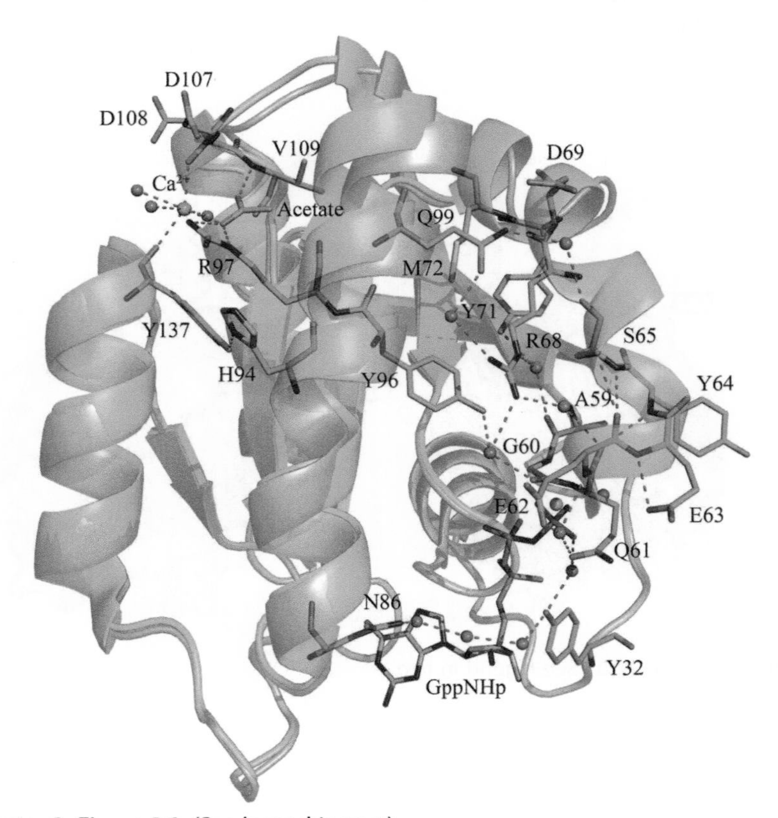

Chapter 3. Figure 3.2 (See legend in text.)

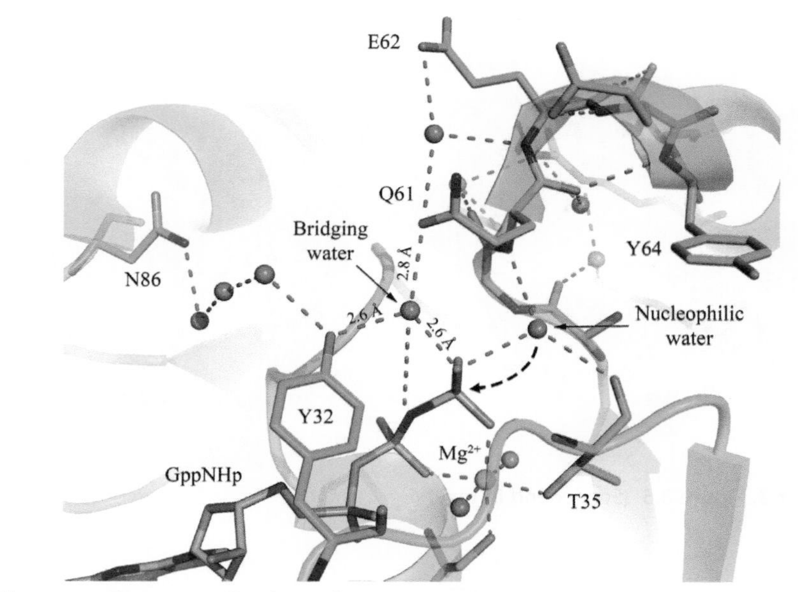

Chapter 3. Figure 3.3 (See legend in text.)

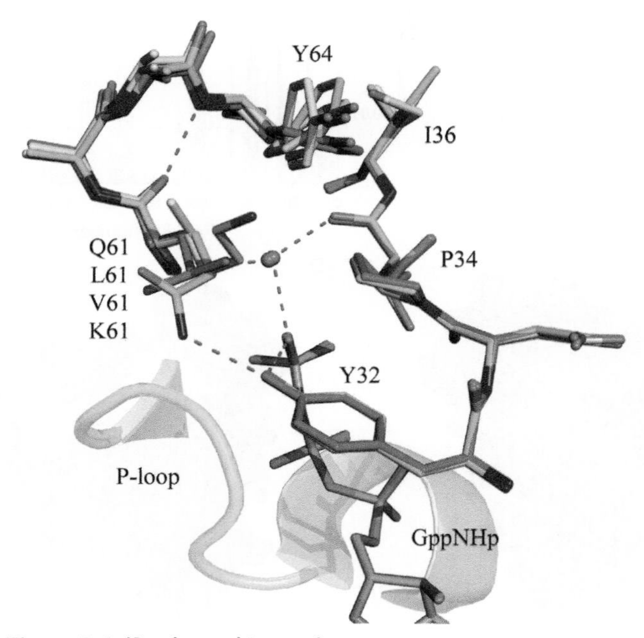

Chapter 3. Figure 3.4 (See legend in text.)

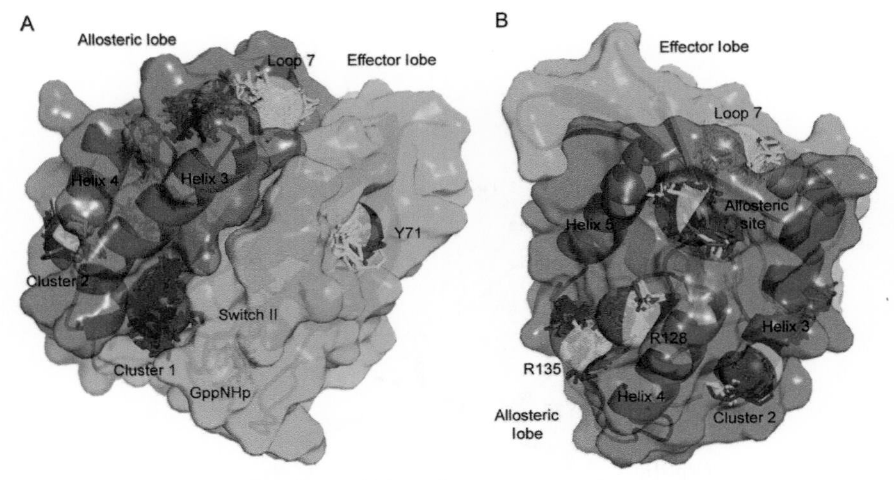

Chapter 3. Figure 3.5 (See legend in text.)

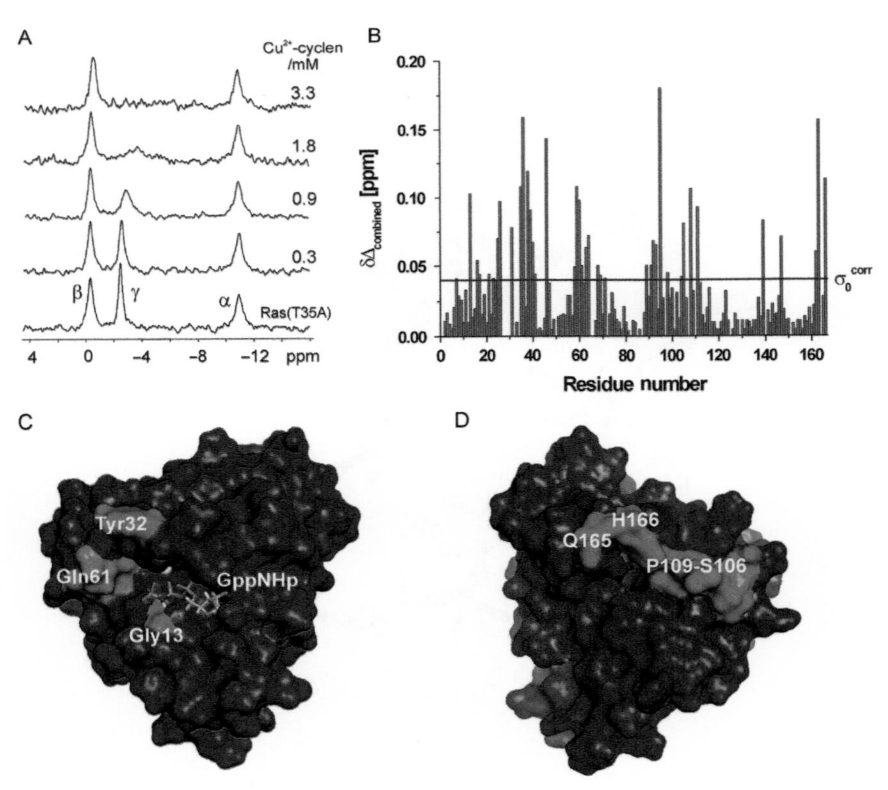

Chapter 4. Figure 4.8 (See legend in text.)

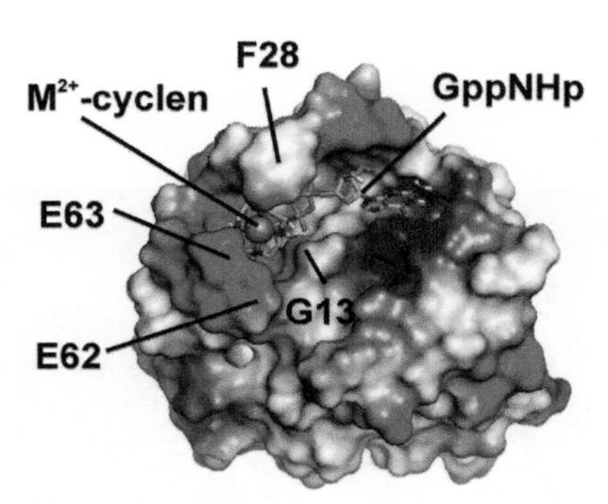

Chapter 4. Figure 4.9 (See legend in text.)

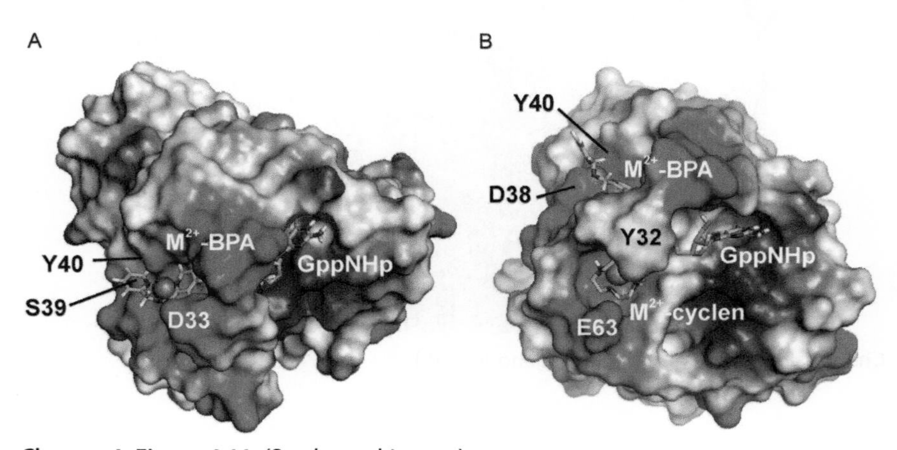

Chapter 4. Figure 4.11 (See legend in text.)

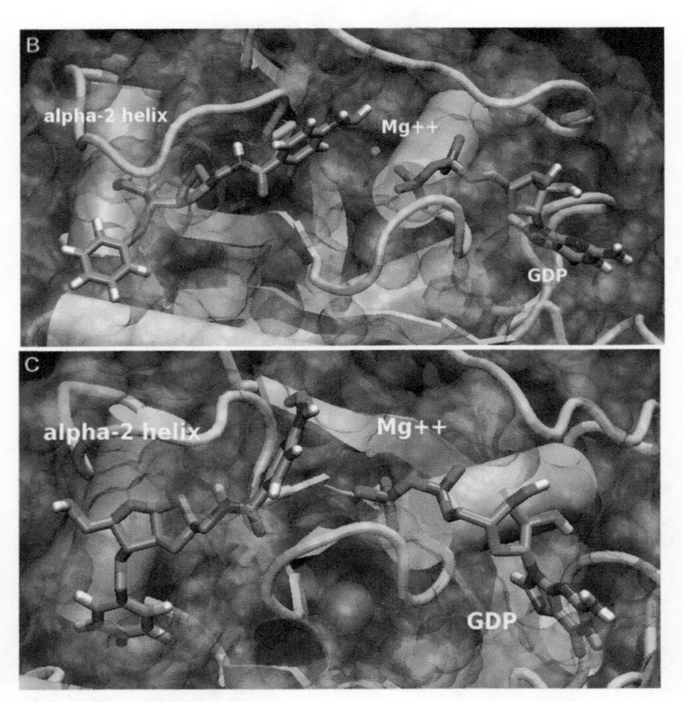

Chapter 5. Figure 5.3B, C (See legend in text.)

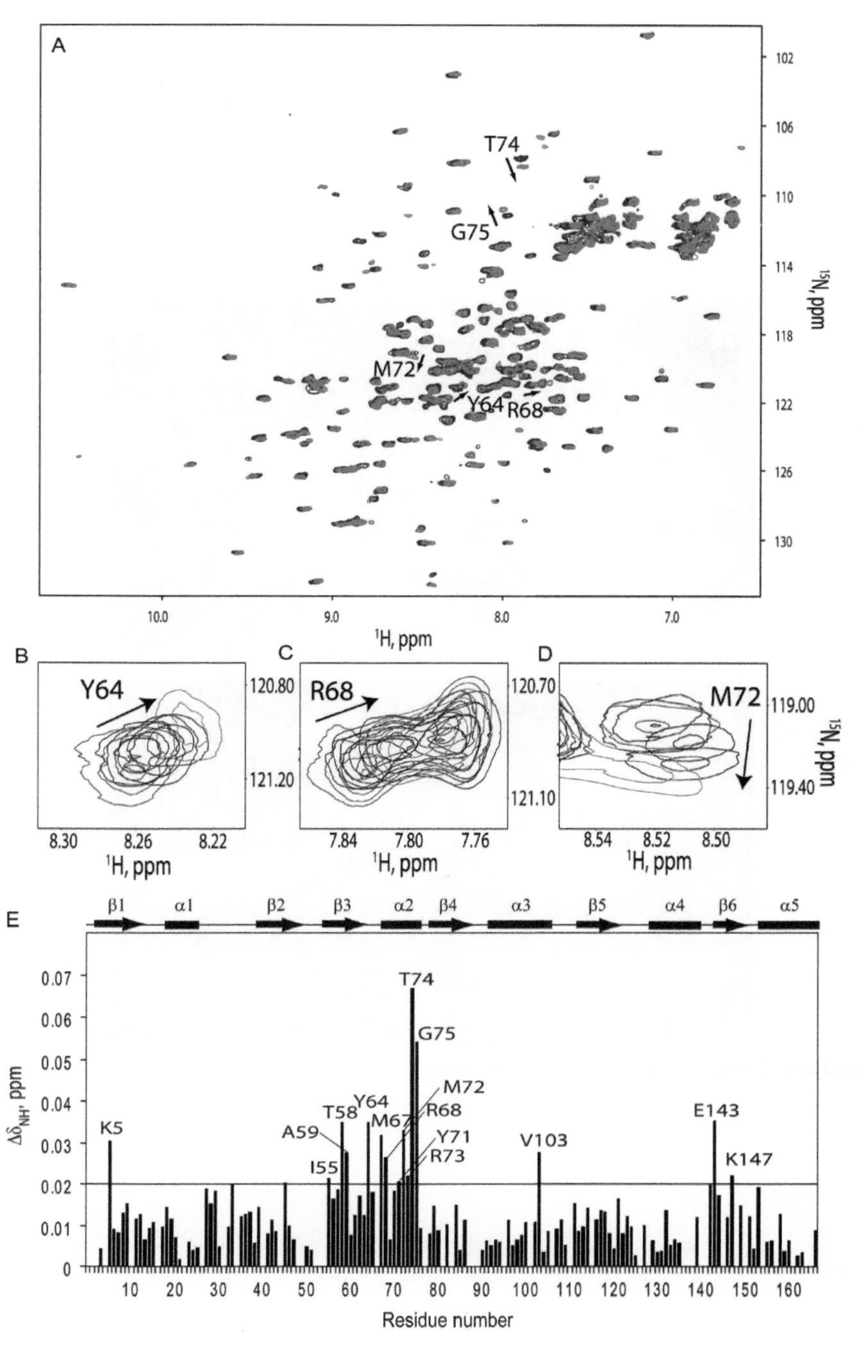

Chapter 5. Figure 5.7 (See legend in text.)

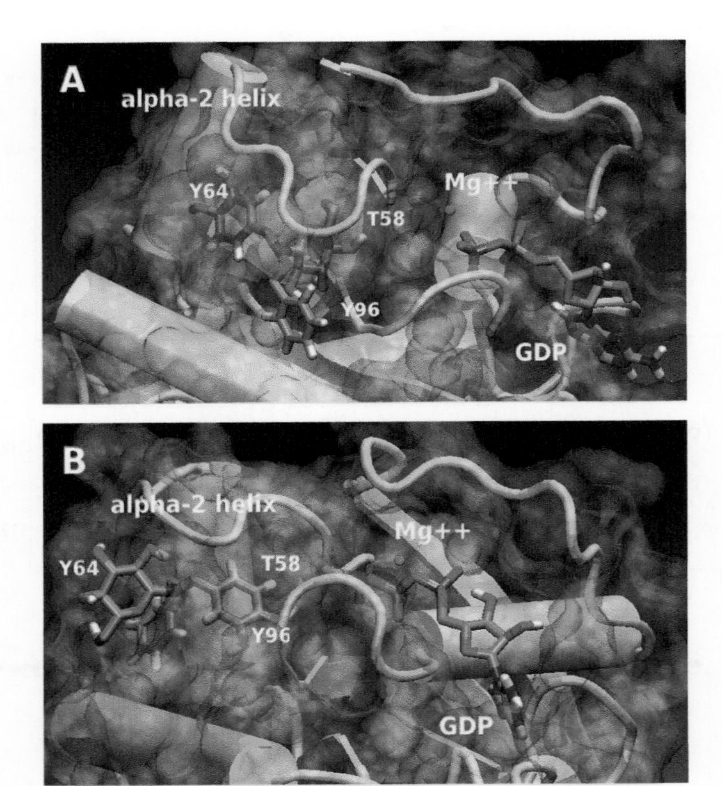

Chapter 5. Figure 5.8 (See legend in text.)

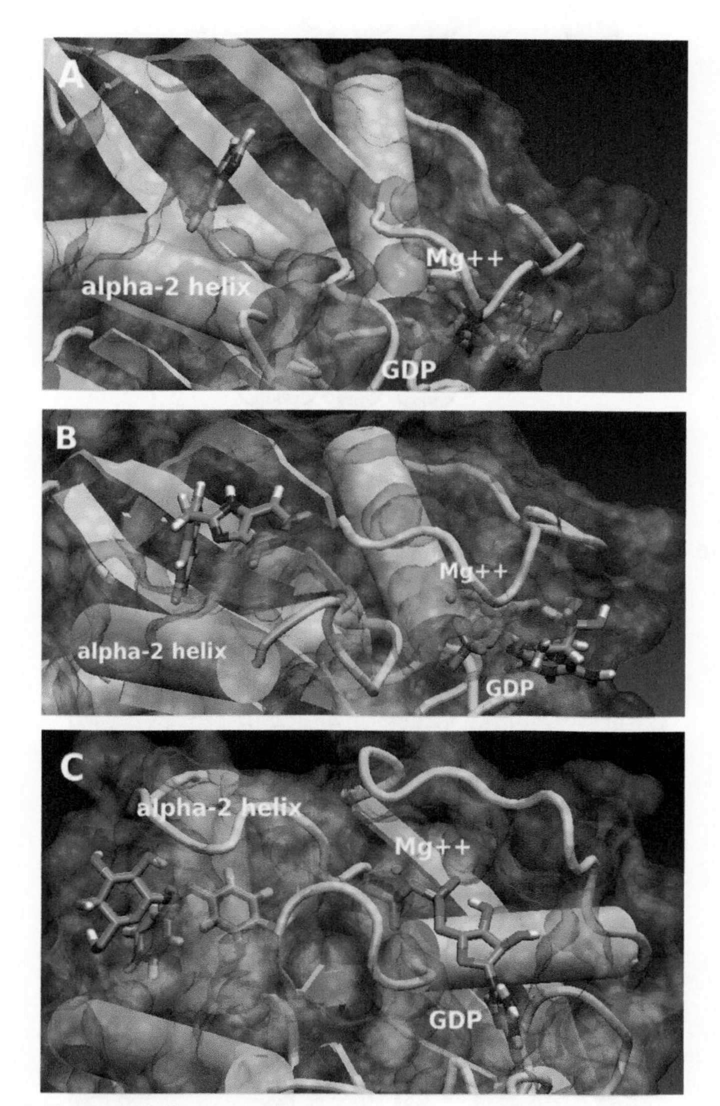

Chapter 5. Figure 5.9 (See legend in text.)

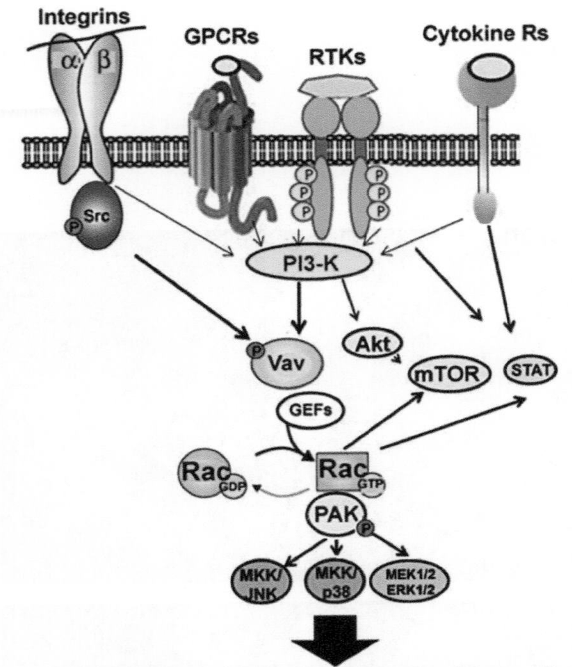

Chapter 6. Figure 6.1 (See legend in text.)

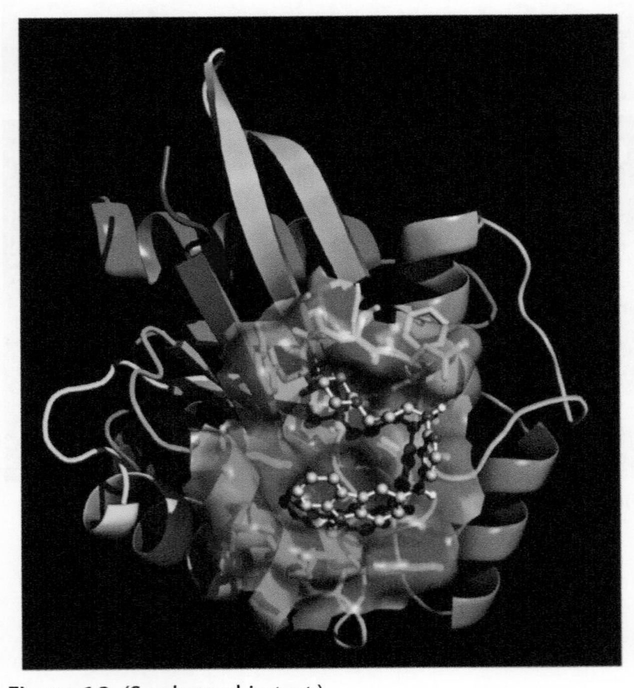

Chapter 6. Figure 6.3 (See legend in text.)

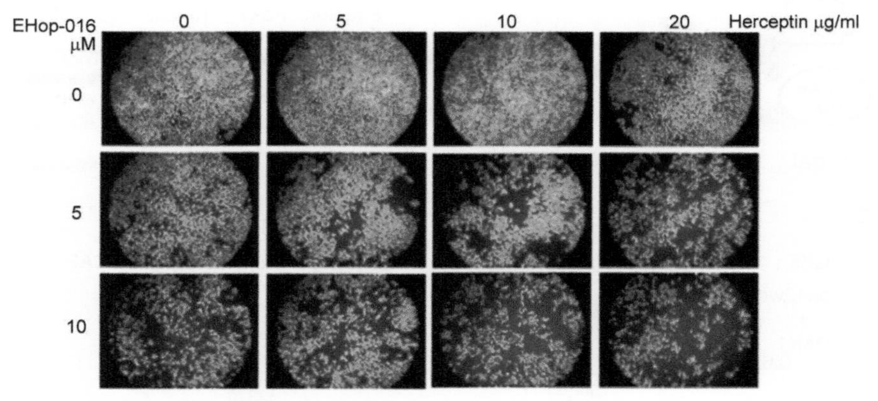

Chapter 6. Figure 6.7 (See legend in text.)

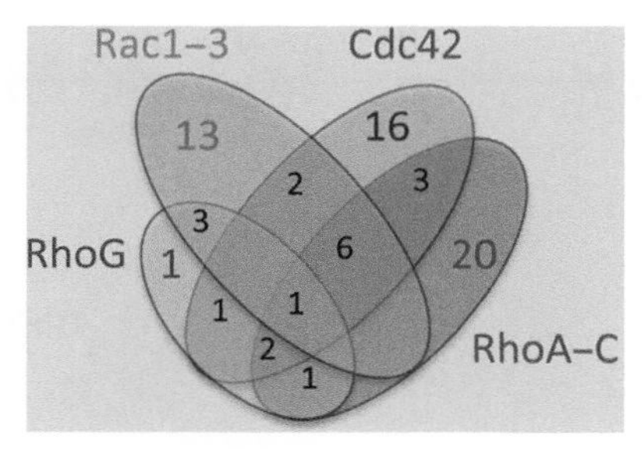

Chapter 8. Figure 8.1 (See legend in text.)

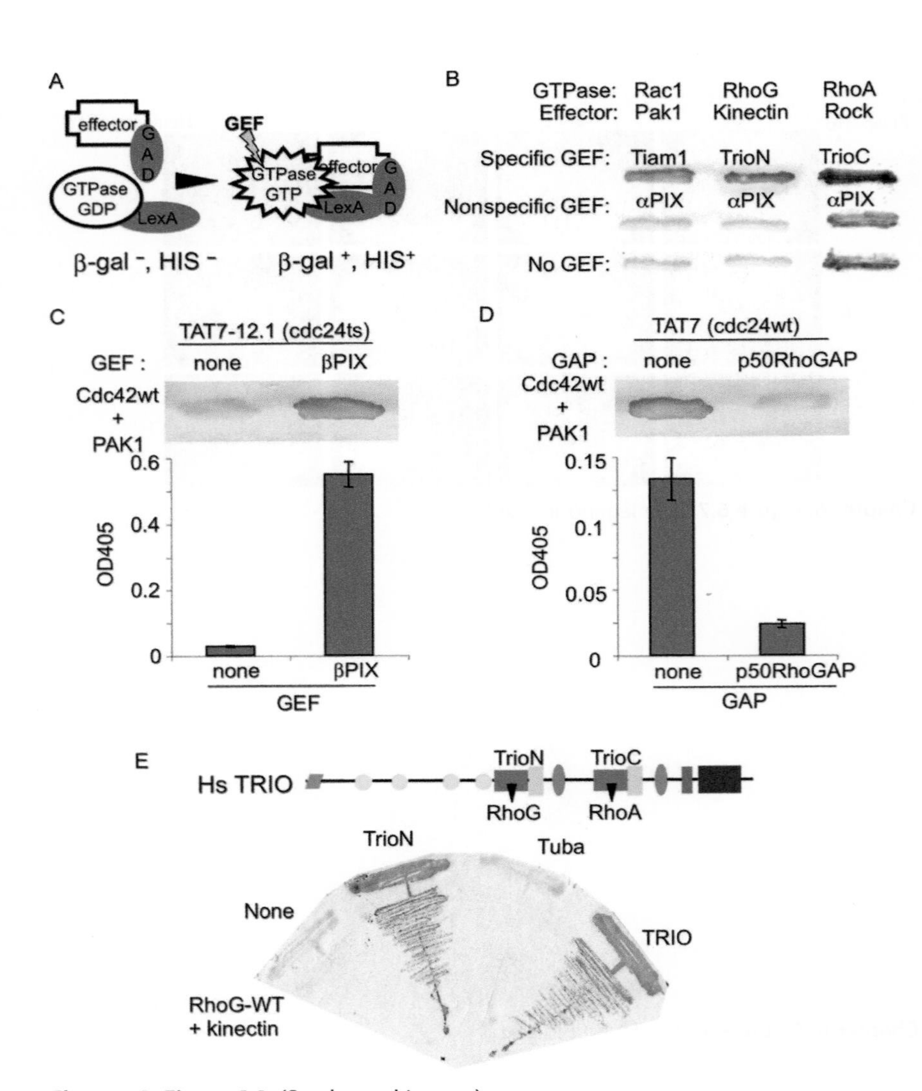

Chapter 8. Figure 8.2 (See legend in text.)

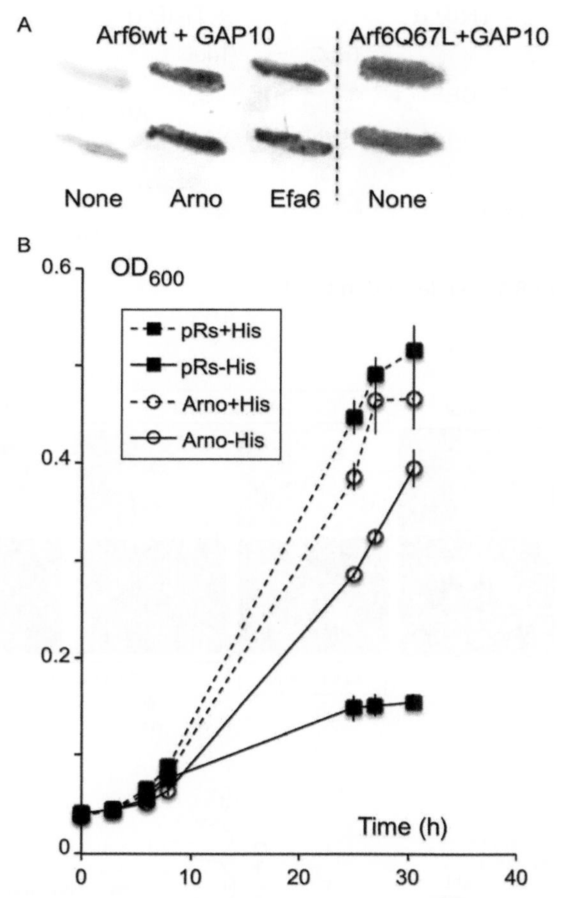

Chapter 8. Figure 8.3 (See legend in text.)

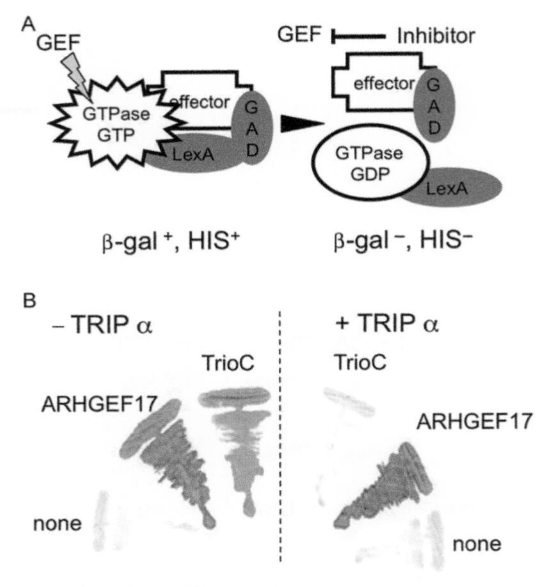

Chapter 8. Figure 8.4 (See legend in text.)

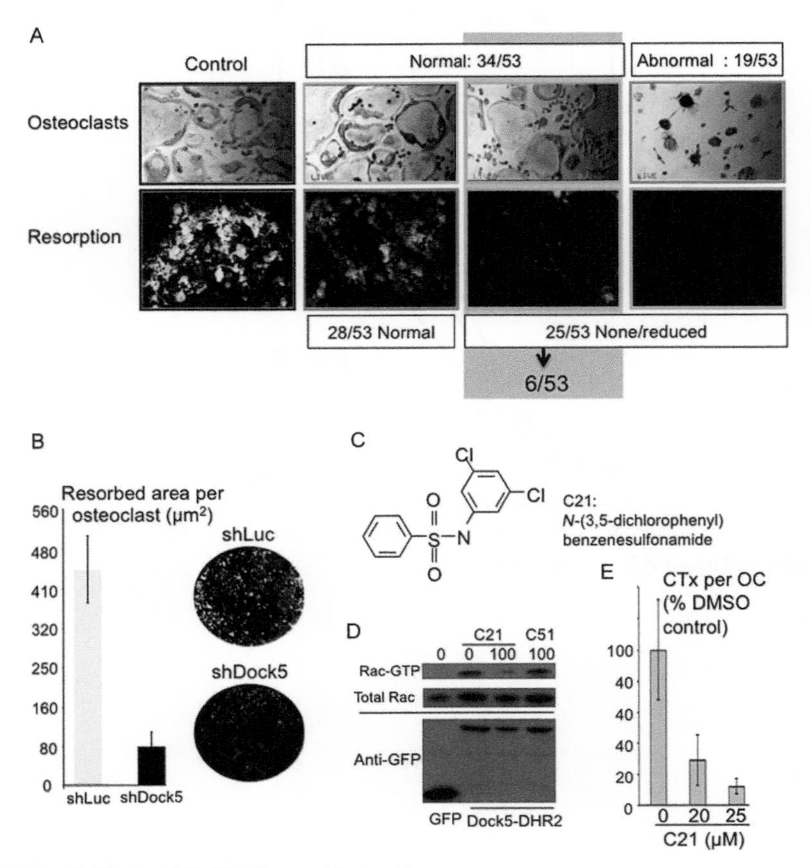

Chapter 8. Figure 8.5 (See legend in text.)

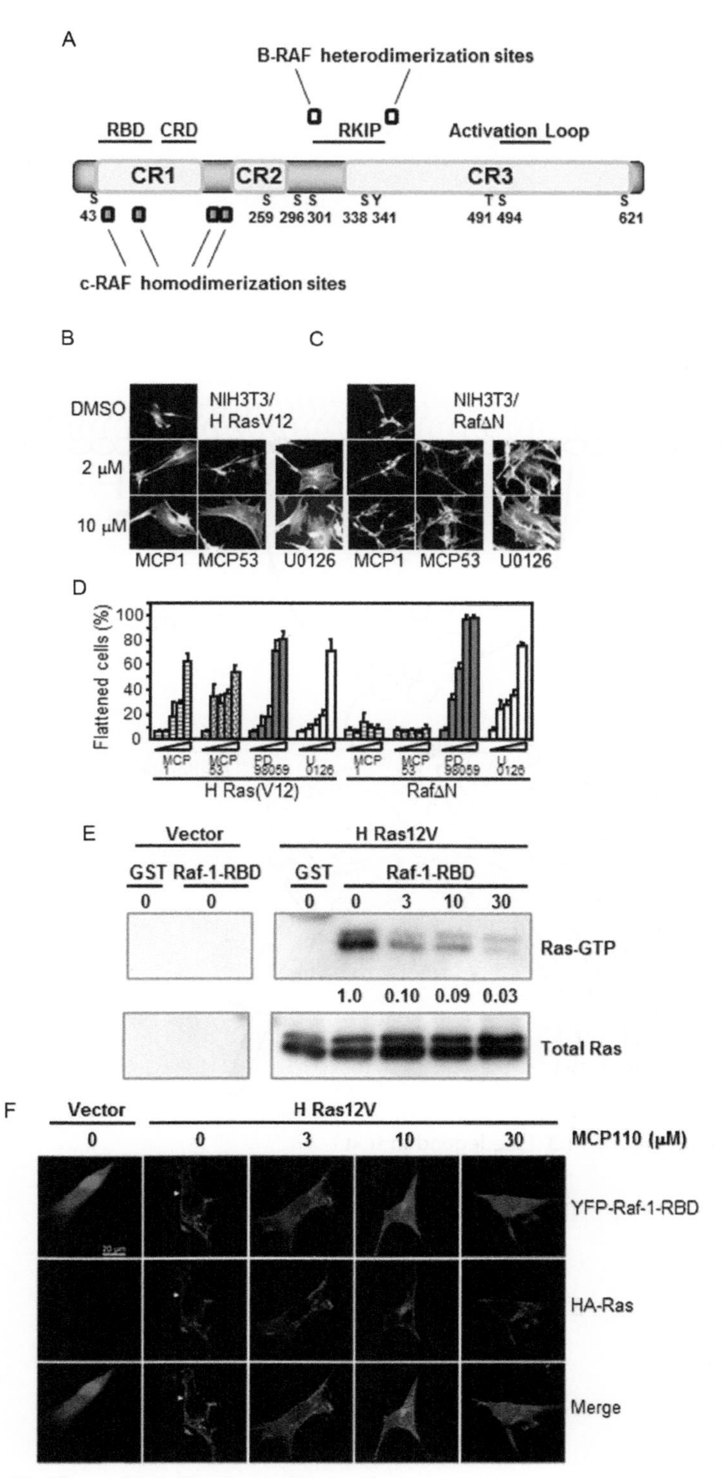

Chapter 10. Figure 10.5 (See legend in text.)

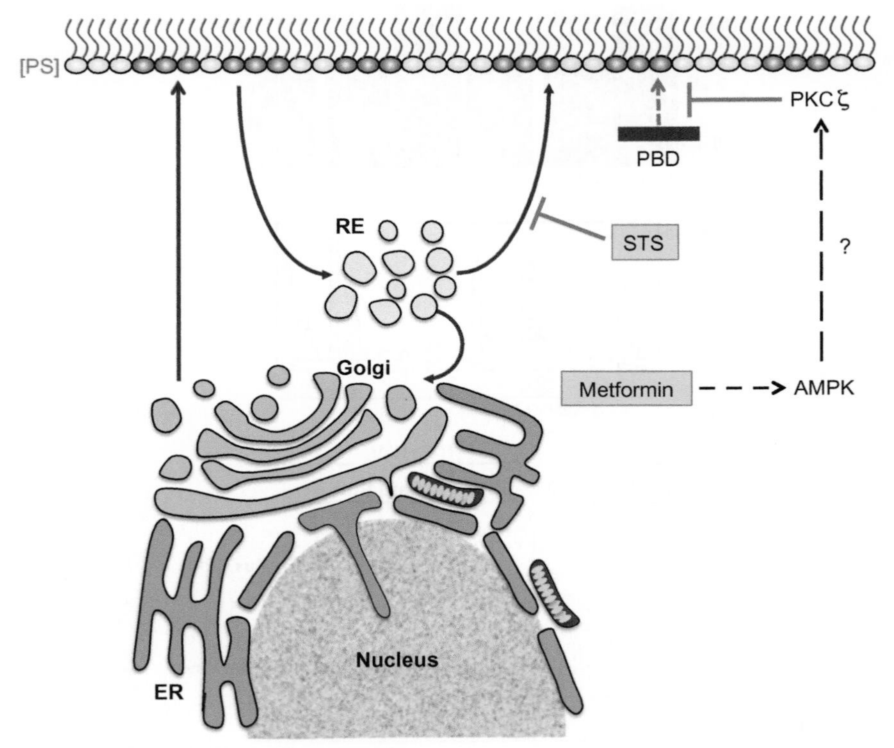

Chapter 11. Figure 11.3 (See legend in text.)

Printed and bound by CPI Group (UK) Ltd, Croydon, CR0 4YY

Printed and bound by CPI Group (UK) Ltd, Croydon, CR0 4YY
08/05/2025
01864961-0001